Meta-analysis and Combining Information in Genetics and Genomics

CHAPMAN & HALL/CRC
Mathematical and Computational Biology Series

Aims and scope:
This series aims to capture new developments and summarize what is known over the whole spectrum of mathematical and computational biology and medicine. It seeks to encourage the integration of mathematical, statistical and computational methods into biology by publishing a broad range of textbooks, reference works and handbooks. The titles included in the series are meant to appeal to students, researchers and professionals in the mathematical, statistical and computational sciences, fundamental biology and bioengineering, as well as interdisciplinary researchers involved in the field. The inclusion of concrete examples and applications, and programming techniques and examples, is highly encouraged.

Series Editors

Alison M. Etheridge
Department of Statistics
University of Oxford

Louis J. Gross
Department of Ecology and Evolutionary Biology
University of Tennessee

Suzanne Lenhart
Department of Mathematics
University of Tennessee

Philip K. Maini
Mathematical Institute
University of Oxford

Shoba Ranganathan
Research Institute of Biotechnology
Macquarie University

Hershel M. Safer
Weizmann Institute of Science
Bioinformatics & Bio Computing

Eberhard O. Voit
The Wallace H. Couter Department of Biomedical Engineering
Georgia Tech and Emory University

Proposals for the series should be submitted to one of the series editors above or directly to:
CRC Press, Taylor & Francis Group
4th, Floor, Albert House
1-4 Singer Street
London EC2A 4BQ
UK

Published Titles

Bioinformatics: A Practical Approach
Shui Qing Ye

Cancer Modelling and Simulation
Luigi Preziosi

Combinatorial Pattern Matching Algorithms in Computational Biology Using Perl and R
Gabriel Valiente

Computational Biology: A Statistical Mechanics Perspective
Ralf Blossey

Computational Neuroscience: A Comprehensive Approach
Jianfeng Feng

Data Analysis Tools for DNA Microarrays
Sorin Draghici

Differential Equations and Mathematical Biology
D.S. Jones and B.D. Sleeman

Engineering Genetic Circuits
Chris J. Myers

Exactly Solvable Models of Biological Invasion
Sergei V. Petrovskii and Bai-Lian Li

Gene Expression Studies Using Affymetrix Microarrays
Hinrich Göhlmann and Willem Talloen

Handbook of Hidden Markov Models in Bioinformatics
Martin Gollery

Introduction to Bioinformatics
Anna Tramontano

An Introduction to Systems Biology: Design Principles of Biological Circuits
Uri Alon

Kinetic Modelling in Systems Biology
Oleg Demin and Igor Goryanin

Knowledge Discovery in Proteomics
Igor Jurisica and Dennis Wigle

Meta-analysis and Combining Information in Genetics and Genomics
Rudy Guerra and Darlene R. Goldstein

Modeling and Simulation of Capsules and Biological Cells
C. Pozrikidis

Niche Modeling: Predictions from Statistical Distributions
David Stockwell

Normal Mode Analysis: Theory and Applications to Biological and Chemical Systems
Qiang Cui and Ivet Bahar

Optimal Control Applied to Biological Models
Suzanne Lenhart and John T. Workman

Pattern Discovery in Bioinformatics: Theory & Algorithms
Laxmi Parida

Spatial Ecology
Stephen Cantrell, Chris Cosner, and Shigui Ruan

Spatiotemporal Patterns in Ecology and Epidemiology: Theory, Models, and Simulation
Horst Malchow, Sergei V. Petrovskii, and Ezio Venturino

Stochastic Modelling for Systems Biology
Darren J. Wilkinson

Structural Bioinformatics: An Algorithmic Approach
Forbes J. Burkowski

The Ten Most Wanted Solutions in Protein Bioinformatics
Anna Tramontano

Chapman & Hall/CRC Mathematical and Computational Biology Series

Meta-analysis and Combining Information in Genetics and Genomics

Edited by

Rudy Guerra
Darlene R. Goldstein

CRC Press
Taylor & Francis Group
Boca Raton   London   New York

CRC Press is an imprint of the
Taylor & Francis Group   an **informa** business

A CHAPMAN & HALL BOOK

Chapman & Hall/CRC
Taylor & Francis Group
6000 Broken Sound Parkway NW, Suite 300
Boca Raton, FL 33487-2742

First issued in paperback 2017

ISBN 13: 978-1-138-11611-5 (pbk)
ISBN 13: 978-1-58488-522-1 (hbk)

Library of Congress Cataloging-in-Publication Data

Meta-analysis and combining information in genetics and genomics / [edited by] Rudy
  Guerra, Darlene R. Goldstein.
      p. cm. -- (Chapman & Hall/CRC mathematical and computational biology series)
    Includes bibliographical references and index.
    ISBN 978-1-58488-522-1 (hardcover : alk. paper)
      1. Meta-analysis. 2. Genetics--Statistical methods. 3. Genomics--Statistical methods.
  I. Guerra, Rudy, 1960- II. Goldstein, Darlene Renee. III. Title. IV. Series.

R853.M48M465 2009
576.5015'1--dc22   2009018853

Visit the Taylor & Francis Web site at
http://www.taylorandfrancis.com

and the CRC Press Web site at
http://www.crcpress.com

Dedication

We dedicate this book to our families: Rudy to Nancy, Amanda, Andrew, and Olivia; and Darlene to Tati.

Contents

Part 0. Introductory Material   **1**

1   A brief introduction to meta-analysis, genetics and genomics   **3**

by Darlene R. Goldstein and Rudy Guerra

    1.1   Introduction   3

    1.2   Combining information   4

    1.3   Genetics   8

    1.4   Genomics   12

    1.5   Combining information in genetics and genomics   16

Part I. Similar Data Types I: Genotype Data   **21**

2   Combining information across genome-wide linkage scans   **23**

by Carol J. Etzel and Tracy J. Costello

    2.1   Introduction   23

    2.2   Meta-analysis of genome-wide linkage scans   23

    2.3   Choice of meta-analysis method   28

    2.4   Discussion   31

    2.5   Appendix A   32

3 Genome search meta-analysis (GSMA): a nonparametric method for meta-analysis of genome-wide linkage studies **33**

by Cathryn M. Lewis

3.1 Introduction 33

3.2 GSMA: Genome Search Meta-Analysis method 34

3.3 Power to detect linkage using GSMA 40

3.4 Extensions of GSMA 42

3.5 Limitations of the GSMA 43

3.6 Disease studies using GSMA 44

3.7 GSMA software 46

3.8 Conclusions 46

4 Heterogeneity in meta-analysis of quantitative trait linkage studies **49**

by Hans C. van Houwelingen and Jérémie J. P. Lebrec

4.1 Introduction 49

4.2 The classical meta-analytic method 51

4.3 Extracting relevant information from individual studies 54

4.4 Example 58

4.5 Discussion 63

5 An empirical Bayesian framework for QTL genome-wide scans **67**

by Kui Zhang, Howard Wiener, T. Mark Beasley, Christopher I. Amos, and David B. Allison

5.1 Introduction 67

5.2 Methods 69

5.3 Results 72

5.4 Discussion 79

Part II. Similar Data Types II: Gene Expression Data   **81**

**6 Composite hypothesis testing: an approach built on intersection-union
   tests and Bayesian posterior probabilities**   **83**

   by Stephen Erickson, Kyoungmi Kim, and David B. Allison

   6.1 Introduction   83

   6.2 Composite hypothesis testing   84

   6.3 Assessing the significance of a composite hypothesis test   86

   6.4 Measuring Bayesian significance evidence in composite hypothesis
        testing   88

   6.5 Combining posterior probabilities in a Bayesian IUT   90

   6.6 Issues and challenges   91

   6.7 Summary   92

   6.8 Software availability   93

**7 Frequentist and Bayesian error pooling methods for enhancing statis-
   tical power in small sample microarray data analysis**   **95**

   by Jae K. Lee, HyungJun Cho, and Michael O'Connell

   7.1 Introduction   95

   7.2 Local pooled error test   97

   7.3 Empirical Bayes heterogeneous error model (HEM)   103

   7.4 Conclusion   112

8 Significance testing for small microarray experiments   **113**

   *by Charles Kooperberg, Aaron Aragaki, Charles C. Carey, and Suzannah
   Rutherford*

   8.1 Introduction   113

   8.2 Methods   114

   8.3 Data   119

   8.4 Results   121

   8.5 Discussion   131

   8.6 Appendix: Array preprocessing   134

9 Comparison of meta-analysis to combined analysis of a replicated microarray study   **135**

by Darlene R. Goldstein, Mauro Delorenzi, Ruth Luthi-Carter, and Thierry Sengstag

9.1 Introduction   135

9.2 Study description   136

9.3 Statistical analyses   136

9.4 Results   141

9.5 Discussion   153

10 Alternative probe set definitions for combining microarray data across studies using different versions of Affymetrix oligonucleotide arrays   **157**

by Jeffrey S. Morris, Chunlei Wu, Kevin R. Coombes, Keith A. Baggerly, Jing Wang, and Li Zhang

10.1 Introduction   157

10.2 Combining microarray data across studies and platforms   158

10.3 Meta-analysis with Affymetrix oligonucleotide arrays   161

10.4 Partial probe sets method   162

10.5 Example 1: CAMDA 2003 lung cancer data   163

10.6 Full-length transcript-based probe sets method   168

10.7 Example 2: Lung cell line data   171

10.8 Discussion   174

11 Gene ontology-based meta-analysis of genome-scale experiments   **175**

by Chad A. Shaw

11.1 Introduction   175

11.2 Ontologies   175

11.3 The Gene Ontology   176

11.4 Statistical methods   182

11.5 Application to stem cell data   189

11.6 Conclusions   197

Part III. Combining Different Data Types   **199**

12 Combining genomic data in human studies   **201**

  by Debashis Ghosh, Daniel Rhodes, and Arul Chinnaiyan

   12.1 Introduction   201

   12.2 Genomic data integration in cancer   202

   12.3 Combining data from related technologies: cDNA arrays   203

   12.4 Combining data from different technologies   206

   12.5 *In vivo/in vitro* genomic data integration   209

   12.6 Software availability   210

   12.7 Discussion   211

13 An overview of statistical approaches for expression trait loci mapping **213**

  by Christina Kendziorski and Meng Chen

   13.1 Introduction   213

   13.2 ETL mapping data and methods   214

   13.3 Evaluation of ETL mapping methods   217

   13.4 Discussion   222

**14 Incorporating GO annotation information in expression trait loci map-
      ping**   **225**

  by J. Blair Christian and Rudy Guerra

   14.1 Introduction   225

   14.2 Expression trait loci mapping   226

   14.3 Data   228

   14.4 Methodology   230

   14.5 Simulations   233

   14.6 Results   238

   14.7 Conclusions   241

15 A misclassification model for inferring transcriptional regulatory networks   **243**

by Ning Sun and Hongyu Zhao

15.1 Introduction   243

15.2 Methods   244

15.3 Simulation results   250

15.4 Application to yeast cell cycle data   253

15.5 Summary   255

16 Data integration for the study of protein interactions   **259**

by Fengzhu Sun, Ting Chen, Minghua Deng, Hyunju Lee, and Zhidong Tu

16.1 Introduction   259

16.2 Data sources   261

16.3 Assessing the reliability of protein interaction data   262

16.4 Protein function prediction using protein interaction data   266

16.5 Discussion   273

17 Gene trees, species trees, and species networks   **275**

by Luay Nakhleh, Derek Ruths, and Hideki Innan

17.1 Introduction   275

17.2 Gene tree incongruence   277

17.3 Lineage sorting   281

17.4 Gene duplication and loss   284

17.5 Reticulate evolution   286

17.6 Distinguishing lineage sorting from HGT   290

17.7 Summary   292

References   **295**

Index   **329**

Contributors

David B. Allison
Department of Biostatistics
Section of Statistical Genetics
University of Alabama
Birmingham, Alabama

Christopher I. Amos
Department of Epidemiology
University of Texas M. D. Anderson Cancer Center
Houston, Texas

Aaron Aragaki
WHI Clinical Coordinating Center
Fred Hutchinson Cancer Research Center
Seattle, Washington

Keith A. Baggerly
Department of Bioinformatics and Computational Biology
and
Department of Biostatistics
University of Texas M. D. Anderson Cancer Center
Houston, Texas

T. Mark Beasley
Department of Biostatistics
Section of Statistical Genetics
University of Alabama
Birmingham, Alabama

Charles C. Carey
Rutherford Lab
Fred Hutchinson Cancer Research Center
Seattle, Washington

Meng Chen
Department of Statistics
University of Wisconsin
Madison, Wisconsin

Ting Chen
Molecular and Computational Biology Program
Department of Biological Sciences
University of Southern California
Los Angeles, California

Arul M. Chinnaiyan
Department of Pathology and Urology
and
Center for Cancer Bioinformatics
University of Michigan Medical School
Ann Arbor, Michigan

HyungJun Cho
Division of Biostatistics and Epidemiology
University of Virginia School of Medicine
Charlottesville, Virginia

Blair Christian
Statistics Department
Rice University
Houston, Texas

Kevin R. Coombes
Department of Bioinformatics and Computational Biology
and
Department of Biostatistics
University of Texas M. D. Anderson Cancer Center
Houston, Texas

Tracy J. Costello
Department of Health Disparities Research
University of Texas M. D. Anderson Cancer Center
Houston, Texas

Mauro Delorenzi
Swiss Institute of Bioinformatics
Lausanne, Switzerland

Minghua Deng
School of Mathematical Sciences and Center for Theoretical Biology
Peking University
Beijing, People's Republic of China

Stephen Erickson
Department of Biostatistics
Section of Statistical Genetics
University of Alabama
Birmingham, Alabama

Carol J. Etzel
Department of Epidemiology
University of Texas M. D. Anderson Cancer Center
Houston, Texas

Debashis Ghosh
Department of Biostatistics
School of Public Health
University of Michigan
Ann Arbor, Michigan

Darlene R. Goldstein
Ecole Polytechnique Fédérale de Lausanne
Institut de Mathématiques
and
Swiss Institute of Bioinformatics
Lausanne, Switzerland

Rudy Guerra
Statistics Department
Rice University
Houston, Texas

Hideki Innan
Graduate University for Advanced Studies
Hayama, Kanagawa, Japan

Christina Kendziorski
Department of Biostatistics and Medical Informatics
University of Wisconsin
Madison, Wisconsin

Charles Kooperberg
Biostatistics and Biomathematics
Fred Hutchinson Cancer Research Center
Seattle, Washington

Jérémie J. P. Lebrec
Department of Medical Statistics and Bioinformatics
Leiden University Medical Center
Leiden, The Netherlands

Jae K. Lee
Division of Biostatistics and Epidemiology
University of Virginia School of Medicine
Charlottesville, Virginia

Cathryn M. Lewis
Department of Medical and Molecular Genetics
Guy's, King's and St. Thomas' School of Medicine

King's College London
London, UK

Ruth Luthi-Carter
Ecole Polytechnique Fédérale de Lausanne (EPFL)
Laboratoire de Neurogénomique Fonctionnelle
Lausanne, Switzerland

Jeffrey S. Morris
Department of Biostatistics
University of Texas M. D. Anderson Cancer Center
Houston, Texas

Luay Nakhleh
Department of Computer Science
Rice University
Houston, Texas

Michael O'Connell
Insightful Corporation
Seattle, Washington

Daniel R. Rhodes
Bioinformatics Program
University of Michigan Medical School
Ann Arbor, Michigan

Suzannah Rutherford
Rutherford Lab
Fred Hutchinson Cancer Research Center
Seattle, Washington

Derek Ruths
Department of Computer Science
Rice University
Houston, Texas

Thierry Sengstag
Swiss Institute of Bioinformatics
Lausanne, Switzerland

Chad A. Shaw
Department of Molecular and Human Genetics
Baylor College of Medicine
Houston, Texas

Fengzhu Sun
Molecular and Computational Biology
Department of Biological Sciences
University of Southern California
Los Angeles, California

Ning Sun
Department of Biostatistics
School of Public Health
Yale University
New Haven, Connecticut

Zhidong Tu
Rosetta Inpharmatics, LLC
Merck & Co., Inc.
Seattle, Washington

Hans C. van Houwelingen
Department of Medical Statistics and Bioinformatics
Leiden University Medical Center
Leiden, The Netherlands

Jing Wang
Department of Bioinformatics and Computational Biology
and
Department of Biostatistics
University of Texas M. D. Anderson Cancer Center
Houston, Texas

Howard Wiener
Department of Epidemiology and International Health
University of Alabama
Birmingham, Alabama

Chunlei Wu
Genomic Institute of Novartis Research Foundation
San Diego, California

Kui Zhang
Department of Biostatistics
Section of Statistical Genetics
University of Alabama
Birmingham, Alabama

Li Zhang
Department of Bioinformatics and Computational Biology
and
Department of Biostatistics
University of Texas M. D. Anderson Cancer Center
Houston, Texas

Hongyu Zhao
Department of Biostatistics
School of Public Health
Yale University
New Haven, Connecticut

Preface

Over the last 20 years biological research has undergone a radical transformation, spawned by earlier developments in recombinant DNA technology. Coupled with parallel advances in computational power, these technological developments provide investigators with the ability to collect a variety of genomic data in a high-throughput fashion. As examples, genotype, gene expression, and structural variation (e.g., copy number) data are now readily obtained on a genome-wide scale. As costs of biotechnology and computing continue to decrease even further, such data sets are becoming more routinely collected by many laboratories. An important aspect of this scenario is that biomedical investigators should be more able to assess replication or synthesis of findings from different groups studying the same diseases or traits.

The formal statistical methods to analyze results from different studies that address the same question traditionally fall under the rubric of meta-analysis. Methodological developments and applications of meta-analysis are increasing in the field of high-throughput biology. With the diversity of data and meta-data (e.g., gene annotation data) now available, there is increased interest in analyzing multiple studies beyond statistical approaches of formal meta-analysis. In this book, we look at the problem of jointly analyzing multiple studies from a broader perspective, considering an extensive range of quantitative information combination methods, not limited to the narrow traditional definition of meta-analysis.

The purpose of this book is to provide a wide-ranging survey of information combination methodology specific to genetic and genomic studies, areas with distinct features that require special accommodation. An introductory chapter presents the basic ideas and tools of meta-analysis, along with some background material to acquaint the reader with the areas of genetics and genomics. After this chapter, we proceed rapidly to more general information combination applications using advanced, specialized techniques, structured in three main parts. Parts I and II address combination of similar data types: genotype data from genome-wide linkage scans in Part I, and data derived from microarray gene expression experiments in Part II. These two parts show some of the data combination problems that can arise even within the same basic framework, and also present approaches to solutions for these problems. The chapters in Part III focus on combined analysis of different data types, giving the opportunity to see data combination approaches in action across a wide variety of genome investigations. Though on occasion there are references to other chapters in the book, all chapters are by and large self-contained and can be read independently of each other.

Part I concerns combination of genome-wide scans for linkage, where in general the aim is to examine the possibility of identifying common disease or quantitative trait loci across the different studies. A number of issues arise in this problem, including different genetic marker maps and study design across the studies. Three chapters discuss different aspects of formal (frequentist) meta-analysis. Chapter 2 by Carol Etzel and Tracy Costello reviews several methods for meta-analysis of genome-wide scans, providing practical advice to guide researchers in their analyses. In Chapter 3, Cathryn Lewis describes the Genome Search Meta-Analysis (GSMA), a widely-used method for meta-analysis of genome scans based on nonparametric rank statistics. Hans van Houwelingen and Jérémie Lebrec (Chapter 4) explore the issue of heterogeneity in the meta-analysis of quantitative trait loci (QTL) linkage studies, giving examples and extending classical methods to the QTL mapping scenario. Chapter 5 by Kui Zhang, Howard Wiener, Mark Beasley, Christopher Amos, and David Allison places the problem of combining QTL linkage scans in an empirical Bayes framework, showing how to update information while accounting for between-study heterogeneity.

Part II also deals with data of the same type, this time emanating from microarray gene expression studies. Chapter 6 by Stephen Erickson, Kyoungmi Kim, and David Allison considers the problem of testing the union of multiple null hypotheses in microarray experiments. A key feature of the approach is the use of an intersection-union test whereby the multiple null hypotheses are replaced by a single (composite) null hypothesis and thus does not require a correction for multiple testing. Two chapters focus on inference for small microarray studies, a very common situation. Jae K. Lee, HyungJun Cho, and Michael O'Connell describe the Local Pooled Error method in Chapter 7. In Chapter 8, Charles Kooperberg, Aaron Aragaki, Charles Carey, and Suzannah Rutherford describe results of a methods comparison.

Chapter 9 by Darlene Goldstein, Mauro Delorenzi, Ruth Luthi-Carter, and Thierry Sengstag compares meta-analysis and analysis based on a combined data set from a replicated microarray study, showing that even in this highly ideal scenario, serious issues still arise in pooling data. In Chapter 10, Jeffrey Morris, Chunlei Wu, Kevin Coombes, Keith Baggerly, Jing Wang, and Li Zhang provide a method for raw data pooling of Affymetrix data from different chip versions. In the final chapter of Part II, Chad Shaw (Chapter 11) gives a comprehensive overview of the Gene Ontology (GO), then discusses methods for using GO annotations to facilitate a combined analysis of microarray experiment gene lists.

In contrast to Parts I and II, Part III is about combining different types. Debashis Ghosh, Daniel Rhodes, and Arul Chinnaiyan start this part of the book with an overview of genomic data integration in studies of human disease (Chapter 12). The remaining chapters consider combination of specific data sources. Chapter 13 by Christina Kendziorski and Meng Chen consider expression trait loci (ETL) mapping, for which the quantitative trait of interest is gene expression, obtained from microarrays. Genotype and gene expression data are the two data types here. Blair Christian and Rudy Guerra extend this by using GO annotations to improve identification of associations between genotype and gene expression.

Two chapters further broaden the types of data examined by considering protein data in addition to DNA (genotype or gene expression) data. Chapter 15 by Ning Sun and Hongyu Zhao describes a Bayesian framework for inferring transcriptional regulatory networks based on combined analysis of gene expression data and protein-DNA interaction data. In Chapter 16, Fengzhu Sun, Ting Chen, Minghua Deng, Hyunju Lee, and Zhidong Tu provide methods for integrating the various data sets available for inferring protein-protein interactions, including protein-protein interaction databases, gene expression, GO, protein localizations, and protein domains.

Moving to the more refined level of DNA sequence information, Luay Nakhleh, Derek Ruths, and Hideki Innan present methods for combining gene trees to give species networks in Chapter 17. An interesting aspect of their work is the notion of non-treelike evolution and its relationship to the coalescent model.

This book is aimed at researchers and students interested in analyzing data from modern biological studies involving multiple data sets, either of the same type or comprising multiple data sources. Statistical analysis of such heterogeneous data will become more common and we fully expect that biological understanding will be significantly aided by jointly analyzing such data using fundamentally sound methodology.

Minimal background for the book includes one to two semesters of statistics; a course in mathematical statistics is preferred. Courses in regression, probability models, and data mining/machine learning would also be helpful. Experienced data analysts with a less mathematically-oriented background can benefit from this book, provided they are willing to gloss over some of the more technical details. A working knowledge of genetics would also be useful, but most biological material is either in the introductory material or self-contained in the chapters.

The chapters have all been peer-reviewed; we have edited them to present a coherent aggregate that we hope the reader will enjoy learning from.

We would like to acknowledge the contributions of several people who helped propel this book from a notion to a reality. First, we are extremely grateful to the contributing authors for sharing their expertise, as well as for their time, effort, and commitment to the project. David Allison (University of Alabama) deserves special thanks as he was highly instrumental in the conception and development of the book and recruitment of contributing authors. Rob Calver, our editor, has been encouraging, supportive, patient, and helpful at every phase of the writing. Sarah Morris, editorial assistant, provided us with excellent guidance through the administrative aspects. Being members of Terry Speed's academic pedigree, we attribute our production of this book in part to his influence on our careers. Finally, we thank our families for their support and patience.

Part 0. Introductory Material

CHAPTER 1

A brief introduction to meta-analysis, genetics and genomics

Darlene R. Goldstein and Rudy Guerra

1.1 Introduction

Synthesizing knowledge from different studies forms the basis of much scientific understanding. Sometimes, the synthesis is qualitative (e.g., literature or case studies review). However, quantitative methods provide a fundamentally sound and reproducible basis for combining information. Quantitative combination of results across genetic or genomic studies, for example using meta-analytic techniques, may aid in assessing overall findings, thereby advancing biological knowledge.

Meta-analysis consists of statistical methods for combining results of independent studies addressing related questions. One aim of combining results is to obtain increased power – studies with small sample sizes are less likely to find effects even when they exist. Putting results together increases the effective sample size, thereby allowing more precise effect estimation and increasing power. The uncovering of a significant effect from a combined analysis, where individual studies have not made positive findings, has been referred to in the microarray meta-analysis literature as "integration-driven discovery" (IDD) (Choi et al., 2003).

Meta-analysis and other information combination methodologies are emerging as an essential tool for modern genetic and genomic analysis. Carrying out such a project requires careful planning and execution of the following six general steps. These generalize the steps of a more traditional meta-analysis in which specific study results (e.g., summary statistics, p-values) are analyzed.

1. Formulate a specific purpose and explicitly define the outcome to be extracted from each study (e.g., effect size, p-value, etc.).
2. Identify relevant primary studies.
3. Establish inclusion/exclusion criteria for primary studies.
4. Detail data abstraction and acquisition.

5. Decide on data analysis methods, including a careful investigation of data quality and between-study heterogeneity.

6. Interpret the results, and decide on appropriate followup steps.

This chapter provides an introduction to the basic ideas and methods of meta-analysis as commonly applied in genetic and genomic studies. More detailed tutorials can be found in Normand (1999) and van Houwelingen et al. (2002).

1.2 Combining information

The options for combining information across studies can be viewed as occurring across a spectrum of possible analyses, moving roughly from combination of least to most "processed" quantities (Wirapati et al., 2009). In general, the more processing that is carried out in the individual studies, the farther the result is from the primary data, with a consequent loss of information. Combination can occur at the level of raw or adjusted (individual patient) data, parameter/effect size estimates, transformed test statistics (including p-values), statistic ranks, or decisions based on a threshold of significance.

1.2.1 Pooling data

One way of combining information across studies is to pool data as a single data set for analysis. This approach is sometimes called a "mega-analysis." For sufficiently homogeneous data, this strategy might be viable. However, this method has a number of drawbacks. It is generally inappropriate to pool raw data from heterogeneous studies (e.g., Simpson's paradox (Simpson, 1951)).

Correction before pooling (e.g., Benito et al. (2004)) consists of applying an adjustment to the data separately for each study, and then combining adjusted values to analyze together as a single set. Unfortunately, without some further modification (such as scaling or some other transformation), even these adjusted observations may well remain too heterogeneous for pooling. Carrying out a stratified analysis is another option (Juo et al., 1978).

1.2.2 Heterogeneity

A major difficulty with synthesizing results is the occurrence of study heterogeneity. Studies which are superficially similar may in fact differ in many ways, some of which can be quite subtle (Sutton et al., 2000). In general, studies carried out by different investigators may vary in scientific research goals, population of interest, study design, quality of implementation, subject inclusion and exclusion criteria, baseline status of subjects (even with the same selection criteria), treatment dosage and timing, management of study subjects, outcome definition or measures, and statistical methods of analysis.

In order for pooling information across studies to make sense, the studies should be sufficiently homogeneous. A standard test of homogeneity (Cochran, 1954a) tests the null hypothesis of homogeneity of effects β_i in k studies (H_0: $\beta_1 = \beta_2 = \cdots = \beta_k$) against the general alternative that at least one β_i is different. The test statistic Q is given by

$$Q = \sum_{i=1}^{k} w_i(\hat{\beta}_i - \bar{\beta}_.)^2, \tag{1.1}$$

where $\hat{\beta}_i$ estimates the treatment effect, w_i is the weight given to study i (most commonly taken as the reciprocal of the variance of the outcome estimate (Cooper and Hedges, 1994)), and $\bar{\beta}_.$ is the weighted average treatment effect

$$\bar{\beta}_. = \frac{\sum_i w_i \hat{\beta}_i}{\sum_i w_i}. \tag{1.2}$$

Under the null hypothesis, the distribution of Q is approximately χ^2_{k-1}.

In the event that the null hypothesis is not rejected, any differences between studies are assumed to be due to chance variation, and it is considered appropriate to combine estimates via a fixed effects model (described in Section 1.2.3 below; see also Section 4.2.2). A major limitation of this approach, though, is the low power of the test, due to small sample sizes or a small number of studies, to detect even substantial heterogeneity. One way to avoid the risk of combining heterogeneous results is to relax the significance criterion, say from 0.05 to 0.10.

If instead the test shows that significant heterogeneity exists between study results, then combination via a random effects model (Section 1.2.3) is typically favored. Where possible, heterogeneity should be scrutinized rather than ignored, with an aim toward explaining important study differences (Bailey, 1987).

1.2.3 Combining parameter estimates

Parameter estimates are typically combined using either a fixed effects or random effects model, depending on the absence or presence of heterogeneity between results of the different studies (Cooper and Hedges, 1994). The overall estimate of the treatment effect is then a weighted average of individual effects, with weights depending on the variances within and between individual study estimates.

Results are often presented by means of a forest plot. A forest plot is a plot showing effect size and precision (usually a 95% confidence interval) for each individual study as well as for the combined effect. A quick, visual impression of the consistency of study results can be obtained via the forest plot. For studies with many outcomes, such as genome-scale experiments in which there is a measure for each gene, forest plots are usually only made for a small set of effect sizes (e.g., for those genes prioritized for follow-up).

Fixed effects meta-analysis

Fixed effects (FE) meta-analysis assumes no heterogeneity between results of the different studies and estimates a common underlying treatment effect. In FE meta-analysis, each individual study estimate $\hat{\beta}_i$ receives weight w_i inversely proportional to its variance. The weighted estimates are combined to yield an overall effect estimate (Equation 1.2). These inverse variance weights are used as they minimize the variance of the combined estimate $\bar{\beta}$. (Cooper and Hedges, 1994). The variance of the weighted estimator is just $1/\sum_{i=1}^{k} w_i$. Under the assumption of normality of $\bar{\beta}$., confidence intervals or p-values can be readily obtained.

Random effects meta-analysis

If the study results do exhibit heterogeneity, then there is assumed to be no single underlying value of the effect but rather a distribution of values. In the presence of heterogeneity, differences among study results are considered to arise from between-study variation of true effect size as well as chance variation. Use of the FE model understates the true degree of variability of $\bar{\beta}$., resulting in p-values which are artificially low – that is, the results appear to be more significant than they in fact are. A more conservative approach is to use a two-stage hierarchical model which accounts for the additional source of variability due to study.

Random effects (RE) meta-analysis assumes that individual studies may be estimating different treatment effects. The aim is to estimate characteristics of the distribution of effects, particularly the mean population effect size and between study variance of effect sizes. As in the FE case weighted estimates are used, but the weights are adjusted to take into account the additional variability between studies:

$$w_i^* = \frac{1}{(1/w_i) + \sigma_B^2},$$

where σ_B^2 measures between-study variability (see Cooper and Hedges (1994) for a derivation). The estimated mean treatment effect is again given by Equation 1.2, but with w_i^* in place of the w_i. Similarly, the variance of the weighted estimator is now given by $1/\sum_{i=1}^{k} w_i^*$. In practice, the variances (and, hence, the weights) are almost always estimated (National Research Council, 1992; DerSimonian and Laird, 1986). When the between-study variance is estimated as 0, the RE model reduces to the FE model. This model is discussed in more detail in Section 4.2.3.

As in the case for the FE model, confidence intervals and p-values from the RE model are obtained assuming normality of the effect distribution.

1.2.4 Combining p-values

In FE and RE meta-analysis, combined estimates of effect size provide the basis for analysis. Other methods of meta-analysis, dating back to at least the 1930s, are based on combining (transformed) p-values.

It is assumed that each of k independent studies tests a common null hypothesis and that each yields a one-tailed p-value in the same direction. Methods for combining p-values are based on testing a consensus or omnibus null hypothesis (i.e., all null hypotheses from the individual studies are true) by combining transformed p-values from each of the individual studies. Methods for combining p-values include those of Fisher (1932), Tippett (1931), and Pearson (1933). A variant method transforms p-values to corresponding standard normal z-statistics (Stouffer et al., 1949; Kulin-skaya et al., 2008). For an early review of these methods see Folks (1984). Erickson et al. (Chapter 6 in this volume) covers composite hypothesis testing more thoroughly.

An extremely popular method is that of Fisher (1932). Under the null hypothesis of no treatment effect, the individual study p-values p_i are independent and uniformly distributed $U(0, 1)$ random variables. Upon rescaling, the test statistic is given by

$$X^2 = -2 \sum_{i=1}^{k} \log(p_i). \qquad (1.3)$$

Under the omnibus null hypothesis that each of the k null hypotheses is true, the distribution of X^2 is χ^2_{2k}.

This method is very widely used as an approach to meta-analysis despite established limitations. It is well known, for example, that the X^2 statistic can be strongly influenced by a single highly significant p-value, which in turn may simply be due to a large sample size. Another concern is the straightforward fact that a single, highly processed term, the p-value, is used to summarize each study, and all the complexities therein, as a basis for meta-analysis.

1.2.5 Combining statistic ranks

In many microarray studies (see Section 1.4), the main "result" consists of an ordered gene list. One way to combine such lists across studies is by averaging percentile ranks. Although order is preserved, the magnitude of differences between genes is lost, entailing at least some, and possibly very great, loss of information.

Combining ranks has been successfully used as a strategy in meta-analysis of genetic linkage studies (Wise et al., 1999; see also Chapter 3 by Lewis), and has also been used in the microarray context (Breitling et al., 2004). For microarrays, the underlying idea is that since the data are noisy, the ranks may be more reliable than the actual measured values. This view is somewhat undermined by the fact that the stability of the ranking depends on the rank, the differences between underlying values and sample size. The measure that is ranked in the original analysis will also affect the ordering, and independent studies may well have used different statistics. Ranking does, however, take care of the problem of different measurement types and scales for different types of microarrays.

1.2.6 *Combining decisions*

At the crudest level, the *conclusions* of statistical hypothesis tests can be combined. That is, rather than combining the p-values for the tests, combine the resulting decisions based on some threshold for significance. This method was proposed over 50 years ago (Wilkinson, 1951). Requiring all tests to individually satisfy a significance criterion has been shown to be inadmissible for testing an exponential family parameter (Birnbaum, 1954), and in general does not appear to have very good power properties (Koziol and Perlman, 1978).

1.2.7 *Bias in meta-analysis*

Bias is a serious concern when carrying out or interpreting a meta-analysis. Bias is generally due to studies selected for inclusion in the meta-analysis being insufficiently representative of the totality of research being carried out. The most commonly discussed is *publication bias*. Publication bias exists when the probability that a result is published depends on the result (typically required to achieve a level of statistical significance). This is also sometimes referred to as the "file drawer problem," since those studies not reaching significant effect sizes are considered to be relegated to a file drawer. Other information dissemination biases include language bias, availability bias, cost bias, familiarity bias, and outcome bias (Rothstein et al., 2005). Bias can be introduced in many ways into the study selection process; see Egger and Smith (1998) for a review.

Examination for the presence of bias is an important part of the meta-analysis process. Sterne et al. (2001) summarizes some of the statistical methods available for detecting and correcting for bias when carrying out a meta-analysis. A funnel plot is a scatter plot of the study effect estimates against a measure of study size (usually SD), and can be used as a diagnostic tool for bias (Egger et al., 1997; Harbord et al., 2006). Sensitivity analyses are also recommended as a strategy to assess and correct for publication bias (Copas and Shi, 2000).

1.3 Genetics

1.3.1 *Linkage analysis*

A central problem in genetic analysis is to determine which gene(s), if any, affect particular *phenotypes* (observable trait characteristics), as well as the chromosomal locations of these genes, their different *alleles* (variant forms of the gene) and, ultimately, their biochemical modes of action. Linkage analysis is an initial step in elucidating the genetic mechanisms affecting a trait of interest. Linkage analysis proceeds by tracking patterns of coinheritance of the trait/phenotype of interest and other traits or genetic *markers* (DNA with an identifiable chromosomal location) in families. Genetic *recombination* is the formation of new combinations of genes that are different from the combinations that are found on the chromosomes of the parents. Linkage analysis relies on the varying degree of recombination between trait and marker loci to map the loci relative to one another. A brief summary of linkage analysis is provided here; more detail is available in Ott (1999).

Mendel's second law of inheritance hypothesizes that different genes segregate to gametes (sperm or egg) independently. In fact, independent assortment of gene pairs only occurs when the genes are unlinked; that is, when they are on different chromosomes or are so far apart on the same chromosome that there is the same chance of recombination as nonrecombination between them. Two genes are *linked* when they do not segregate independently. A measure of the degree of linkage is the recombination fraction θ, the chance of recombination occurring between two loci. For unlinked genes, $\theta = 1/2$; for linked genes, $0 \leq \theta < 1/2$.

Data for linkage analysis consist of sets of related individuals and information on the genetic marker and/or trait genotypes or phenotypes, usually selected on the basis of a disease, such as diabetes, or a quantitative trait, such as glucose tolerance. The recombination fraction is most commonly estimated by the method of maximum likelihood, the likelihood being determined by an appropriate genetic model for coinheritance. The conventional measure of support for the hypothesis of linkage between two loci at recombination fraction θ versus that of no linkage is given by the *lod score*

$$Z(\theta) = \log_{10} \left[\frac{L(\theta)}{L(1/2)} \right], \tag{1.4}$$

where $L(\theta)$ denotes the likelihood for θ given the observed data. Positive values of Z are evidence of linkage, while negative values indicate no linkage. With lod score linkage analysis, the null hypothesis of no linkage to a given locus (H_0: $\theta = 1/2$) is rejected for a sufficiently large value of $Z(\hat{\theta}_{MLE})$, often taken to be 3. Linkage analysis based on the lod score is sometimes referred to in the genetics literature as parametric or model-based, as the mode of inheritance must be specified using some parametric model.

Genetic linkage analysis has been successful at mapping genes for traits following Mendelian inheritance patterns, typically recessive or dominant diseases. Identifying genes affecting complex traits not following these simple modes of inheritance has proved to be more challenging. Lod score linkage analysis for complex traits is difficult to carry out due to many complicating factors. Chief among these is that the mode of inheritance is rarely known. Nonparametric, or model-free, approaches thus have appeal, since they do not require a genetic inheritance model to be specified. Such methods usually focus on *identity by descent* (IBD) allele sharing at a locus between a pair of relatives. DNA at a locus is shared by two relatives IBD if it originated from the same ancestral chromosome. In families of individuals possessing the trait of interest, there is association between allele sharing at loci linked to trait susceptibility loci and the trait (see e.g., Dudoit and Speed (1999) for examples). This association may be used to localize trait susceptibility genes. For loci unlinked to trait susceptibility loci, IBD sharing of DNA is not associated with the occurrence of the trait.

For qualitative traits in humans, there are several methods of linkage mapping based on (affected) relative pairs, most commonly sibling pairs (see Hauser and Boehnke (1998) for a review). A very widely used procedure for quantitative trait locus (QTL) mapping in humans is due to Haseman and Elston (1972); many extensions are also available (Amos, 1994; Amos and Elston, 1989; Amos et al., 1989; Elston et al., 2000; Fulker et al., 1995; Kruglyak and Lander, 1995; Olson et al., 1999; Olson and Wijsman, 1993; Stoesz et al., 1997).

1.3.2 Association mapping

Association mapping aims to find concordance between a trait and a genetic marker. Unlike linkage analysis, which is carried out within families, association mapping is performed at the population level. Association mapping relies on *linkage disequilibrium* (LD) between a marker and the causal locus. In contrast to linkage studies, which generally can detect only relatively large effects, association studies can identify variants with relatively small individual contributions to disease risk (Risch and Merikangas, 1996).

1.3.3 Linkage disequilibrium (LD)

A fundamental notion in association mapping is that of LD between a genetic marker and the locus that affects the trait under study. LD is the nonrandom association of alleles at two (or more) loci. The set of alleles transmitted together on the same chromosome is referred to as a *haplotype*. Loci are in LD when combinations of alleles occur either more or less frequently than expected from random formation of haplotypes based on marginal allele frequencies. LD is not necessarily due to linkage: it may be due to selection, mutation, random genetic drift or co-ancestry, among other possible causes. For this reason, LD is also sometimes referred to as gametic disequilibrium.

1.3.4 Study designs

Several designs are possible for genetic association studies, including case-control, family trios, cohort and, for large marker sets, multi-stage designs. At present the most common design by far is the case-control design.

Case-control studies are retrospective, usually population-based, studies that compare a group with the phenotype of interest (cases) to a group without (controls). This design is popular as it is relatively inexpensive, requiring smaller numbers than a (prospective) cohort study and being faster to complete, and can be used to study rare phenotypes. However, case-control studies are prone to bias (e.g., selection bias, recall bias), and care is needed to avoid confounding.

In a multi-stage design, an initial set of individuals is scanned for a large number of loci. At subsequent stages, smaller numbers of loci are followed up in other sets of individuals. Multi-stage designs are becoming more widely used due to their (potentially large) cost-effectiveness (Pahl et al., 2008).

The essential idea in a genetic association study is that a marker in strong LD with a disease locus is expected to be located nearby. Reasons explaining observed associations are (Cardon and Palmer, 2003):

1. The associated allele directly affects the phenotype (i.e. is causal).
2. There is LD between the tested marker locus and another locus that directly affects the phenotype.
3. The result is spurious or false positive, due to artifact (e.g., confounding, selection bias) or chance variation.

Of particular concern are spurious or false associations. Factors causing false positive associations include population substructure (a type of confounding), differential genotyping error (Clayton et al., 2005), genotype call success rate associated with case/control status, and multiple comparisons, e.g., half a million or more hypothesis tests in a genome-wide study (Section 1.3.6).

1.3.5 Population substructure

In a broad sense, population substructure is any deviation from random mating. Substructure may be attributable to stratification, for instance due to geographic subdivision, admixture of different founder populations, or cryptic relatedness, in which mating pairs have (unknown, possibly remote) relatives in common.

Violation of the assumptions for analyzing case-control study data (genetic homogeneity – i.e. all individuals have the same risk – and statistical independence between individuals) can lead to biased results, spurious associations due to confounding, or false negative associations (e.g., Simpson's paradox due to unaccounted-for admixture). Existence of substructure is therefore problematic.

There are several approaches for handling substructure. The most commonly used techniques to account for population substructure are genomic control (Devlin and Roeder, 1999), structured association (Pritchard et al., 2000), and EIGENSTRAT, a principal components-based correction (Price et al., 2006).

1.3.6 Whole genome studies

Genome-wide scan for linkage

In a genome-wide linkage scan, subjects are genotyped at a large number of highly polymorphic markers located across the entire genome. Parametric and/or nonparametric linkage analyses are carried out at each locus, either one at a time or using a multipoint analysis, in an effort to localize genes influencing the trait of interest.

There are very many methods for analyzing genome-wide linkage scans. Since choice of method depends very heavily on the study aims and type of data available, we do not attempt a general methods outline here. The chapters in Part I of this book describe some of these methods in the context of particular applications.

Genome-wide association studies (GWAS)

The aim of an association study is to find narrow regions of the genome that are associated with the trait of interest. To achieve success, markers must be sufficiently densely spaced so as to be in LD with untyped genomic regions. The advent of high throughput, array-based single nucleotide polymorphism (SNP) genotyping has made genome-wide association studies practical to carry out. Most pedigrees available for linkage analysis include no more than a few hundred meiotic events, limiting localization of a disease mutation to a resolution of about 1Mb (roughly 1cM). On the other hand, LD mapping in association studies takes advantage of thousands of meiotic events over extended pedigrees that relate individuals in a population (Kruglyak, 2008). Population genetics arguments show that LD mapping can provide a resolution on the order of 100Kb, or less.

In a typical GWAS, analysis is carried out on each marker singly, although an emerging area is methodology for interaction analysis in this very high-dimensional setting. The significance of association is determined by calculating an appropriate test statistic. For case-control studies, the test most commonly carried out is the Cochran-Armitage test for trend (Cochran, 1954b; Armitage, 1955; Sasieni, 1997). Haplotype-based tests are sometimes carried out on a reduced set of loci. Amos (2007) cites some factors that contribute to a successful GWAS. The review by Pearson and Maniolo (2008) gives advice on the execution and interpretation of a GWAS.

A major cause of false positive results in GWAS is the high degree of multiple hypothesis testing. Thus, replication is crucial for biologically meaningful progress, although what constitutes a replication is somewhat contentious. The NCI-NHGRI Working Group on Replication in Association Studies (2007) gives an overview of the inherent challenges, along with guidelines for replication.

Despite their advantages and increasing use, GWAS are not a panacea. True associations will be missed or false associations found when causal alleles are not among those genotyped, when the phenotype is genetically heterogeneous, when power is insufficient, when there is uncorrected stratification in the population, or in the presence of high levels of gene-gene or gene-environment interaction.

1.4 Genomics

The genetic information of an organism is contained in its DNA, whose genetic code is used to make proteins. Proteins play a central role in biological activity, with functions in all cellular processes including cell cycle and signaling, and immune response to name just a few. The process of protein synthesis is carried out in a cell via the transfer of information originating in the DNA. According to the central dogma of molecular biology, "information in nucleic acid can be perpetuated or transferred but the transfer of information into protein is irreversible" (Lewin, 2007).

Information is perpetuated by the process of DNA *replication*, which allows for inheritance of genetic information. A protein is built from amino acids based on information in the DNA through the synthesis of an intermediary molecule, messenger RNA (mRNA). The DNA code is copied into mRNA in a process called *transcription*. The protein is built based on the information in the mRNA during the *translation* process. There is, however, no corresponding reverse process that can create nucleic acids (DNA or RNA) from protein.

Genomics is the study of the DNA of an organism, its genome. The term "genomics" is rather broad, encompassing study of sequence variation, gene function and number, interactions between genes, and gene activation and suppression, among other topics. Technological advances, particularly DNA microarrays, have greatly facilitated the study of genomes. Microarrays are used to detect variations in a gene sequence or expression or for gene mapping.

Microarray technologies measure mRNA abundance for thousands of genes in parallel. The high throughput nature of microarrays has contributed to their rise in importance for studying the molecular basis of fundamental biological processes and complex disease traits. The principle underlying microarray technologies is the complementary pairing of nucleic acid bases: specific DNA probe sequences are immobilized on a solid support to detect complementary sequences present in samples.

Several different types of microarray platforms are available. Those currently in common use are cDNA arrays (DeRisi et al., 1997), fabricated in laboratories on-site at many academic and commercial institutions, and high-density short oligonucleotide arrays, such as Affymetrix GeneChips (Lockhart et al., 1996). Other technologies include long oligonucleotide (50-mer to 70-mer probes) arrays, such as those produced by Agilent (Blanchard et al., 1996; Hughes et al., 2001); Illumina BeadChips, in which oligonucleotide probes are attached to silica beads and deposited into wells on glass slides (Kuhn et al., 2004); and digital (count-based) technologies such as serial analysis of gene expression (SAGE) (Velculescu et al., 1995) and variants, and massively parallel signature sequencing (MPSS) (Brenner et al., 2000).

The cDNA and Affymetrix GeneChip technologies are discussed briefly below. Further details on the specifics of microarray data analysis are reviewed by Goldstein and Delorenzi (2004). A bibliography of papers on microarray data analysis is available at http://www.nslij-genetics.org/microarray/.

1.4.1 cDNA microarrays

cDNA microarrays consist of thousands of individual complementary DNA probe sequences printed in a high-density array on a glass microscope slide. The relative abundance of these probes in two samples may be assessed by monitoring the differential hybridization of the two samples to the sequences on the array. Samples are labeled with fluorescent dyes, most commonly cyanine molecules Cy3 ("green") and Cy5 ("red"), whose signal intensity is detected with a fluorescent scanner. The ratio of signals in the two channels is indicative of the relative abundance of the corresponding probe in the two target samples. See The Chipping Forecast (1999) for a more detailed introduction to the biology and technology of cDNA microarrays.

Quantifying expression for cDNA arrays

Primary data for each scanned array consist of pixel-level fluorescence intensities for each channel. The goal is to obtain a single measure of gene expression for each spot, corresponding to a probe, on each array. There are two preprocessing steps to accomplish this: image analysis and normalization.

Image analysis consists of summarizing the pixel level measurements to four numbers for each spot (probe), corresponding to the foreground and background measurements in each of the two (red and green) channels (see Yang et al. (2001) for a comparison of methods). These can be further reduced to two numbers by subtracting the background signal from the corresponding foreground signal.

Relative measures of fluorescence intensity in the two channels are obtained by computing the $\log_2$ ratio of the (background-corrected) foreground intensities R and G: $M = \log_2(R/G)$. Due to artifactual differences, these ratios are generally not comparable between arrays without further preprocessing (normalization).

The purpose of normalization is to remove the effects of systematic variation other than that due to the effect of interest. (The term "normalization" as used here is not related to the normal, or Gaussian, distribution.) Examples of such variation include differences in sample preparation, scanning intensities, and variability among arrays. Ideally, any observed differences in gene expression remaining after normalization are due to biologically meaningful differential expression rather than experimental artifacts.

Many methods have been proposed for normalization of M-values from two-channel arrays. The simplest methods, such as scaling or median-centering, are generally not adequate for removing technical artifacts. A more flexible method in common use that has better performance is *loess* normalization (Yang et al., 2002). Rather than adjusting M-values for each spot by the same amount, the value is adjusted via local regression by an amount depending on the average intensity in the two channels.

Once the data have been preprocessed, the resulting measure of expression for each spot is given by the normalized log-ratio (M-value).

1.4.2 Affymetrix GeneChips

Affymetrix GeneChip arrays use a photolithography approach to synthesize probes directly onto a silicon chip. GeneChips contain several (usually 11 – 20) 25-mer oligonucleotides used to measure the abundance of a given target sequence, the perfect match (PM) probes, as well as an equal number of negative controls, the mismatch (MM) probes. The MM probes differ from the corresponding PM probes by a single (complementary) base substitution at the central (13th) position: A $\Leftrightarrow$ T, G $\Leftrightarrow$ C. The set of probes for a given target sequence is called a *probe set*. Early chips were designed to have the probes within a probe set in close proximity on the array; in more recent chip versions these are scattered across the array, thus allowing some protection against local artifacts.

A single fluorescently labeled sample is hybridized to the array which is then scanned with a laser, yielding absolute (rather than relative) measures of fluorescence intensity. The intensities are indicative of the amounts of mRNAs containing the target sequence in the sample, and thus provide a means of quantifying gene expression.

Quantifying expression for Affymetrix GeneChips

Several methods of quantifying gene expression from probe intensities on Affymetrix GeneChips are in popular use. Two very common measures are the Affymetrix default measure MAS 5.0 (Affymetrix, 2002) and the Robust Multichip Average (RMA) (Irizarry et al., 2003a). For a comparison of many methods see `http://affycomp.biostat.jhsph.edu/` (Cope et al., 2004; Irizarry et al., 2006), where it is easily seen that no method is best under every circumstance.

MAS 5.0 quantification. The MAS 5.0 signal value is calculated from the combined, background-adjusted, PM and MM values of the probe set. The signal is calculated as follows (Affymetrix, 2002):

1. Intensities are preprocessed for global background.
2. An "ideal mismatch" value is calculated and subtracted to adjust the PM intensity.
3. The adjusted PM intensities are $\log_2$-transformed to stabilize the variance.
4. The biweight estimator (Hoaglin et al., 1983) is used to provide a robust mean of the resulting values for each probe set. The signal value S is the antilog of the resulting value.
5. Finally, the signal is scale-normalized to a target value S_c using a trimmed mean removing the upper and lower 2% of the values.

RMA quantification. RMA consists of three steps: a background adjustment, quantile normalization, and probe set summarization. Background is estimated using a model assuming that the observed signal is the convolution of an exponential signal with Gaussian background (noise). Quantile normalization forces equality of quantiles across samples. Such a normalization is appropriate assuming that the true distributions of intensities are the same in all samples (of course, the same probe may occur at different quantiles across samples). For each probe set on the chip, the $\log_2$

background-corrected normalized signal $\log_2 b(PM_{ij})$ is modeled as

$$\log_2 b(PM_{ij}) = \mu_i + \alpha_j + \epsilon_{ij}, \qquad (1.5)$$

where μ_i is the summary measure of expression for the given probe set on chip i, α_j is a probe-specific effect, and ϵ_{ij} are independently and identically distributed mean 0 errors (Irizarry et al., 2003b). For parameter identifiability, it is assumed that $\sum_j \alpha_j = 0$. The model is fit via median polish (Tukey, 1977); the estimated chip effect $\hat{\mu}_i$ is the RMA value of the probe set for chip i.

1.5 Combining information in genetics and genomics

1.5.1 Combining linkage studies

The mega-analysis approach has been used for combined analysis of linkage data (see e.g., McQueen et al., 2005; see also Section 2.3.1 for additional examples), but more commonly it is results rather than primary data that are combined.

Combining information in genetic linkage studies is not new. The lod score method allows combination of lod scores across different studies for recombination fraction estimation or inference (Morton, 1955). An assumption is that there is a common parameter in the studies, i.e. that the trait locus is the same in all pedigrees. This assumption may be assessed with a homogeneity test, and is sometimes loosened to allow for admixture of linked and unlinked pedigrees.

Measures of effect size other than the recombination fraction can be pooled across a collection of genetic studies. For model-free methods, where no natural parameter exists, allele-sharing data may be used to define a genetic effect size. For example, the mean proportion of IBD allele-sharing is an interpretable parameter that can be combined via fixed effects (Goldstein et al., 1999) or random effects (Gu et al., 1998) meta-analysis.

Allison and Heo (1998) consider scenarios in which there is not a common combinable parameter across different studies (due, for example, to related but differing phenotype definitions, lack of a common marker map, or different statistical tests in the primary studies), but for which p-values are available and therefore Fisher's method may be applied. Guerra et al. (1999) discuss application of Fisher's p-value method in the context of genome scans.

There are at present few meta-analytic methods specifically designed for genome scans (see Dempfle and Loesgen (2004), Rao and Province (2001), and Etzel (Chapter 2) for recent overviews). These are generally based on obtaining a meta-analysis test statistic value in each of k_C user-defined segments per chromosome (C) and then testing for significance within each segment. Standard guidelines for genome-wide significance levels are provided by Lander and Kruglyak (1995), but it is unclear how they may apply to meta-analysis, where less stringent significance levels may be appropriate (Lewis, Chapter 3).

The chapters in Part I of this book describe methods for combining results of genome-wide linkage scans. In Chapter 2, Etzel and Costello review several methods for meta-analysis of genome-wide scans and provide practical advice for a variety of typical situations. Lewis (Chapter 3) also gives an overview of the problem and details the widely used GSMA method. van Houwelingen and Lebrec (Chapter 4) explore the issue of heterogeneity in the meta-analysis of quantitative trait linkage studies, giving examples and adapting classical parameter estimate combination methods to the QTL mapping scenario. Chapter 5 by Zhang et al. places the problem of combining QTL linkage scans in an empirical Bayes framework.

1.5.2 Combining GWAS

Although GWAS analysis constitutes a very new field, there have already been a few meta-analyses of GWAS reported in the literature. Prokopenko et al. (2008) consider the leading association signals in 10 GWAS that include information on fasting glucose levels. They base their meta-analysis on a weighted combination of z-statistics derived from individual study p-values. Zeggini et al. (2008) carry out a meta-analysis of three Type 2 diabetes GWAS, identifying additional susceptibility loci. In the primary analysis, they carry out a fixed-effects combination of odds ratios. A secondary analysis is based on weighted z-statistics. Taking a different tack, Ferreira et al. (2008) carry out a collaborative study of GWAS (mega-analysis) examining bipolar disorder.

There is a need for methods to achieve comparability between individual studies to ensure reliability of conclusions based on combined results. The multiple hypothesis testing issue also seems likely to become more problematic. Although combining GWAS is not explicitly covered in this book, it may be possible to adapt some of the methods for combining genome-wide linkage scans to the GWAS setting.

1.5.3 Combining gene expression data

The widespread practice of making microarray data publicly available has fueled an explosion of data combination efforts, spanning all levels of the data acquisition and analysis pipeline. A number of sophisticated techniques for combination of gene expression profiles or signatures are already in use (see e.g., Wirapati et al., 2009, 2008), and others are in development. A general review containing guidelines for meta-analysis of microarray experiments is given in Ramasamy et al. (2008).

Part II of this book contains chapters that review methods and explore new approaches for combining gene expression data. In Chapter 6, Erickson et al. consider inference based on composite hypothesis testing and Bayesian posterior probabilities. Although this methodology is widely applicable in a variety of high-dimensional biological contexts, it is illustrated using gene expression data and is therefore included in this part of the book. In Chapter 7, Lee, Cho, and O'Connell motivate and describe the Local Pooled Error test and extensions as methods for assessing differential expression in small sample studies, while Chapter 8 (Kooperberg et al.) carry out a comparison of several such methods in small experiments. Both of these chapters concern the problem of combination of information across genes within and between microarrays to make microarray-based inference more reliable, particularly in small studies.

In Chapter 9, Goldstein et al. compare data pooling (mega-analysis) with meta-analysis for a microarray experiment replicated under homogeneous conditions. Morris et al. (Chapter 10) propose a new method for combining data from different Affymetrix GeneChips that redefines probe sets based on probe sequence information. The aim of this approach is to make the raw data more comparable across studies and thereby allow more powerful combination, at the level of data rather than results. Chapter 11 by Shaw provides a detailed introduction to the Gene Ontology (GO; http://www.geneontology.org), along with a practical application combining information at the other end of the spectrum, from gene lists obtained from different microarray experiments.

1.5.4 Combining different data types

Many different types of high-throughput biomolecular data are now routinely generated. In addition to the genotype and gene expression data analyzed in Parts I and II, there are now assays for protein-DNA binding, protein-protein interactions, copy number variation (CNV), and sequence determination. It is clear that the standard techniques used in meta-analysis cannot be straightforwardly applied to the problem of combining these diverse, heterogeneous data.

The chapters in Part III address combining different data types in a wide variety of contexts. Ghosh, Rhodes, and Chinnaiyan (Chapter 12) provide a timely review of combining results from genomic studies in humans comprising a diversity of data types. Other chapters examine methods for combining together different data types to address specific problems.

Genotype and gene expression data

The problem of combining genotype and expression data in an analysis is still in its infancy, with new methods being developed at a rapid pace. Since there is as yet no consensus as to what might be "standard" approaches, we do not present an overview of methods here. Instead, the chapters in this volume devoted to these data types focus on specific problems whose solutions may find additional applications. A recent review of some of the earlier approaches may be found in Nica and Dermitzakis (2008). Although not included in this book, GWAS incorporating genotype and gene expression data are starting to become more prominent as well (Schadt et al., 2008). There is also some recent work on network reconstruction using genotype and gene expression data (Zhu et al., 2007). As more GWAS are being carried out, this area is ripe for further development.

In this book, Kendziorski and Chen (Chapter 13) consider a variety of combined analyses of genotype and gene expression data in their study of expression trait loci (ETL) linkage mapping, which involves searching for loci linked to a gene expression phenotype based on microarray data. Christian and Guerra (Chapter 14) describe methods for adding GO data into this mix.

Protein interaction data

DNA functions, such as replication, structure maintenance, repair, and regulation of gene expression, depend on interactions with proteins. One important type of protein is a *transcription factor*. Transcription factors bind to specific DNA sequences to regulate expression. Just as expression can be measured for multiple genes simultaneously, there are technologies for simultaneous profiling of multiple transcription factors.

An area of great interest is the elucidation of the network of interactions between genes and transcription factors. Such networks can be used for hypothesis generation and inference regarding the biological implications of the network. In Chapter 15, Sun and Zhao propose a Bayesian framework for transcriptional regulatory network inference based on the combined analysis of gene expression data and protein-DNA interaction data.

Proteins can also interact with other proteins. These protein-protein interactions are also necessary for a variety of biological functions. Many technologies exist for identifying protein-protein interactions. These vary not only in their approach, but also in their reliability. Sun et al. (Chapter 16) formulate a model to examine fundamental yet crucial aspects of protein interactions: reliability and confidence, based on heterogeneous data.

Sequence data

Sequence data have been used for inferring the phylogeny, or evolutionary history, of a set of organisms. In Chapter 17, Nakhleh, Ruths, and Innan describe different theoretical tree and network models for phylogenetic inference, in which the sequence level data or the phylogenetic trees can be combined. Development of computational tools and statistical methods based on this theory is an active area of research.

Sequence data have until recently only been available on a small scale. With the increasing use of high throughput sequencing machines, sequence data will become more commonly available. The methodological and computational needs for sequence data analysis and integration with other data types will continue to grow.

Other data types

There are other data types for which combination methods are still in early development and are therefore not included in this book. One example is combination of structural variation data, such as copy number variation (CNV) and genotype or gene expression data. CNVs have been shown to be associated with additional risk in several diseases (see Ionita-Laza et al. 2009 for a review). The main technologies used to measure CNV are array comparative genomic hybridization (aCGH) and SNP genotyping arrays, which have the capacity to obtain both genotype and CNV information in a single assay. The most work done in this area so far is in joint analysis of CNV and gene expression (see e.g., Linn et al. (2003); Stranger et al. (2007)). Software facilitating integrated analysis of SNP and CNV data, as might be obtained in a GWAS, is also now available (see e.g., Korn et al. (2008)).

It is also becoming possible to combine genotype data from a GWAS with pathway analysis, as has been done for gene expression data. A promising idea to move beyond single SNP analysis in GWAS is to extend the gene expression data technique of gene set enrichment analysis (GSEA) (Subramanian et al., 2005; Mootha et al., 2003) to GWAS. There is a small but growing literature of approaches based on this concept (Wang et al., 2007; Holden et al., 2008; Peng et al., 2008).

As newer technologies and even more data become available, the need for combining information across high throughput genome-scale investigations will increase. The ideas presented in this book span a wide range of situations, and can provide a basis on which to address the challenges to come in the future.

Part I. Similar Data Types I: Genotype Data

CHAPTER 2

Combining information across genome-wide linkage scans

Carol J. Etzel and Tracy J. Costello

2.1 Introduction

With the formation of international consortia to investigate complex disorders and a variety of cancers, meta-analysis has become a valuable tool to combine linkage results and narrow chromosomal regions of interest. However, between-study heterogeneity, which may include different marker maps, marker informativity, sample sizes, phenotype definition, ascertainment schemes, and statistical tests for linkage, can be problematic in meta-analysis. The presumed etiology of a complex disease is a combination of effects from multiple genes and the environment.

The possibility of identifying some of these genes, which most likely have small effects, from a single study using traditional linkage analysis methods, is small. Instead, pooling raw data across independent studies (mega-analysis) or pooling linkage results across independent studies (meta-analysis) may be the best means to identify such genes. Since current technology allows investigators to perform scans for linkage at the genome-wide level, single marker testing is no longer the standard. Here, we review recent applications and extensions of meta-analytic methods for combining information across independent genome-wide linkage scans. We also provide practical strategies for choosing a method appropriate to the design and available data of a given meta-analysis study.

2.2 Meta-analysis of genome-wide linkage scans

In this section, we review meta-analytic methods that have been applied in genome-wide linkage scans. This coverage is not meant to be exhaustive: we focus on methods whose power and Type I error have been discussed in the published literature, or that are widely used due to their ease of application.

2.2.1 Meta-analytic methods based on p-values and tests of significance

One class of methods is based on combining (transformed) p-values of k independent studies (see Section 1.2.4), each of which has m markers. We let p_{st}, $s = 1, \ldots, k$, $t = 1, \ldots, m$, denote the p-value that provides evidence for linkage at marker t in study s. There are several options for combining these values to arrive at a summary p-value. Fisher's method (Fisher, 1932) has been widely used in genetic linkage studies of single markers, and many extensions have been developed for meta-analyses of genome-wide scans. The Fisher statistic summarizes p-values across studies at locus t as

$$X_t^2 = -2 \sum_{s=1}^{k} \log(p_{st}). \tag{2.1}$$

Guerra et al. (1999) evaluate power and Type I error of Fisher's method in a simulated genome-wide linkage scan, using a per marker α level of 0.1% to account for multiple hypothesis testing. They conclude that although Fisher's method is applicable to genome-wide scans, its power to detect linkage is no greater than that achieved by pooling raw data, at least for the relatively homogeneous data sets generated for the study.

One problem with using Fisher's method is that it is not typically the case that all studies use the exact same marker map. In addition, when relying on published data, it is not guaranteed that results on all markers tested in the studies are provided. Instead, reduced information consisting of local minimum p-values may be all that is available. Therefore, direct application of Fisher's method may not be feasible.

Allison and Heo (1998) combine results from five published studies investigating linkage to body mass index within the human *OB* gene region. The primary studies use different markers as well as different tests for linkage. The combination technique involves obtaining from each study a single p-value within the *OB* region, as opposed to p-values at the same marker locus across studies (which would have been impossible). Fisher's method was then used to combine these p-values across the five studies. This approach illustrates that it is possible to conduct a meta-analysis of multiple linkage studies despite having to work around what they describe as "worst case" conditions.

Badner and Gershon (2002b) introduce the Multiple Scan Probability (MSP) method, a modification of Fisher's method so that meta-analysis can be performed for genomic regions instead of one marker at a time. MSP extends Fisher's method as follows. Let p_{st}^* be defined as the minimum of the observed p-values p_{st} over a specified linkage region t for study s, corrected for the size of the linkage region. The MSP statistic is computed as in Equation 2.1, with p_{st}^* substituting for p_{st}. The correction factor modifying p_{st} is based on the Feingold et al. (1993) estimate of the probability of a p-value being observed in a specified region size, namely

$$p_{st}^* = C p_{st} + 2\lambda G \, \Phi^{-1}\left(p_{st}\right) \phi\left(\Phi^{-1}\left(p_{st}\right)\right) V\left[\Phi^{-1}(p_{st})\sqrt{4\lambda\Delta}\right], \tag{2.2}$$

where C is the number of chromosomes, λ is the rate of crossovers per Morgan (which varies based on the linkage method employed and family structure), G is the size of region t in Morgans, $\Phi^{-1}(\cdot)$ is the inverse of the standard normal distribution function, $\phi(\cdot)$ is the standard normal density function, Δ is the average distance in Morgans between adjacent markers, and the function V is a discreteness correction factor for Δ. Feingold et al. (1993) show that $V(x) \approx \exp(-0.583x)$ for $x < 2$. Under certain conditions, they also show that equation (2.2) is equivalent to the Lander and Kruglyak (1995) p-value correction factor. Badner and Gershon (2002b) show via simulation that the Type I error rate for this modification is at least as low as for any single genome scan study and that the power to detect linkage using this method is equivalent to that of pooling raw data. This method has been applied to studies involving autism (Badner and Gershon, 2002b) and bipolar disorder and schizophrenia (Badner and Gershon, 2002a).

Another problem with applying Fisher's method to genome-wide scans is that many widely used linkage tests truncate negative evidence (i.e., evidence against linkage) into the single lod score value of 0, thus potentially biasing the results given that a lod score of 0 is expected to occur at approximately 50% of the marker loci over the entire genome. Citing the one-to-one correspondence between lod scores and p-values (Ott, 1999)

$$p_{st} = 1 - \Phi \left[\text{sign}(\text{lod}_{st}) \sqrt{2 \ln(10) |\text{lod}_{st}|} \right], \tag{2.3}$$

where $\Phi(\cdot)$ is the standard normal distribution function, Province (2001) recommends that lod scores equal to 0 should be assigned a p-value equal to $1/2\ln(2) \approx 0.72$, instead of either 0.50, as given by equation (2.3), or 1.0, as suggested by maximum-likelihood theory. By doing so, the resulting Fisher test statistic obtained using p-values extracted from published or derived lod scores would roughly follow the assumed χ^2_{2k} distribution under the null of no linkage. This extension has been applied to genome scan studies carried out under the National Heart, Lung and Blood Institute (NHLBI) Family Blood Pressure Program to identify obesity-related genes (Wu et al., 2002), hypertension-related genes (Province et al., 2003), and diabetes (An et al., 2005).

As noted in Section 1.2.4, transformations of p-values other than that of Fisher have also been proposed. The most prominently used is the inverse normal transformation (Stouffer et al., 1949). As noted by Rice (1990), Fisher's method has greater sensitivity to data refuting a common null compared with data that support it, due to the asymmetrical transformation of the p-values. With this method, the p-values, p_{st}, are transformed into a standard normal variate, $z_{st} = \Phi^{-1}(p_{st})$. The individual study z-values may also be weighted. This can be advantageous, since weighting provides a flexible, quantitative way to account for heterogeneity in the (statistical) information that each study supplies. In practice weighting is often based on sample size, since individual study variance estimates are usually unavailable (see Section 1.2.3). A weighted average of the z-values at marker/region t can be calculated as:

$$z_t = \frac{\sum_{s=1}^{k} N_s z_{st}}{\sum_{s=1}^{k} N_s},$$

where N_s represents the sample size (number of pedigrees, number of sib pairs, etc.) for study s. Under the omnibus null hypothesis of no linkage, $z_t/\sqrt{\mathrm{Var}(z_t)}$ follows a standard normal distribution where

$$\mathrm{Var}(z_t) = \frac{\sum_{s=1}^{k} N_s^2}{\left(\sum_{s=1}^{k} N_s\right)^2}.$$

There are also meta-analytic methods for genome scans that use p-values or other outcomes of significance tests but that are not explicit combinations of these quantities. Perhaps the most widely used of these is the Genome Search Meta-Analysis (GSMA; Wise et al., 1999). GSMA is based on a nonparametric ranking of p-values or lod scores across specified genetic regions (or bins). Babron et al. (2003) updated the GMSA method in two ways: replacing individual bin ranks with the average rank of each bin and the ranks of its two flanking bins, and introducing a weighting scheme to account for different information content across studies. Zintzaras and Ioannidis (2005b) introduced a formal heterogeneity analysis for the GSMA method. This method is implemented in the software program HEGESMA (Zintzaras and Ioannidis, 2005a). The GSMA method and extensions are described in great detail by Lewis (Chapter 3).

2.2.2 Meta-analytic methods based on effect sizes

A meta-analysis based on combining results of significance tests can be limited or misleading, especially in cases where the concordance or discordance of significant linkage between two studies may not reflect the existence of true linkage, but rather may be based on the amount of heterogeneity between the studies. Combining effect sizes may provide a more suitable approach, due to the ability to account for between-study heterogeneity. In addition to the methods described here, see van Houwelingen and Lebrec (Chapter 4) for a detailed treatment of heterogeneity in the meta-analysis of quantitative trait loci (QTL) linkage studies.

Loesgen et al. (2001) develop a meta-analytic test statistic that is a weighted average of individual score statistics, $Z_s t$:

$$Z_{MA_t} = \frac{\sum_{s=1}^{k} w_{st} Z_{st}}{\sqrt{\sum_{s=1}^{k} w_{st}^2}}, \tag{2.4}$$

where Z_{st} is the nonparametric lod (NPL) score statistic (Kruglyak et al., 1996) and w_{st} is the assigned weight from study s at position t. They propose several weighting schemes such as sample size, information content and an exponential function based on marker distance. Dempfle and Loesgen (2004) use simulation to compare power and Type I error rates of this method to those of Fisher's method, the GSMA, and other p-value based meta-analytic methods. This study showed that weighted effect size meta-analysis is more powerful for detecting linkage than the p-value methods, with only nominal increases in false positive rates. Further, the method based on effect sizes is more robust and consistent across the simulation parameters compared to the p-value based methods.

Etzel and Guerra (2002) present a meta-analysis technique that combines Haseman and Elston (1972) test statistics across studies that have distinct marker maps. This method produces a combined estimate $\tilde{\beta}$ at any locus l along the genome based on Haseman-Elston slope estimates $\hat{\beta}_{st}$ and corresponding variance estimates S_{st}^2 from each study s at each marker t.

The method is carried out as follows:

1. Define $\{L_q, q = 1, \ldots, l\}$ as the set of analysis points such that L_1 and L_l are at each endpoint of a chromosome segment and the distance between any two adjacent points L_i and L_{i+1} is constant and equal to L/l, where L is the segment length. At each analysis point L_q, calculate the statistics $\hat{\beta}_{stq}$ and S_{stq}^2, utilizing markers within a window of D centimorgans (cM) of L_q:

$$\hat{\beta}_{stq} = \frac{\hat{\beta}_{st}}{[1 - 2\theta_{stq}]^2}, \text{ and } S_{stq} = \frac{S_{st}^2}{[1 - 2\theta_{stq}]^4},$$

where θ_{stq} is the recombination fraction between marker t and analysis point L_q in study s.

2. Next, calculate the weighted least-squares estimate $\tilde{\beta}_q$ at L_q:

$$\tilde{\beta}_q = \frac{\sum_{s=1}^{k} \sum_{t=1}^{n_{sq}} w_{st} \hat{\beta}_{stq}}{\sum_{s=1}^{k} \sum_{t=1}^{n_{sq}} w_{st}}, \text{ where } w_{st} = \frac{1}{\sigma_B^2 + S_{stq}^2}.$$

Here, k is the number of studies, n_{sq} is the number of markers within D cM of L_q in study s, and σ_B^2 is the between-study variance. The estimator $\hat{\sigma}_{B_q}^2$ for σ_B^2 at L_q is given by

$$\hat{\sigma}_{B_q}^2 = \frac{1}{\sum_{s=1}^{k} n_{sq} - 1} \sum_{s=1}^{k} \sum_{t=1}^{n_{sq}} \left(\hat{\beta}_{stq} - \bar{\beta}_{..q}\right)^2 - \frac{1}{\sum_{s=1}^{k} n_{sq}} \sum_{s=1}^{k} \sum_{t=1}^{n_{sq}} S_{stq}^2,$$

where $\bar{\beta}_{..q}$ is the average of the $\hat{\beta}_{stq}$ that are within D cM of L_q. The variance of $\tilde{\beta}_q$ is $1/\sum_{s=1}^{k} \sum_{t=1}^{n_{sq}} w_{st}$.

3. Calculate the test statistic at L_q: $t_{q'} = \tilde{\beta}_q / \sqrt{\text{Var}\left(\tilde{\beta}_q\right)}$.

4. The analysis point $L_{q'}$ that has the smallest (large negative value) significant $t_{q'}$ value over the entire chromosomal segment is the point estimate of location of the QTL. An estimate of the genetic variance at location $L_{q'}$ is given by $\hat{\sigma}_g^2 = \tilde{\beta}_{q'}/(-2)$.

Etzel and Guerra (2002) further describe a bootstrapping procedure to obtain critical values or construct confidence intervals for location of the putative QTL and the genetic variance. Through simulation, they show that the empirical power of this procedure remains high even when power at the individual study level was low. Etzel and Costello (2001) use this procedure on the nine data sets from Genetic Analysis Workshop 12 to assess linkage of human immunoglobulin E (IgE), an asthma-related quantitative trait, finding suggestive linkage for two regions on chromosome 4 and one region on chromosome 11.

The method of Loesgen et al. (2001) allows different linkage tests across the studies but assumes that all studies use the same marker map; the method of Etzel and Guerra (2002) allows for differing marker maps but is limited by the requirement that all studies use the same linkage test. Etzel et al. (2005) proposed a meta-analytic procedure that combines these two methods, resulting in a more flexible procedure to combine effect sizes.

This Meta-Analysis for Genome Studies (MAGS) method (Etzel et al., 2005) is based on a weighted average of effect sizes obtained through reported linkage summary statistics. Let m_s denote the number of markers used in study s. For a specified chromosome, let M_{st} denote the tth marker from study s, $s = 1, \ldots, k$, $t = 1, \ldots, m_s$. Define $\{L_q, q = 1, \ldots, l\}$ as the set of analysis points such that the L_q are equally spaced across the chromosome. For each set of M_{st} on a chromosome, let Z_{st} be an associated score statistic. The statistics Z_{st} can be NPL scores, as in Dempfle and Loesgen (2004), or may instead be derived from other linkage statistics, such as a heterogeneity lod (HLOD) score (see Ott, 1999) or even an appropriately transformed p-value (see Appendix A).

For each analysis point L_q, calculate the weighted normal variate:

$$Z_{MA_q} = \frac{\sum_{s=1}^{k} \sum_{t=1}^{m_k} I_{q\{M_{st}\}} w_{stq} Z_{st}}{\sqrt{\sum_{s=1}^{k} \sum_{t=1}^{m_k} I_{q\{M_{st}\}} w_{stq}^2}},$$

where w_{stq} is the weight given to marker M_{st} and $I_{q\{M_{st}\}}$ is the indicator function defined as 1 if marker is within a set distance D cM from L_q and 0 otherwise. The weight w_{stq} for marker M_{st} can be a function of study sample size, information content at that marker, and/or distance (recombination fraction, θ_{stq}) between marker M_{st} and L_q.

The p-value for each analysis location can then be used to determine areas with combined evidence for linkage. If all studies use the same marker map, then these markers can replace the analysis points L_q and the expression for Z_{MA_t} simplifies to the statistic proposed by Dempfle and Loesgen (2004). Etzel et al. (2005) applied this procedure to the simulated data from the Genetic Analysis Workshop 14, correctly identifying the disease loci on chromosomes 1, 3 and 5; however, there was only low evidence of linkage to the disease modifier genes on chromosomes 2 and 10.

2.3 Choice of meta-analysis method

Data can be obtained from published sources, open-source websites or through consortia group agreements. Choice of meta-analytic method may be limited due to the type of information available: complete data on all studies (e.g., through a consortium), data obtained by contacting corresponding authors from published articles, information published in journals or on web sites, or some combination of these. Here, we describe some scenarios that reflect commonly occurring situations in which a meta-analysis would be performed. For each scenario, we give guidelines regarding the type of meta-analytic method to use.

2.3.1 Scenario 1: Raw data available on all studies

This scenario typically arises for members of a consortium in which all data are freely shared at the individual patient level. This situation is ideal, since the researcher has great latitude in data reanalysis (pooled or separately from each study) using a preferred linkage method. Separate study outcomes can then be combined using any of the above-mentioned meta-analysis methods. To fully account for between-study heterogeneity, one should choose a method that allows for such an adjustment (e.g., methods of Dempfle and Loesgen (2004), Etzel et al. (2005), or Zintzaras and Ioannidis (2005b); see also van Houwelingen and Lebrec (Chapter 4)). Even if the marker maps are different among the studies in the consortium, a scheme for marker map alignment can be developed.

In this case, there is even the option to complete a mega-analysis instead of a meta-analysis. Some notable examples of mega-analysis of genome-wide linkage scans are in multiple sclerosis (The Transatlantic Multiple Sclerosis Genetics Cooperative, 2001; GAMES and The Transatlantic Multiple Sclerosis Genetics Cooperative, 2003), celiac disease (Babron et al., 2003), asthma (Iyengar et al., 2001), diabetes (Demenais et al., 2003), obesity-related phenotypes (Heo et al., 2002), and bipolar disorder (McQueen et al., 2005). A master marker map can be established in a marker location database. It may be desirable to impute missing values, as in Heo et al. (2002). The combined data are then analyzed using a linkage method appropriate to the study design.

Mega-analysis can have more power than meta-analysis to detect linkage (Guerra et al., 1999); however, care must be taken to consider the sources of heterogeneity that may be inherent in the different primary studies. Common sources of such heterogeneity include differing study designs (data from extended pedigrees may not combine well with those from sib pairs, discordant pairs or parent-offspring triads) or genetic heterogeneity (different genes acting in different populations) due to varying ethnic/racial groups across study populations. This heterogeneity may adversely confound or overshadow mega-analysis results.

2.3.2 Scenario 2: All studies use similar linkage tests and similar marker maps

This scenario can also arise when the researcher is a member of a data consortium in which members individually analyze their own data using a common agreed-upon linkage method, then share linkage *results* instead of raw data. Likewise, this scenario could occur when the researcher personally contacts authors of published studies to request the complete linkage analysis results for the data upon which the published results are based. Researchers should in this case collect the most detailed information possible: e.g., score statistics or parameter estimates and their standard errors in preference to p-values, marker information content, recruitment criteria and sample schemes. For this scenario, we once again recommend that the researcher choose a meta-analysis method flexible enough to account for between-study heterogeneity: methods of van Houwelingen and Lebrec (Chapter 4) if effect size estimates and standard errors (or lod scores) are given, Dempfle and Loesgen (2004) or Etzel et al. (2005) if score statistics are available, or Zintzaras and Ioannidis (2005b) if only p-values are provided.

2.3.3 Scenario 3: All studies use similar linkage tests but with different marker maps

This scenario is similar to Scenario 2 except for the commonality of the marker maps between the studies. This scenario occurs in the same circumstances as Scenario 2. The added complexity of differing marker maps does not hinder meta-analysis so long as the method is sufficiently flexible in this respect. Once again, we advise that the researcher request the most detailed linkage information possible and carry out meta-analysis based on the effect size method proposed by Etzel et al. (2005) if score statistics are available, or the GSMA modification proposed by Zintzaras and Ioannidis (2005b) if only p-values are provided.

2.3.4 Scenario 4: Summary results from different linkage tests and different marker maps from published data are available from all studies

In this scenario, the meta-analysis is based on summary linkage results (p-values or lod scores), typically those available from published articles with no additional information obtained from study authors. Although the availability of data in this scenario may seem limited, and can also vary greatly depending on the disease of interest and journal of publication, many fruitful meta-analyses have been based on such data (Allison and Heo (1998) for instance). For this case, the GSMA method (Wise et al., 1999; see also Lewis (Chapter 3)) is the best method to employ as long as the necessary data are available. If possible, the researcher could also employ any of the GSMA modifications if sufficient auxiliary information is available. In cases where application of GSMA is not possible (such as the scenario posed by Allison and Heo (1998)), then application of Fisher's method is still viable.

2.4 Discussion

Here, we have reviewed current meta-analytic techniques for the combination of linkage data across genome-wide studies in order to arrive at a consensus for linkage to a complex disease, offering guidelines for their use. However, we caution that meta-analysis is more than a specific data combination method. It must also be kept in mind that the combination method is only one part of a complete meta-analysis. Just as study design and participant recruitment are important at the beginning of the individual linkage studies, those embarking on a meta-analysis should also develop a study design and participant study plan, including a literature review plan in addition to study inclusion/exclusion criteria. The researcher should also take steps to gather as much information pertaining to the primary studies as possible, e.g., contacting publication authors.

Roadblocks to complete the meta-analysis may exist, including differences among the studies with respect to: marker maps/map density, family structure, environmental factors, population substructure, genetic etiology/disease pathways, marker informativity, sample size, ascertainment scheme, phenotype definition, and linkage analysis method. Additional challenges include publication bias and time-lag bias. Although some of these problems can be accommodated by the methods presented here, no single method exists to handle all such problems.

Two topics not addressed here are determination of an appropriate significance level for a meta-analysis of genome-wide scans and the effect of publication bias (i.e., only positive linkage results published). The topic of genome-wide significance levels for individual studies remains controversial; we advise looking to Morton (1955), Lander and Kruglyak (1995), Feingold et al. (1993), Sawcer et al. (1990), Rao (1998), Rao and Gu (2001), and Levinson et al. (2003), as well as to the substantial multiple hypothesis testing literature (e.g., Ge et al., 2003), for insight into the determination of an appropriate significance level.

Publication bias in a meta-analysis is a factor when the results of the study impact the probability that it will be published in the literature. In the event that the literature is biased in publishing statistically significant results in favor of linkage, a meta-analysis study would also be biased toward linkage. A number of methods for identifying and handling publication bias have been proposed in the meta-analysis literature (e.g., Rosenthal (1979); Iyengar and Greenhouse (1988); see also Section 1.2.7 and references therein). Publication bias is generally thought to be less of a problem in genome-wide scans than in studies using only a small number of markers (Badner and Gershon, 2002a; Wise et al., 1999).

Acknowledgments

This work was partially funded by National Cancer Institute K07 CA093592 and R03 CA110936 (CJE) and R25 CA57730 (TJC).

2.5 Appendix A

Example transformations of a linkage summary statistic to a score statistic:

1. Transform HLOD to a chi-square variate: $X_{st} = 4.6 \times HLOD_{st}$.

2. Obtain p-value for each chi-square variate (Faraway, 1993):
   $$p_{st} = 0.5 \left[1 - P^2(\chi_1^2 < X_{st})\right].$$

3. Transform the resulting p-value to a normal variate by the inverse of the normal distribution: $Z_{st} = \Phi^{-1}(p_{st})$.

CHAPTER 3

Genome search meta-analysis (GSMA): a nonparametric method for meta-analysis of genome-wide linkage studies

Cathryn M. Lewis

3.1 Introduction

Genome-wide linkage studies have been extensively used to identify chromosomal regions which may harbor susceptibility genes for complex diseases. Early enthusiasm for such studies has been replaced by the realization that most complex disease genes have only a minor effect on risk, and consequently many linkage studies have low power to detect such genes (Risch and Merikangas, 1996). This phenomenon was well illustrated by a compilation of 101 genome-wide linkage studies in 31 diseases, which found that few studies achieved significant evidence for linkage, and there was little replication within each disease (Altmüller et al., 2001). Replication of linkage is an important concept in genome-wide linkage studies: that two studies produce high (if not significant) lod scores in the same approximate region lends further weight to the results. This *ad hoc* method of comparing results across studies is formalized in meta-analysis, which provides statistical evidence for the co-localization of linkage evidence across studies. Meta-analysis can also provide a solution to the lack of power in individual studies: combining weak evidence of linkage from several studies may show an overall significant effect.

Several methods for combining linkage studies have been proposed (see Section 1.5.1 and Chapter 2 by Etzel and Costello). What is often considered to be the gold standard is a complete analysis of individual genotype data from all contributing studies (mega-analysis; see Section 2.3.1). However, many investigators are either reluctant or prohibited from sharing raw genotype data, particularly if they are restricted by industrial partnerships or governmental data protection regulations. There are also technical problems of pooling different marker maps, and difficulties in finding an analysis method that would be suitable for all studies. Pooling genotypes in short candidate regions has worked well in many collaborative studies (e.g., Demenais et al., 2003; Levinson et al., 2002). Meta-analysis methods used in epidemiological studies are not always directly applicable to genetic linkage studies. Methods that pool effect sizes across studies are of limited use when linkage studies generally only report results as a test statistic or, most frequently a p-value.

Unless testing for linkage at a strong candidate gene, specifying a single location for the analysis may not be optimal. Simulation studies show that maximum lod scores have poor localization, and can arise as far as 30 cM from a susceptibility gene (Cordell, 2001). Assessing evidence across a region therefore improves the power to detect linkage in a meta-analysis.

3.2 GSMA: Genome Search Meta-Analysis method

3.2.1 Statistical method

The Genome Search Meta-Analysis (GSMA) method (Wise et al., 1999) circumvents some of the common problems of performing meta-analysis on genome-wide linkage studies. GSMA is a nonparametric method, with few restrictions or assumptions, so that any genome-wide linkage search can be included, regardless of study design or statistical analysis method.

GSMA can encompass diverse study designs and analysis methods. The linkage evidence may be extracted from any analysis method, for example: multipoint lod scores calculated at each 1 cM, lod scores calculated at each marker genotyped within a given segment (or bin), or parametric lod scores calculated at a series of recombination fractions for each marker (see also Terwilliger and Ott, 1994). For parametric lod scores, linkage is often tested using a series of models with different modes of inheritance or different penetrance/frequency parameters. The evidence for linkage can be assessed across all models analyzed, provided the underlying distribution of lod scores is approximately equal in each model; an assessment of this assumption can be made based on the distribution of lod scores across the genome. Thus, the maximum evidence for linkage within a bin would be the highest lod score calculated, regardless of the model under which it was obtained.

In GSMA, the entire genome is divided into bins of approximately equal width (measured in cM). For the human genome we conventionally use 120 bins of 30 cM length, so that for chromosome 1, the region between 0 and 30 cM is assigned to bin 1.1, between 30-60 cM to bin 1.2, etc.

Let the number of bins be n, and the number of studies be m. For each study, the maximum lod score (or minimum p-value) within each bin is identified, then the bins are ranked, with the most significant result achieving a rank of n, the next highest result a rank of $n-1$, etc. Across studies, the ranks for corresponding bins are summed; the summed rank for a bin forms the test statistic in that bin. A high summed rank implies that the bin has high lod scores in individual studies, and may contain a susceptibility locus. Under the null hypothesis of no linkage, the summed rank for a bin will be the sum of m ranks, each randomly chosen with replacement from $1, 2, \ldots, n$.

Under no linkage, the probability of attaining a summed rank r in a specific bin, from m studies and n bins is:

$$P\left(\sum_{i=1}^{m} X_i = r \right) = \begin{cases} 0 & \text{for} \quad r < m \\ \frac{1}{n^m} \sum_{k=0}^{j} (-1)^k \binom{m}{k} \left(\frac{r-kn-1}{m-1} \right) & \text{for} \quad m \leq r \leq mn \\ 0 & \text{for} \quad r > mn, \end{cases}$$

where X_i is the bin rank in study i and j is the integer part of $(r - m)/n$ (Wise et al., 1999). Hence the probability of obtaining a summed rank of r or greater (i.e. the p-value) in a bin can be calculated.

Significance levels for each bin can be determined from the distribution function of summed ranks (Wise et al., 1999) or by simulation, by permuting the assigned ranks across bin location labels. First, the ranks within each study are randomly reassigned to bins; then, across studies, the summed rank is calculated for each bin. For d permutation replicates, dn summed rank values are obtained, and the p-value for an observed summed rank r_{obs} associated with a given bin is calculated from r_{sim}, the number of simulated bins with summed rank greater than or equal to the observed summed rank. The p-value is then $p_{SR} = (r_{sim} + 1)/(dn + 1)$, where n is the number of simulated bins (North et al., 2003). Calculating critical values by simulation is particularly appropriate when the assigned ranks depart from the integer values $1, 2, \ldots, n$ assumed in the distribution function (Wise et al., 1999), as happens when there are tied ranks or missing values (see Table 3.1).

The bin-wise summed rank p-value p_{SR} assesses the information in multiple bins and should therefore be corrected for multiple testing. With 120 bins, under no linkage, six bins would be expected to attain $p_{SR} < 0.05$, and 1.2 bins to attain $p_{SR} < 0.01$. Following Lander and Kruglyak (1995), we define genome-wide evidence for linkage as that expected to occur by chance once in 20 GSMA studies, and suggestive evidence for linkage as that expected to occur once in a single GSMA study. Using a Bonferroni correction on 120 bins gives $p = 0.00042$ (= 0.05/120) for genome-wide significance within a study, and $p = 0.0083$ (= 1/120) for suggestive evidence of linkage (Levinson et al., 2003).

For a genome-wide assessment of linkage, an ordered rank (OR) p-value (p_{OR}) may be used (Levinson et al., 2003). This p-value can be thought of as evidence determining the extent to which the bins ranked highest on average tend to be ranked higher than the bins ranked highly in most studies. To compute p_{OR}, simulations of the complete GSMA are created to compare the summed rank of the observed k^{th} highest bin with the simulated distribution of summed ranks of the kth highest bin,

Table 3.1 *Common sources of incomplete data in the GSMA, and possible solutions.*

Missing data problem	Possible solutions
Many bins with a maximum lod score of zero	Use tied ranks. For example, 20 bins with a maximum lod score of zero would be assigned ranks 10.5.
Bins with no genotyped markers or no linkage data	Assign the median rank (i.e. $(n+1)/2$ for n bins), or assign a rank which is the weighted average of flanking bins (since multipoint lod scores are correlated in adjacent bins).
Results only reported for regions with the strongest evidence for linkage	Contact study authors for full information, and carry out the study collaboratively. Alternatively, if the observed results fall into b bins, assign these ranks $n, n-1, n-2, ..., n-(b+1)$, and assign all remaining bins the average remaining rank. For many missing bins, or bins missing in several studies, this method is not advisable, as the assumed distribution function for the summed rank no longer provides a good fit.
Different chromosomes have been included (e.g., some studies have not tested the X chromosome)	Analyze all relevant subsets of studies to obtain maximum information, and for each bin/region, report results from the analysis with most complete data. For example, if chromosome X is missing for r studies (out of m), analyze the remaining $m-r$ studies for the whole genome, and report these results from this analysis for chromosome X.
Two-stage genome wide study, with some regions genotyped on additional families	Use only the first stage analyses: the distribution of the maximum lod score per bin depends on the number of families included, and a consistent study design should be used across the genome.
High-density genotyping in previously identified candidate regions	Obtain original lod scores from markers used in the genome search. The maximum evidence for linkage within a bin increases with denser genotyping, thus inflating the evidence for linkage in more densely-genotyped bins.

i.e. compares the "place" of the bins in the full listing of results. Therefore, in a simulation of 5,000 complete GSMAs, the bin with the highest observed summed rank is compared to all 5,000 bins with highest summed rank, and the ordered rank p-value p_{OR} calculated. Similarly, the summed rank of the bin in the kth place is compared to summed ranks of all bins lying in k^{th} place. This test can identify evidence for many bins with increased evidence for linkage, although the evidence for linkage within each bin may be modest. In the study of 20 genome-wide searches for schizophrenia, 12 bins in the weighted analysis had significant summed rank and significant ordered ranks ($p_{SR} < 0.05$, $p_{OR} < 0.05$). Our simulations based on these studies showed that this combination of significant results was not consistent with that occurring by chance (not observed in 1,000 GSMA simulations of an unlinked study). The combination of a significant p_{SR} and p_{OR} is therefore highly predictive of a linkage within a bin, however empirical criteria for linkage for an arbitrary number of studies have not yet been developed (Levinson et al., 2003).

In assessing linkage we recommend proceeding sequentially as follows:

1. A genome-wide significant summed rank p-value ($p_{SR} < 0.05/\#\text{bins}$),
2. Nominal evidence for linkage in both statistics ($p_{SR} < 0.05$, $p_{OR} < 0.05$)
3. Nominal evidence for linkage in the summed rank ($p_{SR} < 0.05$).

No evidence for linkage should be declared where bins do not have a significant summed rank p-value. Within bins with a significant summed rank, a significant ordered rank p-value can be considered to enhance the evidence for linkage. Clearly, if the kth bin has nominal evidence for linkage under both statistics, then any bin with higher summed rank must also be considered significant. By plotting the observed summed ranks by size, with the distribution of ordered ranks, a "scree slope" may be seen where the summed ranks decrease rapidly and the ordered ranks become non-significant (see van Heel et al., 2004, Figure 2).

In regions where the $p_{SR} > 0.05$ but $p_{OR} < 0.05$, one interpretation is that the power to identify linkage in these bins is low, and a larger meta-analysis might increase significance of p_{SR}, while retaining the significance of the ordered rank statistic.

3.2.2 Using published information vs. collaborative study

There are two major ways in which GSMA analysis can be carried out: using published information or collaboratively. A GSMA can be based on published information, for example extracting relevant linkage statistics from graphs and tables. In some cases, investigators may have posted detailed genome-wide results or original genotype data on a website. In papers, genome-wide studies are frequently displayed as line graphs of linkage statistics along each chromosome. This may be used in the GSMA by dividing each chromosome into the required number of equal length bins, and reading off the maximum statistic attained in each bin. Inaccuracies in the method arise from different marker maps used in each study, or different chromosome lengths (so that bins will not be exactly compatible across studies). If marker names are given, bins may be designated more accurately by mapping the bin boundary markers relative to the genotyped markers. In some studies, tables of linkage statistics attained at each marker genotyped are given. These markers may be placed into relevant bins, and the maximum linkage statistic for each bin identified.

A more satisfactory method of performing a meta-analysis study is to form a collaboration of relevant research groups, where there is access to the full information on the location and magnitude of linkage statistic. This strategy should improve the accuracy of the resulting study. In practice, meta-analyses of genetic studies have been widely supported by researchers (e.g., schizophrenia (Lewis et al., 2003), bipolar disorder (Segurado et al., 2003), and inflammatory bowel disease (van Heel et al., 2004)).

3.2.3 Bin width

GSMA is heavily dependent on the chosen bin width. The original description of GSMA listed 120 bins, defined by specific boundary markers (see table at `http://www.kcl.ac.uk/depsta/memoge/gsma/` for full marker-bin information). The exact bin width depends on both chromosome length (to give equal width bins on each chromosome) and marker location. Other studies have chosen different bin widths (see Table 3.2). Although intuition suggests that narrow bins may provide more information (see Figure 3.1), localization through linkage information is broad. Adjacent bins may show evidence for linkage (see, for example, GSMA studies of rheumatoid arthritis (Fisher et al., 2003) and inflammatory bowel disease (van Heel et al., 2004)) and simulation studies have shown that the strongest information for linkage may arise in the bin flanking the true location (Levinson et al., 2003). In a study of age-related macular degeneration (Fisher et al., 2005), the original 120 bins (of length 30 cM) were bisected, and ranks for the resulting 240 bins re-assigned to determine whether more bins would improve localization information or identify novel loci. The results were disappointing, with similar evidence for linkage spreading across several 15cM-width bins, and no novel regions were identified. The relative advantages of narrow or wider bins are listed in Table 3.3.

3.2.4 Weighted analysis

The original formulation of the GSMA assumes that all studies contribute equally. However, a study of 500 affected sibling pairs (ASPs) has higher power to detect a true locus than a study of 100 ASPs. This aspect can be reflected in the meta-analysis by weighting the studies by sample size. Weighting by the square root of the number of genotyped affected individuals has been widely used (see Table 3.2), increasing the power to detect linkage by approximately 7% compared to unweighted analyses in a simulation study based broadly on studies in the schizophrenia GSMA (Levinson et al., 2003). The optimal weighting function is unclear, particularly when some studies have used extended pedigrees and others have used ASPs. The power

Table 3.2 Examples of published GSMA studies (geno: genotyped individuals; aff: affecteds; arp: affected relative pairs; asp: affected sib pairs; Significance – Nom: nominal; Sugg: suggestive; Gen: genome-wide).

Disease/Trait	Publication	# studies	# families	# bins	Weights	# bins with SR Nom./Sugg./Gen.	# bins with $p_{SR} < 0.05$ and $p_{OR} < 0.05$
Multiple sclerosis	Wise, 1999	4	257	120	–	8/2/1	–
Type 2 diabetes	*Demenais, 2003	4	1,127	120	–	6/1/0	–
Schizophrenia	*Lewis, 2003	20	1,208	120	$\sqrt{\#aff}$	12/4/1	12
Bipolar disorder[a]	*Segurado, 2003	18	370	120	$\sqrt{\#aff}$	9/2/0	2
Celiac disease	*Babron, 2003	4	442[b]	115	# ped	5/5/2	–
Rheumatoid arthritis	Fisher, 2003	4	570	120	# asp	10/3/1	–
Coronary heart disease	Chiodini, 2003	4	807	124	$\sqrt{\#asp}$	4/3/1	–
Inflammatory bowel disease	Williams, 2003	5	709	117	–	8/4/1	–
Crohn's disease	Williams, 2003	5	472	117	–	9/4/0	–
Inflammatory bowel disease	*van Heel, 2004	10	1,253	105	$\sqrt{\#arp}$	8/5/1	6
Crohn's disease	*van Heel, 2004	10	711	105	$\sqrt{\#arp}$	10/5/0	8
Ulcerative colitis	*van Heel, 2004	7	314	195	$\sqrt{\#arp}$	5/1/0	0
Hypertension/blood pressure	*Koivukoski, 2004	9	1,992	120	$\sqrt{\#aff}$	9/3/1	2
Psoriasis	†Sagoo, 2004	6	493	110	–	5/2/2	–
Cleft lip/palate	†Marazita, 2004	13	574	120	$\sqrt{\#geno}$	12/3/1	12[c]
Body mass index	*Johnson, 2005	5	505	121	$\sqrt{\#geno}$	–/1/0	–
Age-related macular degeneration	*Fisher, 2005	6	908	120	$\sqrt{\#aff}$	15/2/1	11

* = collaborative study; † = partially collaborative; [a]very narrow phenotype definition; [b]based on fine-scale mapping; [c]maximum number, including candidate region follow-up

Table 3.3 *Comparison of properties affecting choice of bin width.*

Property	Narrower bins (e.g., 120 x 30 cM bins)	Wider bins (e.g., 60 x 60 cM bins)
Bin width	Little variability	Unequal bin widths for different length chromosomes
Correlation in ranks in adjacent bins	Highly correlated, particularly for multipoint linkage analysis. May violate distributional assumptions for test statistic.	Low correlation
Localization	Reasonable, although adjacent bins may be significant	Poor
Power to detect linkage	High, except where maximum lod scores occur in different bins	Lower, except where wider bins substantially increase the study rank in linked regions
Consistency of bin definition across studies	Poor, especially based on published information	More overlap between bins in adjacent studies, even when poorly defined

to detect linkage will depend on the locus effects (mutation frequency, penetrance). Extended pedigrees may have higher power to detect linkage for some loci, while ASPs may be the optimal sampling unit for other genes. Defining a single weighting parameter is therefore somewhat unsatisfactory.

The chosen weighting function can be standardized by its average value for all studies, so that the mean weight is 1. Using a narrow range of weights (e.g., 0.9 – 1.1) will give an analysis that is very close to the unweighted analysis. However, using one study with a very high weight (e.g., four studies with weights 3.0, 0.4, 0.3, 0.3) will give results close to those obtained in this single study. Both these situations should be avoided, and alternative weighting functions may need to be tested.

3.3 Power to detect linkage using GSMA

An extensive simulation study of the GSMA was carried out by Levinson et al. (2003) based on genome scans contributed to the meta-analyses of schizophrenia (Lewis et al., 2003) and bipolar disorder (Segurado et al., 2003). For the simulation, a number of sib pairs with broadly equivalent information to the pedigrees from the original studies were used, with 1,625 ASPs for schizophrenia, 1,017 ASPs for bipolar disorder (narrow phenotype definition), and 501 ASPs for bipolar disorder (very narrow phenotype definition). These three studies therefore give a wide range of study sizes covering those seen in many GSMA studies (Table 3.2). Disease loci were placed halfway between adjacent markers. Full details of other assumptions required in the simulation, including the number of genotyped parents, marker density, and number of loci simulated are given in Levinson et al. (2003). Power in this context is defined as the proportion of linked bins achieving genome-wide suggestive or nominal significance levels.

The schizophrenia study had high power to detect linkage with a locus conferring a sibling relative risk (λ_s) of 1.3 at a significance level of $p < 0.01$. For a significance level of 0.05, a power of at least 70% was attained in the following situations:

- 1,625 ASPs (schizophrenia), for a locus with $\lambda_s = 1.15$,
- 1,017 ASPs (bipolar disorder, narrow phenotype) for a locus with $\lambda_s = 1.3$,
- 501 ASPs (bipolar disorder, very narrow phenotype) for a locus with $\lambda_s = 1.4$.

The power of a study to detect linkage depends on the number of studies m and the number of bins n, in addition to the genetic effect size in each study. The average rank threshold for declaring genome-wide suggestive or nominal linkage changes with the number of studies ($m = 4, 7, 10, 15, 20$) and the number of bins ($n = 60, 120$), as shown in Figure 3.1. We note again here that the (Bonferroni-corrected) thresholds for genome-wide (p_{GW}) and suggestive (p_{SUG}) linkage depend on the number of bins used: $p_{GW} = 0.00042$ and $p_{SUG} = 0.0083$ for 120 bins, and $p_{GW} = 0.00056$ and $p_{SUG} = 0.017$ for 60 bins; nominal evidence for linkage was fixed at $p = 0.05$ throughout.

With 120 bins, an average rank threshold for nominal linkage is 32 for 4 studies, but over 48 for 20 studies – so the average rank is not even within the top third of reported ranks. An average rank of 32 gives nominal evidence for linkage with 4 studies, but provides genome-wide evidence for linkage with 20 studies. For a given study size, relative to 120 bins an analysis with 60 bins requires smaller average ranks for linkage (Figure 3.1). Thus, the evidence must be stronger by pooling smaller correlated bins into wider ones. Provided the maximum lod scores for a locus localize to a narrow region, using the narrower bins yielded more evidence for linkage: with 10 studies, an average rank of ≈ 20 gives genome-wide evidence for linkage if this is obtained using 120 bins, but only nominal significance with 60 bins. Reducing the number of bins could, however, increase the power to detect linkage if the peaks in the lod scores are too widely spread to be contained in a single bin (for example if the locus lies close to a bin boundary), so that the average ranks decrease using fewer bins.

One critical issue is the loss of information arising when the GSMA divides the genome into discrete bins. Simulation studies have compared the power of the GSMA to the power of "mega-analysis," based on pooling the raw genotype data from each study. Dempfle and Loesgen (2004) showed that the power of the GSMA was less than the mega-analysis approaches tested, but they applied the Lander and Kruglyak criteria for genome-wide significance, which is much more stringent than using a Bonferroni multiple testing correction (0.05/#bins). Using the less stringent Bonferroni correction, Levinson et al. (2003) showed that the power of the GSMA to detect linkage was higher than for the analysis of pooled genotypes.

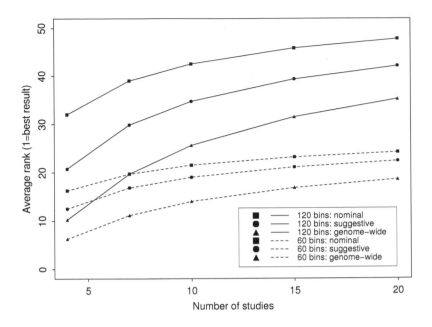

Figure 3.1 *Critical values of average rank required for genome-wide, suggestive, and nominal evidence for linkage, by number of bins.*

3.4 Extensions of GSMA

Many different diseases have been studied using the GSMA method, but little further methodological development has been carried out. Some authors have proposed minor enhancements to the method. For example in their study of celiac disease, Babron et al. (2003) used a summed rank function that was a weighted average of the ranks of a bin and two flanking bins. This extends the potential area in which evidence for linkage can be shown, since high linkage statistics in a flanking bin will be included. However, it will also increase the correlation between summed ranks in adjacent bins. An alternative approach to the problem of maximum lod scores being attained in adjacent bins in different studies is "pooled bins" strategy used in the rheumatoid arthritis study by Fisher et al. (2003). Here, adjacent bins are pooled, and the original analysis of n bins is redone as two analyses of $n/2$ bins each, where bins 1+2, 3+4, ... are pooled in the first analysis, and 2+3, 4+5 ... are pooled in the second analysis. This analysis would be valuable where a true locus lies close to a bin boundary, and the bin-location of maximum linkage evidence is inconsistent across studies. However, as Figure 3.1 shows, reducing the total number of bins from 120 to 60 reduces the power to detect linkage.

In a study of cleft lip/palate, Marazita et al. (2004) use a series of overlapping bins from 0–30 cM, then 10–40 cM, 20–50 cM, etc. and assess the maximum evidence for linkage across each possible bin. This should give better localization information, and may determine whether two linkage peaks exist in one region. However, unresolved problems of multiple testing remain in this approach.

Zintzaras and Ioannidis (2005b) provide a major extension to GSMA in developing methods to test for homogeneity of linkage evidence within a bin. Homogeneity testing (see Section 1.2.2) is a component of standard meta-analysis, but had not previously been implemented in the GSMA. Three (highly correlated) homogeneity statistics are used to assess homogeneity of the study rank statistics. The significance of each statistic is assessed by simulation, randomly reassigning the ranks to bins within each study, and recalculating each statistic. Zintzaras and Ioannidis (2005b) applied the methods to published ranks in GSMA studies of rheumatoid arthritis (Fisher et al., 2003) and schizophrenia (Lewis et al., 2003). They identify several bins in each study that show evidence for high heterogeneity (different evidence for linkage across studies) or low heterogeneity (consistent linkage evidence). Zintzaras and Ioannidis (2005b) acknowledge that the distribution of the homogeneity statistics may depend on the summed rank statistic attained within the bin. The test is therefore carried out under two scenarios: where the observed statistic is compared to all simulated bins, and where the observed statistic is compared only to those simulated bins with similar summed rank values (± 2).

3.5 Limitations of the GSMA

Three classic sources of error in meta-analysis studies are the so-called file drawer problem, the garbage in-garbage out phenomenon, and the comparison of apples to oranges. These are discussed here with their relevance to the GSMA.

3.5.1 File drawer problem

This error arises when unpublished studies are not included in the meta-analysis, as their existence is unknown to the investigators. For linkage studies of candidate regions, a publication bias exists because negative studies are less likely to be published, thus biasing the results of the meta-analysis. For genome-wide studies this is not a major concern: these studies are large and expensive to perform, and therefore should be publishable regardless of the significance of lod scores obtained.

3.5.2 Garbage in, garbage out

Any meta-analysis is reliant on the quality of both the data and the results from the individual studies. Errors due to genotyping problems, inaccurate phenotype definition, incorrect pedigree reconstruction, or poor analysis methods will be carried through to the meta-analysis, and will reduce power to detect evidence for linkage. We assume, verifying where possible, that each study has a high quality of phenotype and genotype data, and that standard quality control checks have been performed (e.g., testing for non-paternity, genotyping error detection). The most challenging problem in GSMA is ensuring a consistent bin definition, particularly when studies have used marker maps that differ in order or distance.

3.5.3 Apples and oranges

Pooling data from many different studies is statistically appealing, but it is only of value if a common effect is occurring across the studies. There are several sources of heterogeneity that can limit the value of a meta-analysis of genetic linkage studies. Potential sources of heterogeneity are population, family sampling units (extended pedigrees or affected sibling pairs), and clinical characteristics (diagnostic criteria, age of diagnosis, severity of disease). Heterogeneity of linkage evidence can be tested using the methods of Zintzaras and Ioannidis (2005b). A subset analysis can also be performed to analyze a more homogeneous set of studies. We have little understanding of how the distribution of genetic variants contributing to complex disease may be affected by these features, although the "common disease, common variant" hypothesis for complex diseases implies that a variant would be present across a wide range of study designs. Some GSMA studies have detected linkage to several genetic regions (schizophrenia, inflammatory bowel disease), suggesting that at least some common disease genes can be detected across diverse studies.

3.6 Disease studies using GSMA

The GSMA has been applied in several studies of complex diseases, some of which are summarized in Table 3.2. Most of these studies have analyzed qualitative diseases, but quantitative traits (blood pressure, body mass index) have also been studied. The average number of linkage studies included per meta-analysis was 7.9 (range 4–20), and the average number of families was 736 (range 257–1992). Of 14 studies, 8 were full collaborations, while others relied at least partially on published information. All studies found at least one suggestive result and in 12 studies, at least one result of genome-wide significance was found.

In the auto-immune diseases, genome-wide significance was found in the HLA region on chromosome 6 (multiple sclerosis (Wise et al., 1999), rheumatoid arthritis (Fisher et al., 2003), psoriasis (Sagoo et al., 2004), inflammatory bowel disease (van Heel et al., 2004)), confirming findings of the original linkage studies. In other studies, a region of genome-wide significance was observed on chromosome 2 for schizophrenia (Lewis et al., 2003), which had not previously been highlighted as a strong candidate region for schizophrenia (O'Donovan et al., 2003). Similarly, regions of genome-wide significance were detected on chromosome 4 for psoriasis (Sagoo et al., 2004), on chromosome 3 for coronary heart disease (Chiodini and

Lewis, 2003), on chromosome 2 for cleft lip/palate (Marazita et al., 2004), on chromosome 3 for hypertension (Koivukoski et al., 2004), and on chromosome 10 for age-related macular degeneration (Fisher et al., 2005). No susceptibility genes have yet been localized in these regions. The identified regions provide strong candidates for follow-up linkage or association studies. Genome-wide significance is an extremely stringent criteria (occurring only once in 20 GSMAs by chance), as illustrated by the results for Crohn's disease in the region of CARD15 on chromosome 16. This region attained a (weighted analysis) p-value of 0.003 (van Heel et al., 2004), despite the presence of this confirmed susceptibility gene. Across the diseases, there is no correlation between the number of bins with nominal or suggestive significance and the number of studies included. Only five studies had used the Ordered Ranks test to assess clustering of linkage results, but the easy availability of this method in the GSMA software package (Pardi et al., 2005) should make this analysis more widely used.

These results show that GSMA can play an important role in synthesizing data across genome-wide linkage studies and directing follow-up studies. The number of significant regions arising from GSMA studies has raised enthusiasm for the potential utility of linkage studies. These studies suggest that provided sample sizes are sufficient, susceptibility genes for complex diseases are detectable using linkage studies.

3.6.1 Comparison of GSMA and the Multiple Scan Probability (MSP) method

The Multiple Scan Probability (MSP) method (Badner and Gershon, 2002b) extends Fisher's p-value method, using the minimum p-values attained in a region with a correction to the overall p-value for the total region length and genotyping density (Section 2.2.1 has further details). A replication analysis excluding the original linkage finding is also recommended.

This method has been applied to autism (Badner and Gershon, 2002b), schizophrenia and bipolar disorder (Badner and Gershon, 2002a). In schizophrenia, significant evidence for linkage was detected on chromosome 8p, 13q and 22q. These regions on chromosome 8p and 22q were also detected in the GSMA study of schizophrenia (Lewis et al., 2003), but the 13q region was absent. Linkage to 13q and 22q were also found in bipolar disorder, neither of which was detected in the GSMA study (Segurado et al., 2003), however for both schizophrenia and bipolar disorder, the studies included in the GSMA and the MSP differed substantially.

The major contrast between the GSMA and MSP methods is in the test statistic. MSP uses a p-value, and therefore retains the magnitude of the significance of the original study. In contrast, the GSMA is a nonparametric rank method, so the maximum contribution from any study is the maximum number of bins (i.e. rank 120 in a study of 120 bins). MSP should therefore have higher power to detect regions which have strong evidence for linkage in some studies, but with genetic heterogeneity present. Interestingly, the analysis of heterogeneity in the schizophrenia GSMA showed significant genetic heterogeneity on chromosome 13q, which may contribute to the different GSMA and MSP meta-analysis results in this region (Zintzaras and Ioannidis, 2005b). MSP would have lower power to detect regions where linkage evidence is moderate in all studies, as this would not trigger the investigation of a region.

3.7 GSMA software

Software to perform GSMA on genome-wide linkage studies is available at `http://www.kcl.ac.uk/depsta/memoge/gsma/` (Pardi et al., 2005). This program is written in C++ and available for Windows, Mac, and Unix/Linux platforms. The input is simply a table of maximum linkage statistics for each bin in each study. The program allows for an arbitrary number of bins and studies. Missing values in bins are permitted, and are replaced with the median linkage statistic for that study. Since by convention larger statistic values are considered stronger evidence for linkage, the entry values for studies reporting p-values should be $1 - p$ to ensure correct ranking of results. The program calculates the summed rank, then determines the summed rank and ordered rank p-values (p_{SR}, p_{OR}) by simulation. The program is rapid, completing 10,000 simulations in under 3 seconds on a typical desktop PC. Both unweighted and (user-defined) weighted analyses are performed. Output consists of three results files: (1) results for the most significant bins only, (2) a full genome listing of bin, summed rank, p_{SR}, p_{OR} (weighted and unweighted analyses), and (3) ranks assigned to each study, for data checking.

3.8 Conclusions

Millions of dollars have been spent on individual linkage studies of complex genetic disorders, but the results have been overwhelmingly disappointing. In hindsight, many of these studies are under-powered to detect linkage to genes that confer only a modest increase in risk for a complex disease. However, the utility of linkage studies has been demonstrated by the localization of a few genes (e.g., CARD15 in inflammatory bowel disease, NRG1 in schizophrenia, CAPN10 in Type 2 diabetes) following fine-mapping of regions detected in linkage analysis. Linkage studies still have an important role in localizing disease genes: genotyping of many large cohorts is in progress, and linkage studies are still widely published. Meta-analysis of linkage studies is therefore a timely approach. It provides a rapid and cost-effective method to ensure that maximum information is extracted from the many linkage studies already performed. Chromosomal regions identified through meta-analysis of linkage studies can be used to prioritize future gene localization studies, whether these are based on fine-scale linkage, on association studies of candidate genes, or on follow-up of genome-wide association studies (GWAS).

Acknowledgments

I thank Douglas Levinson, Sheila Fisher and Lesley Wise for invaluable discussions on the GSMA and meta-analysis in genetics, Fabio Pardi for developing the GSMA software, and all researchers of genome-wide linkage studies for generously contributing their data to meta-analysis studies.

CHAPTER 4

Heterogeneity in meta-analysis of quantitative trait linkage studies

Hans C. van Houwelingen and Jérémie J. P. Lebrec

4.1 Introduction

In complex diseases, where many genes might be causally involved, individual loci influencing a quantitative trait are most likely to explain only a small proportion of the total variance. Consequently, there is a huge problem of lack of statistical power to detect such effects. Most linkage studies published to date consist only of a few hundred pedigrees, with only a small number of individuals each and, therefore, have little power to detect linkage of any but the "largest" quantitative trait loci (QTL). In order to enhance power, it is now common practice to retrospectively pool evidence for linkage from several different studies. However, in combining information from different studies, one should be aware of possible heterogeneity between studies. The aim of this chapter is to present statistical models for describing this heterogeneity, along with approaches to analyze heterogeneous data.

We distinguish two types of heterogeneity: locus and size heterogeneity. Individual study populations often have different genetic backgrounds, so that a locus affecting the trait of interest in one population might have no effect in another one; we refer to this type as *locus heterogeneity*. In other instances, the same locus may influence the trait in all populations, but there may be reason to believe that the size of the effect will vary across studies. For instance, the frequency of the causal allele may be much smaller in some populations or it may interact with other loci, or with environment and other risk factors. We refer to this type as *size heterogeneity*.

Besides these biological sources of heterogeneity, some common logistical sources of variation often arise: typically, genotyping will have been carried out on different marker maps (and even when identical markers are used, their allele frequencies may vary across populations), and families may have been sampled according to different schemes. More simply, the phenotypes measured may vary in their method of collection from study to study.

When raw data are available, one obvious way to combine evidence from several studies is to pool the data and proceed with an overall (mega-)analysis. In the case of linkage studies with different marker maps, the data manipulations involved are very tedious. Moreover, the data sets become unnecessarily large because of the artificially created missing data on markers that are used in other studies. Furthermore, carrying out standard analyses on such large data files usually requires uncommon computing capacities. Therefore, we advocate the meta-analytic approach that uses summary information from each study as the starting point for further analysis. A more simple reason for favoring meta-analysis is that (outside of consortium agreements) researchers usually are unable to access the raw data and must instead be content with individual test statistics or p-values, and, at best, with parameter estimates.

Most meta-analysis methods for linkage studies are in the spirit of the classical meta-analysis. An interesting, widely applicable, alternative is provided by rank-based methods such as the GSMA (Wise et al., 1999; see also Lewis (Chapter 3)). These might be sub-optimal compared to approaches based on the combining estimates of a common linkage parameter, but are much more robust because of the built-in genomic control. There are also associated methods that assess heterogeneity (Zintzaras and Ioannidis, 2005b; see also Section 3.4). Interestingly, Gu et al. (1998) explicitly adjust for the (study-specific) marker to locus distance and allow for heterogeneity across studies by means of a random effects model (see also Zhang et al., Chapter 5). Unfortunately, they do not seem to efficiently take into account the within-study dependence structure between markers.

For meta-analysis of linkage studies by combining parameter estimates (see Section 1.5.1), the sufficient statistics consist of a measure of effect on a common grid of putative locations and its associated standard error. In the case of quantitative traits, a natural estimate of common linkage effect is the proportion of total variance explained by a putative location.

We first describe the meta-analytic tools, assuming that QTL effect estimates and standard errors are available for all studies on a *common grid* of locations. In Section 4.2 the traditional meta-analytic approach in the context of linkage is reviewed, including how to test and allow for *size heterogeneity*. We also introduce a simple finite mixture model to account for potential *locus heterogeneity*. A complication that arises in both approaches for heterogeneous data is that variance components are by definition nonnegative. We discuss the consequences of this for estimation and testing. In Section 4.3, we quickly review the methods which should be used for the analysis of individual studies in order to yield the relevant statistics required for the meta-analysis we advocate. All methods are illustrated in Section 4.4 with four genome-wide scans for lipid levels.

4.2 The classical meta-analytic method

In this section, we recall briefly how meta-analysis is classically carried out and introduce some refinement that is specific to the variance component model used in linkage studies. We assume that at a given *common putative position*, each study (indexed by $i = 1, \ldots, k$) provides a consistent estimate $\widehat{\gamma}_i$ of the true QTL effect γ_i of that locus and an associated standard error s_i. The link with the traditional lod score is given by $\mathrm{lod}_i = (\widehat{\gamma}_i{}^2/s_i^2)/(2 \times \ln(10))$. The definition of the variance component and details of its estimation are given in Section 4.3.

4.2.1 Analysis under homogeneity

The simplest approach corresponds to a (normal) fixed effects model meta-analysis (Section 1.2.3), which assumes that the effects γ_i are all equal to a common value γ, with $\widehat{\gamma}_i \sim N(\gamma, s_i^2)$. This is known as the *homogeneity assumption*. In this situation, the corresponding maximum likelihood estimator of γ is given by the weighted average

$$\widehat{\gamma}_{\mathrm{hom}} = \frac{\sum_i \widehat{\gamma}_i/s_i^2}{\sum_i 1/s_i^2} \quad \text{with standard error} \quad SE_{\mathrm{hom}} = \frac{1}{\sqrt{\sum_i 1/s_i^2}}.$$

The null hypothesis of no effect, that is $\gamma = 0$, versus the alternative $\gamma > 0$, can be tested by means of the one-sided statistic

$$\left(z_{\mathrm{hom}}^+\right)^2 = \begin{cases} (\widehat{\gamma}_{\mathrm{hom}}/SE_{\mathrm{hom}})^2 & \text{if } \widehat{\gamma}_{\mathrm{hom}} > 0 \\ 0 & \text{if } \widehat{\gamma}_{\mathrm{hom}} \leq 0, \end{cases}$$

which follows the mixture distribution $\frac{1}{2}\chi_0^2 + \frac{1}{2}\chi_1^2$ under the null hypothesis (χ_0^2 denotes the degenerate density with all mass at 0). The corresponding $\mathrm{lod}_{\mathrm{hom}}$ score can be calculated as $\left(z_{\mathrm{hom}}^+\right)^2/(2 \times \ln(10))$. Observe that we do not truncate the estimated $\widehat{\gamma}_i$ at zero, if negative, because that would complicate the combination considerably. However, truncation is no problem in the final stage.

4.2.2 Test for heterogeneity

Even when the same locus is affecting a trait in different populations, it seems difficult to believe, for reasons given in Section 4.1, that the QTL effects γ_i are all equal. This situation of *size heterogeneity* can be tested:

$$H_0: \gamma_1 = \gamma_2 = \cdots = \gamma_K \equiv \gamma_{\mathrm{hom}}$$
$$H_1: \text{at least one } \gamma_i \text{ is different ;}$$

the null hypothesis of homogeneity H_0 can be tested using the statistic

$$Q = \sum_{i=1}^{k} \frac{(\widehat{\gamma}_i - \widehat{\gamma}_{\mathrm{hom}})^2}{s_i^2},$$

whose approximate null distribution is χ^2_{k-1} (Cochran, 1954a). In practice, any test for homogeneity is likely to have little power because individual studies tend to have low precision. Nonetheless, the test can formally suggest heterogeneity in some instances, as will be seen in Section 4.4. Note that the Q-statistic has an appealing interpretation (at least for researchers with experience in parametric linkage). Indeed, it can be re-written as

$$
\begin{aligned}
Q &= \sum_{i=1}^{k} \frac{\widehat{\gamma}_i^2}{s_i^2} - \frac{\widehat{\gamma}_{\text{hom}}^2}{(\sum_i 1/s_i^2)^{-1}} \\
&= 2 \times \ln 10 \times \left(\sum_{i=1}^{k} \text{lod}_i - \text{lod}_{\text{hom}} \right).
\end{aligned}
$$

In other words, the individual lod scores add up only when the effect is perfectly homogeneous.

4.2.3 Modeling size heterogeneity

The classical way to allow for heterogeneity between studies is to introduce an additional layer to the earlier homogeneous model by assuming that the true study-specific effects γ_i, $i = 1, \ldots, k$, themselves arise from some distribution. The usual model is a normal distribution with common mean γ and between-study variance σ_B^2. This is a normal hierarchical model with a conjugate normal mean, which we refer to as a normal mixture model (or random effects model; see Section 1.2.3), and results in marginal distributions for the observations given by $\widehat{\gamma}_i \sim N(\gamma, s_i^2 + \sigma_B^2)$. If the between-study variance σ_B^2 were known, the estimate of γ would be

$$
\widehat{\gamma}_{\text{het}}(\sigma_B^2) = \frac{\sum_i w_i \widehat{\gamma}_i}{\sum_i w_i}, \quad \text{with standard error} \ \ \text{SE}_{\text{het}} = 1/\sqrt{\sum_i w_i},
$$

with $w_i = 1/(\sigma_B^2 + s_i^2)$. One way to carry out estimation of σ_B^2 is by maximization of the profile log-likelihood pl $(\sigma_B^2) = l(\widehat{\gamma}_{\text{het}}(\sigma_B^2), \sigma_B^2)$.

In the context of linkage, where the actual effects γ_i are themselves standardized variance components, this model only makes sense if the probability $\Phi(-\gamma/\sigma_B)$ of negative values of γ is negligibly small. In practice, this condition is achieved if the coefficient of variation $\sigma_B/\gamma < 1/2$. Similarly, the null hypothesis of no locus effect requires that all γ_i, $i = 1, \ldots, k$, should be equal to 0 with probability 1. Hence, the null hypothesis specifies both $\gamma = 0$ and $\sigma_B^2 = 0$, which is different from the usual situation in meta-analysis of clinical trials. The test for linkage is then given by the corresponding log-likelihood difference

$$
2 \times \left[pl(\widehat{\sigma}_B^2) - l\left(\gamma = 0, \sigma_B^2 = 0 \right) \right],
$$

so that evidence for heterogeneity potentially contributes to the rejection of the null hypothesis of no linkage. The use of the usual mixture $\frac{1}{2}\chi_0^2 + \frac{1}{2}\chi_1^2$ for the null distribution of this non-standard likelihood is anti-conservative; the correct asymptotic distribution is given by a mixture $(\frac{1}{2}-p)\chi_0^2+\frac{1}{2}\chi_1^2+p\chi_2^2$, where p represents the probability that both estimates are positive (Self and Liang, 1987). However, asymptotic results are unlikely to be useful since typically only very few studies are available for pooling. In practice, we use the anti-conservative limits dictated by the $\frac{1}{2}\chi_0^2 + \frac{1}{2}\chi_1^2$ mixture as a screening tool and resort to parametric bootstrapping for a refined estimate of significance level once interesting positions have been identified.

4.2.4 A two-point mixture model for locus heterogeneity

In some cases, the previous model will not be adequate to model differences between studies because heterogeneity is qualitative rather than quantitative; that is, the locus influences the trait in some studies/populations and not at all in others. Such heterogeneity is indicated when the normal mixture model yields a large coefficient of variation σ_B/γ allowing negative γ values under the normal mixture. In analogy to the admixture model, used routinely at the family level in parametric linkage (e.g., Ott (1999); see also Holliday et al. (2005), Papaemmanuil et al. (2008), and Everett et al. (2008) for recent applications), and to what can be done in the variance components setting (Ekstrom and Dalgaard, 2003), one can fit a two-point mixture model at the study level as follows: $\widehat{\gamma}_i|\gamma_i \sim N(\gamma_i, s_i^2)$ with

$$\gamma_i = \begin{cases} \gamma, & \text{with probability } \alpha \\ 0, & \text{with probability } 1 - \alpha, \end{cases}$$

so that the marginal distribution of $\widehat{\gamma}_i$

$$\widehat{\gamma}_i \sim \alpha N(\gamma, s_i^2) + (1 - \alpha)N(0, s_i^2) \,.$$

The basic idea is that only a proportion α of the studies show linkage to the putative locus and γ is the QTL effect among those studies only. Hence, γ is no longer the mean value of the γ_i as in the normal mixture model. Care is therefore needed when comparing the models.

For estimation purposes, this mixture of normal distributions naturally lends itself to the EM algorithm (Dempster et al., 1977). Let $\phi(x; \mu, \sigma^2)$ denote the normal density function with mean μ and variance σ^2. The E (estimation) step at stage $m + 1$ of the iterative procedure consists in calculating the posterior probabilities $\tau_i^{(m+1)}$ that the $\widehat{\gamma}_i$ have arisen from a normal distribution with mean $\gamma^{(m)}$ given the prior mixing proportion $\alpha^{(m)}$, i.e.

$$\tau_i^{(m+1)} = \frac{\alpha^{(m)}\phi(\widehat{\gamma}_i; \gamma^{(m)}, s_i^2)}{\alpha^{(m)}\phi(\widehat{\gamma}_i; \gamma^{(m)}, s_i^2) + (1 - \alpha^{(m)})\phi(\widehat{\gamma}_i; 0, s_i^2)} \,.$$

The M (maximization) step gives the updated parameters $\alpha^{(m+1)}$ and $\gamma^{(m+1)}$ as

$$\alpha^{(m+1)} = \sum_{i=1}^{k} \tau_i^{(m+1)} / k$$

$$\gamma^{(m+1)} = \frac{\sum_{i=1}^{k} \widehat{\gamma}_i \tau_i^{(m+1)} / s_i^2}{\sum_{i=1}^{k} \tau_i^{(m+1)} / s_i^2}.$$

The model parameters α and γ are constrained to lie in $[0, 1]$ and $[0, +\infty)$, respectively. Although the EM estimation procedure described above ensures that $\alpha \in [0, 1]$, the estimate of γ will sometimes be negative. In this case we set $\widehat{\gamma} = 0$ and $\widehat{\alpha} = 0$, too. Under the usual regularity conditions, the corresponding likelihood ratio test would be asymptotically distributed as a $\frac{1}{2}\chi_0^2 + \frac{1}{2}\chi_1^2$ under the null (Self and Liang, 1987). However, here the situation is further complicated by the fact that the model parameters are not identifiable under the null hypothesis: indeed, if $\gamma = 0$, any choice of α will give the same likelihood. One way to tackle this problem is to slightly modify the likelihood as done by Chen et al. (2001) and derive the corresponding simple asymptotics. Again though, since asymptotics are unlikely to be useful when combining small numbers of studies (see Section 4.2), we prefer using parametric bootstrapping techniques to assess significance of the likelihood ratio test.

The models for size heterogeneity and locus heterogeneity could be combined into a single model where either $\gamma = 0$ with probability $1 - \alpha$ or γ follows a normal distribution with probability α.

4.3 Extracting relevant information from individual studies

As stated above, the basic ingredients for a classical meta-analysis of k linkage studies are study specific QTL effect estimates $\widehat{\gamma}_i$ and their associated standard errors s_i on a *common* fine *grid* of genome locations. In this section, we explain how to obtain these estimates in practice and how to adjust for varying information across studies.

4.3.1 General approach

For random samples of the trait values, the variance components method (Almasy and Blangero, 1998; Amos, 1994) is the standard way of testing for linkage to a quantitative trait. Unfortunately, most computer programs implementing the variance components method have placed the emphasis on testing rather than estimation and rarely provide both QTL effect estimates and associated standard errors. In the context of linkage, two exceptions are the MENDEL (Lange et al., 2001) and Mx (Neale et al., 1999) software programs. However, in principle, this is not so much of a problem because asymptotic standard errors s can be obtained provided both the QTL effect estimate $\widehat{\gamma}$ is present (and differs from 0) as well as its statistical significance, using the approximate relation $(\widehat{\gamma}/s)^2 \simeq \text{lod} \times 2 \times \ln(10) \sim \chi_1^2$. At positions where the QTL estimate is 0, one could interpolate values of s at neighboring positions where $\hat{\gamma} \neq 0$. One problem with the variance components method, as far as pooling of estimates is concerned, is that $\hat{\gamma}$ is constrained to remain nonnegative and pooling of several imprecise estimates $\hat{\gamma}_i$ could result in a positively biased estimate of the true QTL effect γ. Whenever possible, we favor adequate regression or score test approaches to linkage in which the slope is equal to $\hat{\gamma}$ and is allowed to be negative (Lebrec et al., 2004). As shown by Putter et al. (2002), such approaches are equivalent to the variance components method.

Most often, study subjects are not obtained through random (population-based) sampling, but are selected based on phenotype values (e.g., affected sib pairs, extremely discordant pairs, etc.). In this case, the variance components method is no longer valid and appropriate methods that take into account the sampling scheme need to be employed. These so-called inverse regression methods, first introduced by Sham and Purcell (2001), have been implemented in `MERLIN-regress` (Sham et al., 2002). A typical output from the software will provide a signed estimate of the QTL effect $\hat{\gamma}$ and associated standard error s at a grid of arbitrarily spaced positions. This software can also be used as an alternative to the variance components modules in case of random samples. Because of its very convenient output, we advocate the use of `MERLIN-regress` when analyzing linkage data whenever suitable. One problem with `MERLIN-regress` is the use of an imputed covariance for IBD sharing; this can lead to bias in estimation especially in genome areas where markers information is very low. One clear indication that the imputed covariance is not a good approximation is when the software produces either QTL estimates larger than 1 with huge associated lod scores or no estimates at all (NA). In practice, marker maps and densities vary widely and it is frequently the case that there will be areas of the genome with scarce information. In this case, we advocate the use of a more reliable IBD covariance matrix calculated by Monte Carlo simulations (Section 4.3.2).

4.3.2 Inverse regression approach for sib pair designs

To show how we adjust for differing marker maps (or different allele frequencies on the same map), we outline the inverse regression approach in the simple, widely used sib pair study design. The trait values for the pair $x = (x_1, x_2)'$ are assumed to have been standardized and to follow the usual additive variance components model; i.e. the vector x is assumed to follow a bivariate normal distribution with mean 0 and covariance matrix Σ (Amos, 1994)

$$\Sigma = \begin{bmatrix} 1 & \rho + \gamma(\pi - \frac{1}{2}) \\ \rho + \gamma(\pi - \frac{1}{2}) & 1 \end{bmatrix},$$

ρ is the marginal sib-sib correlation for the trait of interest and π is the proportion of alleles shared IBD measured exactly at the QTL position; γ therefore represents the proportion of total variance explained by the QTL. An extension of a relation shown in Putter et al. (2003) under complete information gives an approximate regression (valid for small values of γ) between excess IBD sharing and a function of the phenotype trait values. This relation forms the basis of the inverse regression approach:

$$E\left(\widehat{\pi} - \frac{1}{2}|\boldsymbol{x}, \gamma\right) \simeq \gamma \, \text{var}_0(\widehat{\pi}) \, C(\boldsymbol{x}, \rho),$$

where

$$\widehat{\pi} = \frac{1}{2} \times P_0(\pi = \frac{1}{2}|M) + 1 \times P_0(\pi = 1|M)$$

is the usual estimate of IBD sharing given marker data M available, while

$$C(\boldsymbol{x}, \rho) = \left[(1 + \rho^2)x_1 x_2 - \rho(x_1^2 + x_2^2) + \rho(1 - \rho^2)\right]/(1 - \rho^2)^2,$$

sometimes referred to as the optimal Haseman-Elston function. For a sample of N sib pairs, the method of least squares provides an approximately consistent estimate of γ, given by

$$\widehat{\gamma} = \frac{\sum_{j=1}^{N}(\widehat{\pi}_j - \frac{1}{2})C(\boldsymbol{x}_j, \rho)}{\text{var}_0(\widehat{\pi}) \times \sum_{j=1}^{N} C^2(\boldsymbol{x}_j, \rho)},$$

with standard error

$$s = \left(\text{var}_0(\widehat{\pi}) \times \sum_{j=1}^{N} C^2(\boldsymbol{x}_j, \rho)\right)^{-1/2}.$$

Here $\text{var}_0(\widehat{\pi})$ represents the variance of $\widehat{\pi}$ under the null hypothesis, equal to $\frac{1}{8}$ under complete information. Although exact calculation is extremely tedious, it can be closely approximated by simple Monte Carlo simulations. For example, one can use the options `--simulate` and `--save` in MERLIN (Abecasis et al., 2002) to generate a large number of pedigrees with a given structure (sib pairs here), marker characteristics (i.e. allele frequencies and inter-marker distances), and pattern of missing genotypes. Then $\text{var}_0(\widehat{\pi})$ can be accurately approximated by the sample variance of $\widehat{\pi}$. Figure 4.1 shows how widely this measure of marker information may vary within and between studies. It is therefore crucial to appropriately account for this variation when estimating γ, as failure to do so may introduce bias in the QTL estimates. If no such information is available, it is possible in principle to calibrate a linkage scan by comparing mean or median QTL variance components over the whole genome between studies, but in small studies this method might be prone to error.

4.3.3 Retrieving information on the common grid from an individual study

For the meta-analysis we need to define a common grid of locations and obtain QTL estimates on that grid for each study. However, it can happen that in the individual studies, the only data at hand are QTL estimates ($\widehat{\gamma}$) and their standard errors (s) on an original grid of locations which is not the common one we wish to use. Typically, this original grid would be a set of M marker positions. We show how to obtain QTL estimates and associated standard errors on the common grid of locations, if the characteristics of the original map are available, from the IBD distribution under the null hypothesis for the original map.

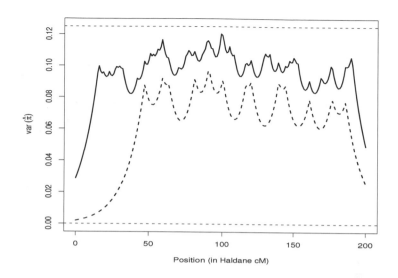

Figure 4.1 *Marker information* $(\mathrm{var}_0(\widehat{\pi}))$ *in the Australian (continuous line) and Dutch (broken line) data sets vs. position (Haldane's cM) – chromosome 6.*

For the sake of simplicity, we stick to sib pair designs as in the previous section. Given the $M \times 1$ vector of original QTL effect estimates $\widehat{\gamma} = (\widehat{\gamma}_t)_{t=1,\ldots,M}$ and associated standard errors $(s_t)_{t=1,\ldots,M}$, the best linear approximation of the QTL effect $\widehat{\gamma}_q$ at an arbitrary position q is given by a weighted least squares estimate

$$\widehat{\gamma}_q = \frac{\omega_q' V^{-1} \widehat{\gamma}}{\omega_q' V^{-1} \omega_q},$$

$$\text{with standard error } s_q = \left(\omega_q' V^{-1} \omega_q\right)^{-1/2}.$$

Here the symbol $'$ denotes the transpose of a vector. The matrix V is proportional to the variance-covariance matrix of the vector $\widehat{\gamma}$ under the null hypothesis of no linkage and is given by

$$V_{kl} = \begin{cases} \mathrm{var}_0(\widehat{\pi}_k)^{-1} & \text{if } k = l \\ \mathrm{Cov}_0(\widehat{\pi}_k, \widehat{\pi}_l) \, (\mathrm{var}_0(\widehat{\pi}_k) \, \mathrm{var}_0(\widehat{\pi}_l))^{-1} & \text{if } k \neq l; \end{cases}$$

furthermore, ω_q is the $M \times 1$ vector whose k^{th} element is given by

$$\omega_{q,k} = \frac{\mathrm{Cov}_0(\widehat{\pi}_k, \widehat{\pi}_q)}{\mathrm{var}_0(\widehat{\pi}_k)}.$$

All the var_0 and Cov_0 terms can in principle be calculated by Monte Carlo simulations, provided the map characteristics and pedigree structure are known.

In the idealized case of a saturated map, i.e., one that would supply perfect IBD knowledge at any location on a chromosome, all var_0 terms are equal to $\frac{1}{8}$ and

$Cov_0(\widehat{\pi}_{t_1}, \widehat{\pi}_{t_2}) = \frac{1}{8}(1 - 2\theta_{t_1,t_2})^2$, where θ_{t_1,t_2} is the recombination fraction between loci at t_1 and t_2 (Risch, 1990). Taking the off-diagonal terms in V to be equal
to 0 (i.e. assuming that markers are not linked), the estimate of QTL effect is that
advocated by Etzel and Guerra (2002) (in the special case that between-study variance $\sigma_B^2 = 0$). In the context of meta-analysis, it is important to properly account for
differences in marker information between studies, unless the marker maps are close
to saturated in all studies. Remarkably, the elements needed to calculate $\widehat{\gamma}_q$ and s_q at
any arbitrary location are just the corresponding estimates at M marker locations and
map characteristics; none of the subject-specific data (traits values, individual IBD
estimates $\widehat{\pi}_i$) are needed.

4.4 Example

We have applied these homogeneity (fixed effects), heterogeneity (random effects),
and two-point mixture methods to four data sets examining lipid levels. These orig-
inate from Australia (aus), the Netherlands (nlj and nlo), and Sweden (swe).
Full results are reported in Heijmans et al. (2005); we select only one endpoint (LDL
cholesterol levels) for illustration purposes. The data available for linkage analysis
consist almost entirely of sib pairs (371, 83, 110 and 36 pairs in the aus, nlj, nlo
and swe data sets, respectively), with the exception of the Australian data set, which
also had one family with three siblings and three families with four siblings. Genotyping was carried out using a common marker map for the nlj, nlo and swe data
sets but with a different denser map for the aus data set. We had access to the raw
data sets and could therefore easily obtain QTL estimates and standard errors on a
common grid of positions.

Prior to linkage analysis (using MERLIN-regress), raw phenotypic data were adjusted for sex and age, within country. The analysis of these data revealed little difference between the three methods described in Section 4.2. This is partly due to the
small sample sizes in the individual data sets, which does not allow heterogeneity between studies to be clearly established. We present the original results for two interesting chromosomes: chromosome 2 (Figure 4.2) and chromosome 13 (Figure 4.3).
In order to make all curves visible, the QTL variance estimates and lod scores of
the "Combined" analyses have been multiplied by 0.95 and 1.05 for the random effects model (labeled "het") and the two-point mixture model (labeled "2-p mixt"),
respectively.

For chromosome 2, the three combined estimates of QTL variance coincide every-
where apart from the 20–60 cM region where the two-point mixture model gives a
higher estimate with corresponding estimate of proportion of study linked $\widehat{\alpha} = 0.75$
(i.e. the nlo data set is not linked) at 32 cM where the maximum lod score is attained. The corresponding combined lod score is roughly the same as the maximum
lod score obtained in the aus data set and therefore there seems to be no gain in
pooling the three linked data sets in this case. On chromosome 13, combining results
leads to a very slight increase in lod score in the region around 20 cM compared to

EXAMPLE   59

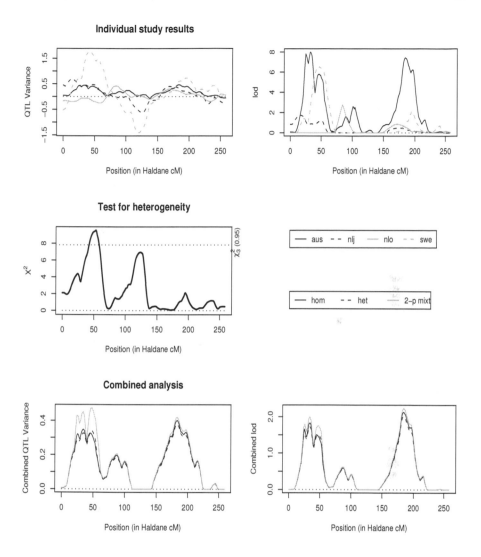

Figure 4.2 *Original data – chromosome 2 – LDL cholesterol level.*

the maximum of the individual lod scores; the three methods give the same score. Note the sudden rise and fall in the estimate of QTL variance $\widehat{\gamma}$ for the two-point mixture at 52 cM; the result corresponds to a decrease in $\widehat{\alpha}$ from 1.0 to 0.36. The fitting algorithm of the two-point mixture actually gave negative values for $\widehat{\gamma}$ to the right of 54 cM, so the estimates were truncated to 0.

Given these unconvincing real-life examples, one can legitimately ask two questions:

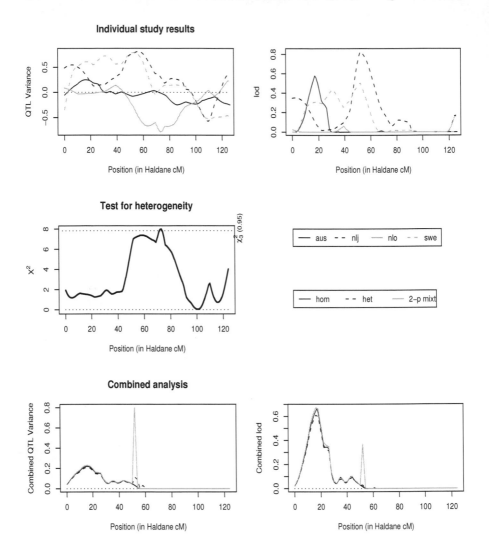

Figure 4.3 *Original data – chromosome 13 – LDL cholesterol level.*

1. In practice, is there any gain in combining estimates at all? That is, can we obtain higher lod scores than the maximum of the individual lod scores?

2. Does allowance for heterogeneity help in enhancing statistical significance? That is, are the lod scores for the random effects model and/or the two-point mixture model ever higher than the lod score of the homogeneity model?

The answer to Question 1 is "Yes" even when individual studies are small, provided that the QTL effects are more or less the same in all studies (i.e. the assumption of ho-

EXAMPLE   61

mogeneity holds). The answer to Question 2 is also "Yes" but only when the sample size in the individual studies are large enough, as we show here by means of a simulated example inspired from the original lipid levels data. We artificially increased the sample size of each of the four data sets by a factor 4 (i.e. the standard errors were divided by 2). The corresponding results are displayed in Figure 4.4 for chromosome 2 and in Figure 4.5 for chromosome 13. In the 20–70 cM region of chromosome 2, studies aus and swe both show clear linkage signals, QTL estimates vary quite widely across studies, now unambiguously shown by the homogeneity test. There is probably both quantitative and qualitative heterogeneity here since study nlo shows no QTL effect at all. As a result, the significant signals observed in the aus and swe studies (maximum lod score $\simeq$ 8) weaken in the homogeneous model (maximum lod score $\simeq$ 7), while both the heterogeneity model and the two-point mixture enhance it further (maximum lod score $\simeq$ 10). Accounting for heterogeneity therefore contributes to the evidence that a linkage effect is present. Similar outputs are displayed for chromosome 13 in Figure 4.5. In the 40–70 cM region, heterogeneity of QTL effects is now clearly qualitative (both nlj and swe have similar QTL effects with corresponding suggestion for linkage) and the homogeneous analysis is dominated by the large aus study with QTL variance estimates close to 0, which entirely obliterates the individual linkage signals of nlj and swe. The two-point mixture works best here in pooling evidence from the two positive studies, enhancing the lod score beyond 4 in a much narrower region (maximum lod score $\simeq 3.5$ in individual studies).

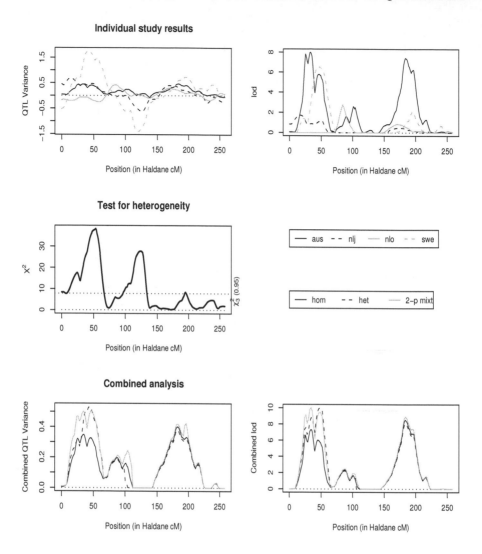

Figure 4.4 *Artificial data – chromosome 2 – LDL cholesterol level.*

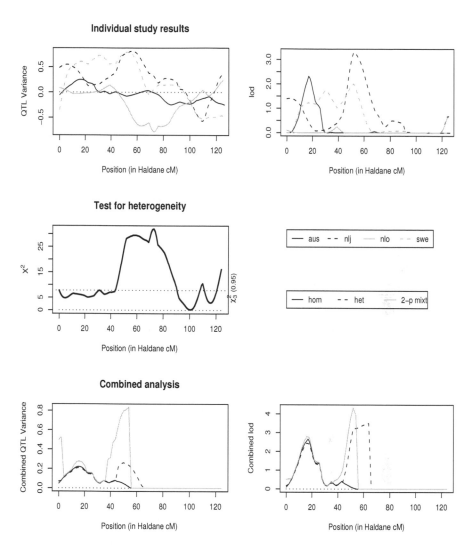

Figure 4.5 *Artificial data – chromosome 13 – LDL cholesterol level.*

4.5 Discussion

We have detailed how classical meta-analytic methods can be adapted to linkage studies provided consistent estimates of QTL effects along with standard errors are available for each study on a common grid of positions. The methods required to obtain such summary statistics are now well developed, with software implementations publicly available for a number of years. We realize, however, that most studies published to date will not have sufficient information to carry out the methods detailed here. Indeed, it is still common practice nowadays in the literature, even for QTL mapping where the effect to be estimated is fairly uncontroversial, to publish statistics conveying statistical significance only (i.e. lod scores) without any indication of the actual effect estimate. This practice substantially hinders powerful combining of the many small linkage studies available in the community. Gu et al. (1998) presented guidelines on how to report linkage studies that would enable future meta-analysis using IBD sharing as a common linkage parameter. Since the analysis tools are available (e.g., `MERLIN-regress`) and disk space plentiful, it should be expected by journals that researchers publish QTL effects and associated standard errors on a grid of locations, at least as supplementary information on a web site.

Given the small individual study sizes typically encountered, any test to detect heterogeneity of QTL effects across studies is bound to suffer from a lack of power. This was reflected in the test for homogeneity of the real lipid levels data as well as in the estimate of the between-study variance component σ_B^2, which only rarely differed from 0 (Heijmans et al., 2005). Another way to test for homogeneity in the random effects model setting is to test whether $\sigma_B^2 = 0$. This approach is known to be asymptotically equivalent to the Q-test presented above (Andersen et al., 1999). Note that the classical (normal) random effects model is probably not the most appropriate in the case of linkage studies. The fact that the QTL effect is a variance component precludes it from being negative (yet which is not impossible under the normal mixture model) and suggests that the random effects γ_i could be more appropriately modeled as arising from a gamma distribution, although estimation is then less straightforward.

The idea of applying finite mixture models to meta-analysis is not new (Böhning et al., 1998), although it does not yet seem to have been widely used for meta-analysis of linkage studies. The mixture model is based on the simple idea that only studies with a positive effect should be able to contribute to the evidence for linkage. Instead of making this choice "by hand," we let the data decide which studies exhibit positive linkage. Note that it is also possible to formally test for locus homogeneity by assessing whether α differs from 1. Ultimately, given a sufficiently large number of studies with acceptable levels of precision, it would be possible to fit a model that adapts to both locus and size heterogeneity by combining the random effects and the two-point mixture models.

Acknowledgments

We wish to thank Bas Heijmans, Marian Beekman, Hein Putter, Nico Lakenberg, Henk Jan van der Wijk, John Whitfield, Danielle Posthuma, Nancy Pedersen, Nick Martin, Dorret Boomsma and Eline Slagboom for allowing us to use their lipid levels data as an example. Gathering these data was a tremendous effort of many individuals, and in particular we would like to acknowledge the work of Kate van Duijn, Eka Suchiman and Dennis Kremer for expert technical assistance; Harry Beeby, Scott

Gordon and Megan Campbell for assistance with cleaning Australian genotype data; Dr. Daniel O'Connor for his effort to complete the Australian genotype data set; and the Mammalian Genotyping Service of the Center for Medical Genetics, Marshfield, WI, USA (director Dr. James Weber) for genotyping part of the Australian twin sample. This work was partially funded by the GENOMEUTWIN project which is supported by the European Union Contract No. QLG2-CT-2002-01254.

CHAPTER 5

An empirical Bayesian framework for QTL genome-wide scans

Kui Zhang, Howard Wiener, T. Mark Beasley, Christopher I. Amos, and David B.

Allison

5.1 Introduction

Genome scan studies for linkage analysis have been widely used to search candidate regions containing quantitative trait loci (QTLs). However, most genome scan studies for QTL mapping are analyzed without formal consideration of information provided by other genome scan studies of the same trait. When multiple genome scan studies of the same trait are available, we may increase the power to detect linkage between markers and QTLs by using information provided from all these studies. Methods that can formally integrate data from multiple genome scan studies are emerging as useful and powerful tools in the field of linkage analysis for QTL mapping.

Marked heterogeneity can exist in multiple genome scan studies and pose daunting challenges in such analysis (see Chapter 4 by van Houwelingen and Lebrec). Different genome scan studies can use different genetic marker loci and marker maps, different statistical methods to test for linkage, and different sampling schemes. Furthermore, the QTL effect can vary across studies because of disparate environmental effects and population substructures. The combination of raw data from all studies with a well-designed pre-analysis procedure would be a preferred approach to overcome such difficulties. However, in many situations this is not feasible because only summary statistics, rather than the raw data, are available.

For these reasons, two closely related but distinct groups of statistical methods can be used to test and locate QTLs by integrating the same type of summary statistics or estimates obtained from multiple genome scans: meta-analysis and empirical Bayesian (EB) methods.

The first group of methods is meta-analysis, which can be viewed as a set of statistical procedures designed to summarize statistics across independent studies that address similar scientific questions. Several meta-analysis methods have been developed to detect linkage between genetic markers and QTLs: Allison and Heo (1998); Etzel and Guerra (2002); Gu et al. (1998); Guerra (2002); Guerra et al. (1999); Hedges and Olkin (1985); Li and Rao (1996); Rice (1997); Wise et al. (1999). For example, Allison and Heo (1998) used Fisher's method (Fisher, 1932; see also Section 1.2.4) to show strong evidence of linkage in obesity gene (OB) regions by combining p-values from five published linkage studies on these regions, illustrating this method's applicability in the presence of marked heterogeneity across studies. However, it is difficult to use this technique to estimate the parameters of interest, such as the location and effect of a QTL, because of the method's nonparametric nature. Several meta-analysis methods can combine parameters of interest across studies. For example, methods have been developed to combine estimates of Haseman-Elston regression slopes and associated variances at marker loci (Haseman and Elston, 1972; Etzel and Guerra, 2002; Gu et al., 1998; Li and Rao, 1996). For a more detailed review of these meta-analysis methods, see Chapter 2.

The second group of methods is based on the EB framework (Beasley et al., 2005; Bonney et al., 1992; Zhang et al., 2006).

The EB approach as applied here makes use of linkage statistics (e.g., Haseman-Elston regression slopes and associated variances at marker loci) obtained from each individual genome scan study. A key feature of the EB method is that linkage parameter estimates of each study are updated by incorporating the linkage statistics from other studies. The final linkage parameter estimates can be used both to detect and to estimate the location of the QTL.

It is worth emphasizing a key difference between the EB methods and the meta-analysis methods. In EB analysis, an individual genome scan of interest is identified as the primary study; the remaining scans are considered as background studies. Theoretically, any individual study could be considered to be the primary study. However, the study of primary interest to an investigator would typically be the study conducted by the investigator; presumably the investigator would be able to obtain further genotypes from the individuals in the primary study for fine mapping, while this type of information would not necessarily be available from the background studies. In the meta-analysis methods, each individual study is equivalent to the others and investigators are typically interested in the overall results.

In this chapter, we describe EB methods for genome-wide scans carried out using sib pairs. We also provide a simulation-based assessment of the performance of the EB methods for detecting linkage between markers and the QTL and mapping the location of the QTL. We conclude by giving some possible EB method extensions.

5.2 Methods

5.2.1 Haseman-Elston regression analysis for a single genome scan

Haseman-Elston regression (Haseman and Elston, 1972) has been widely used to detect linkage between genetic markers and QTLs using sib pairs. For sib pair i, $i = 1, \ldots, n$, denote the trait values by $y_i = (y_{i1}, y_{i2})$, the squared trait difference by $Y_i^D = (y_{i1} - y_{i2})^2$, and the estimated proportion of alleles shared identical by descent (IBD) at a marker locus by π_i. The Haseman-Elston method can be represented by a simple regression of Y_i^D on π_i:

$$Y^D = \beta_0 + \beta\pi + \varepsilon.$$

The regression slope β has expectation $E(\beta) = -2(1 - 2\theta)^2\sigma_g^2$, where θ is the recombination fraction between the marker locus and the QTL, and σ_g^2 is the phenotypic variance explained by the additive effect of this QTL. Thus, the regression slope β is 0 under the null hypothesis of no linkage, and is negative under the alternative hypothesis. Specifically, if there are m markers with slope estimates $\widehat{\beta}_j$ and associated variance estimates $\widehat{S}_j^2$, $j = 1, \ldots, m$, then under the null hypothesis of no linkage at marker j, the t-statistic

$$t_j = -\widehat{\beta}_j / \sqrt{\widehat{S}_j^2}$$

asymptotically follows a standard normal distribution. In the original Haseman-Elston method, the regression slope and its variance are estimated only at the genotyped marker loci with determined genotype. Due to the coarseness of most marker maps in linkage analysis, this method is more suitable for detecting linkage between markers and the QTLs rather than estimating the QTL location and effect.

The Haseman-Elston method assumes the normality of trait values, but for a sufficiently large sample size ($n > 100$ sib pairs) is robust even when this assumption is violated (Allison et al., 2000). However, the original Haseman-Elston regression tends to have low power. Several modified Haseman-Elston regression methods were subsequently developed (Amos, 1994; Drigalenko, 1998; Elston et al., 2000; Feingold, 2002; Sham et al., 2002; Xu et al., 2000). For example, additional power can be acquired by regressing the mean corrected squared sums of trait values $Y_i^S = (y_{i1} - \bar{y} + y_{i2} - \bar{y})$ (Drigalenko, 1998), the mean corrected cross product of trait values $Y_i^P = (y_{i1} - \bar{y})(y_{i2} - \bar{y}) = (Y_i^S - Y_i^D)^2$ (Drigalenko, 1998; Elston et al., 2000), or a weighted combination of Y_i^D and Y_i^S (Xu et al., 2000) on π_i, where $\bar{y}$ is the mean trait value over all sib pairs.

5.2.2 The interval mapping (IM) method

Fulker et al. (1995) developed an interval mapping (IM) method to detect linkage between markers and QTLs and to estimate the QTL location and effect. They first used the estimated proportion of alleles shared IBD at all marker loci on a single chromosome and the genetic distance between these markers to estimate the proportion of

IBD sharing at virtually any location on the chromosome. Haseman-Elston regression can then be carried out for any location, whether genotyped or not. Denote the estimates of the regression slope and its variance at each analysis point q along the chromosome by $\widehat{\beta}_q$ and $\widehat{S}_q^2$, respectively. The null hypothesis of no linkage is tested with the test statistic $\widehat{t}_q = -\widehat{\beta}_q / \sqrt{\widehat{S}_q^2}$, whose asymptotic distribution is also standard normal. The analysis point $\widehat{q}$ that gives the maximum value $\widehat{t}_q$ over all location points examined is taken as the estimate for the QTL location. The point estimate for the variance of the QTL effect, σ_g^2, is given by $\widehat{\sigma}_g^2 = -\widehat{\beta}_{\widehat{q}}/2$.

5.2.3 Empirical Bayesian method with a hierarchical normal model

In Bayesian analysis, the choice of a suitable prior distribution for parameters is sometimes not obvious. However, if data from several independent studies are available, the prior information can be extracted from the data. Such approaches are called empirical Bayes methods (Carlin and Louis, 2000b). These methods are not Bayesian but can be viewed as approximations to a complete hierarchical Bayesian analysis, or as hybrids between classical frequentist methods and fully Bayesian methods. There are both parametric and nonparametric EB methods (Carlin and Louis, 2000a), but even the parametric varieties typically do not depend on strong distributional assumptions (Efron and Morris, 1973).

The empirical Bayes approach as proposed by Efron and Morris (1973, 1975) can be described in terms of a two-level hierarchical normal model. Suppose β is the parameter of interest and there are k populations available to estimate it. Let β_i denote the value of β in population i. At the first level, the maximum likelihood estimators $\widehat{\beta}_i$, $i = 1, \ldots, k$, for β_i can be obtained; it is assumed that $\widehat{\beta}_i | \beta_i$ asymptotically follows a normal distribution, $N(\beta_i, S_i^2)$. At the second level, β_i is specified by a normal model with an r-dimensional predictor x_i, a common regression coefficient μ, and an unknown variance $A \geq 0$; i.e., $\beta_i | \mu \sim N(x_i'\mu, A)$. Using Bayes' rule, it is easy to compute the marginal distribution of $\widehat{\beta}_i$ (given μ and A) and the conditional distribution of β_i (given $\widehat{\beta}_i$, μ, and A):

$$\widehat{\beta}_i | \mu, A \sim N(x_i'\mu, S_i^2 + A), i = 1, \ldots, k \tag{5.1}$$

and

$$\beta_i | \widehat{\beta}_i, \mu, A \sim N((1 - B_i)\widehat{\beta}_i + B_i x_i'\mu, S_i^2(1 - B_i)), \quad i = 1, \ldots, k, \tag{5.2}$$

where $B_i = S_i^2/(S_i^2 + A)$ is an unknown shrinkage factor. Generally, S_i^2 is unknown and is replaced by $\widehat{S}_i^2$, the estimates of variance for the $\widehat{\beta}_i$. A and μ can be estimated by the maximum likelihood methods or by more advanced techniques developed by Tang (2002). Then we can use $\widetilde{\beta}_i = (1 - \widehat{B}_i)\widehat{\beta}_i + \widehat{B}_i x_i'\widehat{\mu}$ and $\widetilde{S}_i^2 = \widehat{S}_i^2(1 - \widehat{B}_i)$ as the final estimators for β_i and S_i^2.

5.2.4 Empirical Bayes application

Empirical Bayes methods have been used in many contexts, including genetic research (Bonney et al., 1992; Li and Rao, 1996; Lockwood et al., 2001; Witte, 1997). We can tailor the general EB procedure for use in linkage analysis. Here, we apply the EB method in the case of k different sib pair genome-wide scans for detecting linkage to the same QTL. We assume here that each study has m markers based on the same marker map. For each marker locus j in each study i, the parameter of interest is the regression coefficient, β_{ij}, which describes the effect of the putative QTL on the phenotype. The expected value of β_{ij} equals $-2(1 - 2\theta_{ij})^2\sigma_{gi}^2$, where θ_{ij} is the recombination fraction between the QTL and the marker j in study i and σ_{gi}^2 is the total genetic variance of the QTL in study i. From the Haseman-Elston regression, we obtain the estimator $\widehat{\beta}_{ij}$ for β_{ij} along with its estimated sampling variance $\widehat{S}_{ij}^2(i = 1, \ldots, k; j = 1, \ldots, m)$. First, all k studies are used to estimate parameters μ_j and A_j $(j = 1, \ldots, m)$; then, the empirical Bayes estimators $\widetilde{\beta}_{ij}$ and $\widetilde{S}_{ij}^2$ are easily obtained for each study using Formulas 5.1 and 5.2.

5.2.5 The IM-EB method

We now give a detailed description of the Interval Mapping-Empirical Bayes method (IM-EB) to detect linkage between markers and QTLs and estimate QTL location and effect from multiple genome scan studies using sib pairs. We assume that the data are from k genome scans on sib pairs for the same trait and, by convention, consider the first study as the primary study. Within each study, the markers are within the same chromosomal region and are denoted as M_{ij}, $i = 1, \ldots, k$, $j = 1, \ldots, m$. We note here, though, that the method does not restrict the marker map to be common across the studies since the interval mapping estimates are obtained for the analysis points q, which are not necessarily at marker locations.

For the IM-EB method, the estimates of the regression slope and variance, $\widehat{\beta}_{iq}$ and $\widehat{S}_{iq}^2$, $i = 1, \ldots, k$, at each analysis point q on the chromosome are obtained using the IM method described in Section 5.2.2 (Fulker et al., 1995). Then, the empirical Bayes estimates, $\widetilde{\beta}_{iq}$ and $\widetilde{S}_{iq}^2$ $(i = 1, \ldots, k)$, are obtained from each of the k studies by using the Gaussian Regression Independent Multilevel Model (GRIMM; see Tang (2002)). GRIMM is independently applied to each analysis point along the chromosome. The test statistic for the primary study is then calculated on the basis of $\widehat{t}_{1q} = -\widehat{\beta}_{1q}/\sqrt{\widehat{S}_{1q}^2}$ and $\widetilde{t}_{1q} = -\widetilde{\beta}_{1q}/\sqrt{\widetilde{S}_{1q}^2}$ at the analysis point q. The analysis point $\widehat{q}$ having a maximum value $\widehat{t}_{1\widehat{q}}$ over the entire chromosome is considered as the IM estimate of QTL location and consequently, the IM estimate of σ_{1g}^2 is given by $\widehat{\sigma}_{1g}^2 = -\widehat{\beta}_{1\widehat{q}}/2$. The same procedure can be applied to $\widetilde{t}_{1q}$ to obtain $\widetilde{q}$ and $\widetilde{\sigma}_{1g}^2$, the IM-EB estimates of QTL location and variance, respectively.

5.2.6 Simulation designs

To investigate the performance of the empirical Bayes method to incorporate data from multiple genome scans of sib pairs, we conduct the following simulations. We assume that there is only one QTL with no background polygenic variation and no shared sib environment effect, or equivalently that such effects are subsumed into the residual variance. There are two alleles at the QTL with the high-risk allele having a frequency of 0.05. We use 15 microsatellite markers on chromosome 11 that were used for a genome scan of Alzheimer's disease (Blacker et al., 2003) because it provides realistic simulation parameters, including the location and the allele frequencies at each marker locus. The trait value y of each individual is generated according to the genetic model, $y = \mu + g + \varepsilon$, where μ is the overall trait mean across the population, g is the additive effect of the high-risk allele, and ε is a normally distributed random error term. We set $E(\varepsilon) = 0$, $\text{cov}(g, \varepsilon) = 0$, $\mu = 70$, and set the total variance of g and ε, $\sigma_g^2 + \sigma_\varepsilon^2$ equal to 1 for all studies.

For each simulation, 5, 10, or 15 studies are generated, corresponding to a single study of interest with 4, 9, or 14 background studies, respectively. We generate genotype and phenotype data for 500 independent sib pairs in each study, each using the same marker map. For the primary study, the QTL is positioned 65 cM from the p-terminus of the chromosome. QTL heritability is set either to 0 (without QTL effect) or 15% (with non-zero QTL effect). QTL location in the background studies is set either at 35 cM or at 65 cM from the p-terminus of the chromosome. Marker locations along with the QTL location are shown in Figure 5.1. The heritability of the QTL in each background study varies between 0 and 25% in increments of 5%. The number of background studies with non-zero QTL effect varies but all background studies having a non-zero QTL effect are given the same value of heritability. This simulation strategy can accommodate different degrees of heterogeneity among the primary study and the background studies. It can also include a variety of combinations of weak to strong linkage signals among the primary and background studies. For example, we can set the QTL heritability in the primary study to 15%, the heritability of half of the background studies to 0, and the heritability of the other half of background studies to 25% to represent the situation that the primary study has a moderate linkage signal while some background studies show small to no QTL effect and some of background studies have stronger QTL effect. Other situations are readily accommodated by varying the number of background studies with non-zero QTL effect and the heritability levels.

Once genotypic and phenotypic data are generated, Haseman-Elston regression slope estimates and variances at each marker or analysis point in each study are determined by regressing the weighted combination of Y^D and Y^S on π (Xu et al., 2000).

5.3 Results

Here, we use simulation to assess the performance of IM-EB in terms of its power to detect linkage between markers and QTLs and its accuracy to estimate the QTL location and effect.

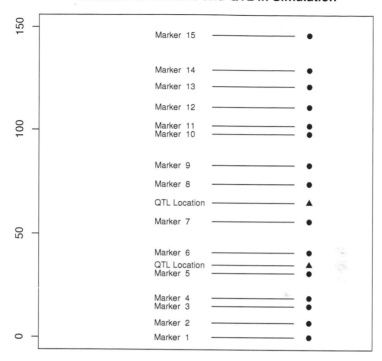

Figure 5.1 *Map for 15 micro-satellite markers (circles) from the National Institute of Mental Health Alzheimer's Diseases Genetics Initiative and the locations of two hypothetical QTLs (triangles) used in simulations. Minimum distances between the marker and two QTLs, 65 cM and 35 cM from the p-terminus of the chromosome, are 9 cM and 4 cM, respectively.*

5.3.1 The Type I error and power of the IM-EB method

We first investigate the Type I error rate of the IM-EB method. It is important to understand that a null model in this context refers only to the study of interest, whether or not the background studies contain a linked QTL. We generated 1,000 data sets with 5, 10, and 15 studies. In all studies, the QTL is positioned 65 cM from the p-terminus of the chromosome. In the primary study, QTL heritability is set to 0 (representing the null). In the background studies, the heritability is set either to 0 or to some value between 5% and 25% in increments of 5%. The number of background studies with non-zero QTL effect is varied. Under any particular condition, all background studies with non-zero QTL effect have the same heritability. In the primary study, the null hypothesis is rejected at level $\alpha = 5\%$ when the IM-EB statistic at 65 cM from the p-terminus of the chromosome exceeds 1.645. Figure 5.2 shows the

Type I error rate of the IM-EB method, estimated as the proportion of simulations in which the null hypothesis is rejected.

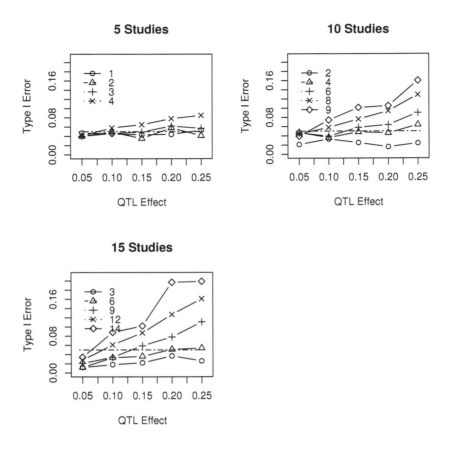

Figure 5.2 *Type I error rates of the IM-EB estimator at 65 cM from the p-terminus of the chromosome with 5, 10, and 15 studies. The QTL in all studies is at 65 cM. In the primary study, QTL heritability is 15%; the number of background studies having non-zero QTL effect varies.*

For 5 studies, the number of background studies with non-zero QTL effect is varied to be 1, 2, 3, or 4. It can be seen when three or fewer background studies have non-zero QTL effect, the Type I error rate stays below the nominal 5% error rate. When all 4 background studies have a heritability of 10%, the Type I error rate can be greater than the nominal 5% error rate. The highest Type I error rate is 8.5% for all 4 background studies having a heritability of 25%. For 10 studies, the number of background studies with non-zero QTL effect was set to 2, 4, 6, 8, or 9. When there are fewer than 4 background studies having non-zero heritability as high as 20%, the Type I error rate is below the nominal 5% error rate. The Type I error rate is inflated when 8 or more background studies have a heritability greater than 10%. The highest Type I error is 16%. For 15 studies, the number of background studies with non-zero QTL effect was set to 3, 6, 9, 12, or 14. When fewer than 6 background studies have a heritability as high as 25%, the Type I error rates do not exceed the nominal 5% rate. Again, the Type I error rate is inflated when 9 or more background studies have a heritability greater than 10%. The Type I error rate is 20% when all 14 background studies have a heritability of 25%. In summary, the Type I error rate of the IM-EB stays below the nominal 5% rate when most of the background studies have a heritability less than 10%. At the same time, we did find some inflated Type I error rates of the IM-EB method when most of background studies have a higher heritability. This is expected because the empirical Bayes-based method borrows the information from the other studies. If a large number of studies have a large QTL effect, the empirical Bayes-based method will detect a QTL even if the results from the primary study show small to no effect. However, from an EB perspective, it is debatable whether this truly represents a "null" situation.

We then investigate the power of the IM-EB method. Again, we generate 1,000 data sets with 5, 10, and 15 studies. In the primary study, the heritability of QTL is 15%. In the background studies, the heritability is either 0 or some value between 5% and 25% in increments of 5%. The number of background studies with non-zero QTL effect is varied and all background studies with non-zero QTL effect are assigned the same heritability.

Figure 5.3 shows the power of the IM-EB method, estimated as the proportion of simulations in which the null hypothesis is rejected. In these simulations, the QTL is positioned 65 cM from the p-terminus of the chromosome. Above, we used 1.645 as the 95% cutoff value to reject the null hypothesis of no linkage between the marker and the QTL. This value is valid only for one single study. When the empirical Bayes method was used, this cutoff value tends to be conservative. Therefore we follow the simulation method proposed by Beasley et al. (2005) to determine the cutoff value. We simulate 1,000 data sets with 5, 10, and 15 studies, none of which has a QTL effect. For the IM-EB method, the 95% cutoff values are determined to be 1.464, 1.406, and 1.224 for 5, 10, and 15 studies, respectively. These simulated cutoff values are used as critical values to reject the null hypothesis at the nominal 5% level.

It can be seen in Figure 5.3 that the power of the IM-EB estimator can be substantially increased when a majority of background studies have the same or higher QTL effect. When all 4, 9, and 14 background studies have a heritability of 15%, the power of the IM-EB estimator increases from 0.191 (the power of the IM estimator for an individual study) to 0.266, 0.343, and 0.466, respectively. When all 4, 9, and 14 background studies have a heritability of 25%, the power of the IM-EB estimator increases to 0.322, 0.525, and 0.688, respectively. The power of the IM-EB estimator also increases even when some of the background studies disagree with the primary study. For example, when about half of the background studies have no QTL effect and half of the background studies have the same heritability of 15%, the power of the IM-EB estimator is 0.224, 0.221, and 0.345 for 4, 9 and 14 background studies,

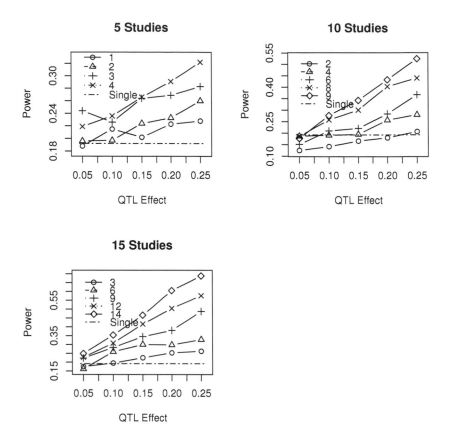

Figure 5.3 *Power of the IM estimator (Single) and IM-EB estimator with 5, 10, and 15 studies at 65 cM from the p-terminus of the chromosome. The QTL in all studies is at 65 cM. In the primary study, QTL heritability is 15%; and the number of background studies having non-zero QTL effect varies.*

respectively. As would be expected, the increase in power is slightly less than the situation when all of the studies agreed.

To see how the existence of other QTLs along the same chromosome affects the power of the IM-EB method, we simulate data sets in which all background studies have a heritability of 15% but half of them have the QTL positioned 35 cM from the p-terminus of the chromosome. The power of the IM-EB estimator at each marker locus is shown in Figure 5.4. We find that the IM-EB estimator increases the power to detect linkage near the QTL of interest at a very small cost of inflated Type I error rates.

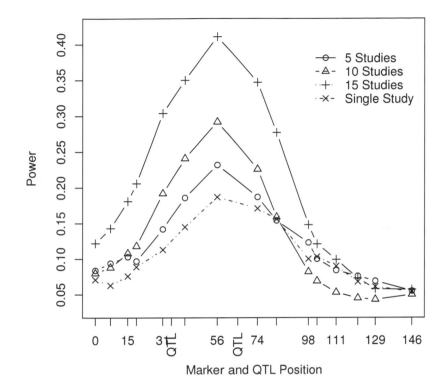

Figure 5.4 *Power of the IM estimator (Single Study) and IM-EB estimator with 5, 10, and 15 studies at the marker loci. In all studies, including the primary study and background studies, QTL heritability is 15%. In the primary study and half of the background studies, the QTL is 65 cM from the p-terminus of the chromosome. For the other half of the background studies, the QTL is at 35 cM.*

5.3.2 Accuracy of the IM-EB estimates for QTL location and effect

To investigate the accuracy of the IM-EB estimates for QTL location, we recorded their mean value (MEAN), standard error (STD), and the square root of the mean squared difference between the estimates and the true value (MSE). The simulations here are carried out with the same parameter values as the power simulations in Section 5.3.1.

The mean and MSE of the IM and IM-EB point estimates for the QTL location and effect under several different simulation strategies are presented in Tables 5.1 and 5.2. Several general conclusions emerged from these two tables.

Table 5.1 *The mean and MSE (in parentheses) for the point estimates of QTL location.*

Number of QTLs in Background Studies	Number of Studies	Number of Background Studies with Non-zero QTL Effect	Method	The Heritability in Background Studies		
				5%	15%	25%
	5	2	IM	71.8 (43.4)	69.2 (43.6)	69.6 (43.9)
	5	2	IM-EB	72.2 (44.2)	70.6 (42.1)	68.7 (42.5)
	5	4	IM	65.4 (41.7)	66.8 (41.9)	69.5 (41.4)
	5	4	IM-EB	64.9 (40.6)	67.8 (40.7)	69.3 (39.0)
	10	4	IM	71.1 (41.8)	68.2 (42.2)	69.6 (42.4)
One QTL	10	4	IM-EB	72.0 (41.4)	68.1 (40.2)	70.0 (40.1)
at 65 cM	10	9	IM	70.8 (43.5)	71.0 (41.8)	69.9 (42.2)
	10	9	IM-EB	68.9 (41.9)	69.5 (37.5)	67.2 (34.9)
	15	6	IM	72.4 (42.6)	70.4 (42.6)	70.8 (41.9)
	15	6	IM-EB	71.9 (43.3)	68.4 (40.4)	70.8 (38.3)
	15	14	IM	69.8 (41.2)	68.3 (42.7)	71.0 (43.2)
	15	14	IM-EB	69.9 (39.3)	69.4 (38.5)	68.7 (33.2)
	5	2	IM	70.8 (42.7)	70.8 (42.6)	66.0 (42.2)
	5	2	IM-EB	69.4 (42.3)	70.2 (42.4)	64.8 (41.9)
	5	4	IM	67.8 (41.5)	70.0 (42.9)	70.6 (41.9)
	5	4	IM-EB	66.3 (41.1)	68.6 (41.2)	65.8 (40.2)
One QTL	10	4	IM	72.0 (42.2)	69.1 (42.5)	72.0 (43.0)
at 65 cM	10	4	IM-EB	71.6 (41.7)	68.4 (42.9)	67.7 (41.2)
One QTL 65 cM	10	9	IM	70.2 (42.6)	68.7 (42.5)	68.2 (42.4)
at 35 cM	10	9	IM-EB	65.8 (42.0)	65.2 (38.3)	62.7 (36.8)
	15	6	IM	69.9 (41.3)	69.7 (41.9)	69.8 (42.5)
	15	6	IM-EB	70.6 (41.8)	66.0 (40.7)	64.8 (39.5)
	15	14	IM	69.8 (43.1)	70.7 (42.7)	69.7 (41.4)
	15	14	IM-EB	67.4 (41.5)	63.6 (37.9)	60.3 (34.2)

First, as expected, in most of the simulated situations the IM-EB method, using multiple studies, estimates the QTL location and effect more closely and with a smaller MSE than does the IM method using a single individual study in most situations we simulated. This improvement becomes more notable with more independent studies having larger QTL heritability included in the analysis. For 5 studies, the MSE of the estimates for the QTL location is reduced 5% (from 41.4 to 39.0) when all 4 background studies have a heritability of 25%. For 10 studies, the MSE of the estimates for the QTL location is reduced 18% (from 42.4 to 34.9) when all 9 background studies have a heritability of 25%.

Second, the heterogeneity among background studies and the disagreement between the primary study and background studies only slightly affect the accuracy of the IM-EB estimates. In addition, we do not observe a large bias for the estimates of either

Table 5.2 *The mean and MSE (in parentheses) for the point estimates of QTL effect.*

Number of QTLs in Background Studies	Number of Studies	Number of Background Studies with Non-zero QTL Effect	Method	The Heritability in Background Studies		
				5%	15%	25%
	5	2	IM	0.27 (0.18)	0.27 (0.18)	0.26 (0.18)
	5	2	IM-EB	0.22 (0.14)	0.23 (0.14)	0.23 (0.14)
	5	4	IM	0.27 (0.18)	0.27 (0.17)	0.26 (0.18)
	5	4	IM-EB	0.22 (0.14)	0.22 (0.14)	0.23 (0.14)
	10	4	IM	0.27 (0.19)	0.27 (0.18)	0.26 (0.18)
One QTL	10	4	IM-EB	0.17 (0.10)	0.17 (0.10)	0.18 (0.10)
at 65 cM	10	9	IM	0.27 (0.18)	0.27 (0.18)	0.27 (0.18)
	10	9	IM-EB	0.17 (0.10)	0.19 (0.10)	0.19 (0.10)
	15	6	IM	0.27 (0.18)	0.27 (0.18)	0.27 (0.18)
	15	6	IM-EB	0.14 (0.08)	0.15 (0.09)	0.16 (0.09)
	15	14	IM	0.27 (0.18)	0.27 (0.18)	0.27 (0.18)
	15	14	IM-EB	0.15 (0.08)	0.16 (0.09)	0.17 (0.08)
	5	2	IM	0.26 (0.17)	0.26 (0.17)	0.27 (0.18)
	5	2	IM-EB	0.21 (0.13)	0.22 (0.13)	0.22 (0.14)
	5	4	IM	0.27 (0.18)	0.27 (0.18)	0.26 (0.17)
	5	4	IM-EB	0.22 (0.14)	0.22 (0.14)	0.23 (0.14)
One QTL	10	4	IM	0.27 (0.18)	0.27 (0.17)	0.27 (0.18)
at 65 cM	10	4	IM-EB	0.16 (0.09)	0.17 (0.09)	0.17 (0.10)
One QTL 65 cM	10	9	IM	0.27 (0.18)	0.27 (0.18)	0.27 (0.18)
at 35 cM	10	9	IM-EB	0.17 (0.09)	0.18 (0.10)	0.19 (0.10)
	15	6	IM	0.27 (0.18)	0.27 (0.18)	0.27 (0.18)
	15	6	IM-EB	0.14 (0.08)	0.15 (0.08)	0.15 (0.08)
	15	14	IM	0.27 (0.17)	0.27 (0.18)	0.28 (0.19)
	15	14	IM-EB	0.14 (0.07)	0.16 (0.08)	0.17 (0.08)

the QTL location or effect in the presence of other QTLs and different QTL effects in the background studies, as observed in Zhang et al. (2006).

5.4 Discussion

With availability of multiple genome-wide scans aimed at detecting linkage between the same QTL and a marker, there is a need to develop novel methods that can borrow or combine information from all available studies. In this chapter, we have summarized empirical Bayes methods (Beasley et al., 2005; Zhang et al., 2006) and assessed their performance using extensive simulations. We found that the empirical Bayes-based methods have more power to detect the QTL and provide more precise estimates of QTL location and effect than do methods using an individual study.

To assess the effect of between-study heterogeneity, we have assumed that the background studies could have no QTL effect, have a non-zero QTL effect different from that of the primary study, or have the QTLs different from that of the primary study. Although the influence of these factors varies, the empirical Bayes methods were generally robust under all simulated situations. That is, they had more power to detect the QTL and yielded more precise estimates for QTL location and effect, with Type I error increased only under extreme situations. These simulations were carried out assuming that all studies used identical marker maps. This condition, however, is not required by the empirical Bayes method (IM-EB) described here. Zhang et al. (2006) found that varied marker maps across studies had only slight impact on the performance of empirical Bayes methods, and could even be helpful in a few situations.

We did not compare empirical Bayes methods with meta-analytic methods in this chapter. Zhang et al. (2006) compared several empirical Bayes-based methods with a weighted least-square methods developed by Etzel and Guerra (2002). These results showed that no method was superior to the others under all simulation situations. Although it is of great interest to conduct such a comparison of combination procedures, it is important to point out that the empirical Bayes methods introduced here are not meta-analysis methods. In meta-analysis methods, results from several studies of the same relationship are combined to obtain an overall inference or estimate of that relationship. In such an analysis, the results of the studies are combined with equal regard (though possibly weighted by their relative precisions). In these empirical Bayes methods, we assume that there is one study of primary interest, with the rest of studies regarded as background studies. The results obtained from the background studies are incorporated in an empirical Bayes step as prior information with the aim of improving inference or estimates in the primary study.

In summary, we conclude that empirical Bayes methods can account for between-study heterogeneity. They can also be more powerful for detecting linkage between markers and a QTL, as well as provide more precise estimates of QTL location and effect.

Acknowledgments

This paper is supported by grant R01ES09912 from National Institutes of Health.

Part II. Similar Data Types II: Gene Expression Data

CHAPTER 6

Composite hypothesis testing: an approach built on intersection-union tests and Bayesian posterior probabilities

Stephen Erickson, Kyoungmi Kim, and David B. Allison

6.1 Introduction

Combining information from multiple domains can provide meaningful answers to important biological questions. The domains (or "experiments," a term which we use interchangeably with "domains" throughout this chapter) involved might address the same scientific hypothesis, or a range of different but related hypotheses. We restrict attention to situations in which the multiple experiments aim to determine common genetic variants across multiple stimuli, such as cross-stimulus and cross-species studies. For example, one might wish to find:

1. genes that are differentially expressed in response to a particular stimulus across several experimental crosses or different tissue types within the same cross,
2. genes that are differentially expressed in a similar pattern in response to multiple stimuli, or
3. genes that independently influence two or more phenotypic traits (i.e., pleiotropy).

These examples involve testing the union of multiple null hypotheses.

There are two arms of multiple hypothesis testing: one that concerns composite hypotheses (Rhodes et al., 2002, 2004a; Kim et al., 2004) and the other which concerns combining the results of individual hypothesis tests (Choi et al., 2003; Moreau et al., 2003; Parmigiani et al., 2004; Stevens and Doerge, 2005). In this chapter, we focus on the simultaneous testing of a composite hypothesis based on intersection-union tests in the context of genetic studies. Although the methodologies are illustrated in gene expression studies, the rationale can be extended to allow for tests to identify quantitative trait loci (QTLs) which affect multiple traits from the same experiment or a particular trait in multiple experiments, and many other similar situations which occur in high-dimensional biology.

6.2 Composite hypothesis testing

To introduce our theoretical framework and its notation, suppose that n experiments are independently conducted to determine whether a gene g is differentially expressed in a condition of interest. We consider one-sided hypotheses: for gene g, we wish to test the n hypotheses $H_0^{(i)} : \theta_i = 0$ against the one-sided alternatives $H_A^{(i)} : \theta_i > 0$ for experiments $i = 1, 2, \ldots, n$, where θ_i is the true expression change of gene g in the experiment i. By "true" we mean a population mean expression change. For hypothesis i, a statistical test is performed on the basis of the test statistic T_i, a random variable. Large values of T_i result in small p-values, which lead to rejection of the null hypothesis $H_0^{(i)}$ in favor of the alternative hypothesis $H_A^{(i)}$. Let t_i be the observed value of the test statistic T_i; the p-value of the test for experiment i is calculated as

$$p_i = P(T_i > t_i | H_0^{(i)}).$$

If the null hypothesis $H_0^{(i)}$ is indeed true and the test statistic T_i has a continuous distribution under the null, then p_i is uniformly distributed on the interval $[0, 1]$. The individual hypotheses need not have the same substantive meaning across all the experiments.

There are two types of composite hypothesis tests: the union-intersection test and the intersection-union test. A *union-intersection test* (UIT; Roy, 1953) can be used to test whether a gene is differentially expressed in *at least* one of the experiments. In other words, we wish to test the composite null hypothesis that is the intersection of all nulls against the union of alternatives:

$$H_0 = \bigcap_{i=1}^{n} H_0^{(i)} \quad \text{vs.} \quad H_A = \bigcup_{i=1}^{n} H_A^{(i)}.$$

The composite null hypothesis is rejected if any of the individual null hypotheses is rejected, but at a multiplicity-adjusted threshold that controls the overall Type I error rate.

Alternatively, an *intersection-union test* (IUT; Berger, 1982, 1996) is used to test if a gene is differentially expressed in *all* of the experiments. We wish to test the composite null hypothesis that is the union of all nulls against the intersection of alternatives,

$$H_0 = \bigcup_{i=1}^{n} H_0^{(i)} \quad \text{vs.} \quad H_A = \bigcap_{i=1}^{n} H_A^{(i)}.$$

Table 6.1 *A layout of two-component hypothesis for an intersection-union test.*

	Null $H_0^{(1)}$: $\theta_{gA} = 0$	Alt. $H_A^{(1)}$: $\theta_{gA} > 0$
Null $H_0^{(2)}$: $\theta_{gB} = 0$	Null H_0	Null H_0
Alt. $H_A^{(2)}$: $\theta_{gB} > 0$	Null H_0	Alternative H_A

In this case, the composite null hypothesis is rejected only if all the individual null hypotheses are rejected, and the IUT does not require a multiplicity adjustment. The IUT is well-suited for testing whether genes are differentially expressed in the same way in response to a stimulus of interest across multiple experiments.

As a motivating example, suppose we wish to determine which of 36,000 assayed genes are differentially expressed in response to caloric restriction (CR) in two experiments, A and B. For a given gene g, the null hypothesis of interest would be that the gene is not differentially expressed in experiment A, experiment B, or both. Define $\theta_g = |\mu_{gA} - \mu_{gB}|$ for $g = 1, \ldots, 36,000$ as the absolute mean difference in expression levels of gene g, between the caloric restriction (CR) group and the placebo group, in experiment i ($i = A$ or B). Consider testing the composite null hypothesis for gene g

$$H_0 = H_0^{(1)} \bigcup H_0^{(2)},$$

as the union

$$H_0^{(1)}: \theta_{gA} = 0 \quad \text{or} \quad H_0^{(2)}: \theta_{gB} = 0,$$

against the composite alternative hypothesis

$$H_A = H_A^{(1)} \bigcap H_A^{(2)},$$

being the intersection

$$H_A^{(1)}: \theta_{gA} > 0 \quad \text{and} \quad H_A^{(2)}: \theta_{gB} > 0,$$

This two-component composite hypothesis is illustrated in Table 6.1. Gene g is judged to be differentially expressed in response to caloric restriction in both experiments A and B if the composite null hypothesis H_0 is rejected (i.e., if both $H_0^{(1)}$ and $H_0^{(2)}$ are rejected at level α). In other words, H_A is true if and only if both $H_A^{(1)}$ and $H_A^{(2)}$ are true. Hence, individual tests for the multiple experiments can be combined by means of an IUT to yield a single, overall test of the consistent gene effects across the different experiments.

Zhang et al. (1998) give an example of IUTs in linkage analysis, although the authors do not explicitly use the terminology of composite hypothesis testing. Their study maps QTLs for milk production and health in dairy cattle under two competing models: one model assumes a single QTL per chromosome-trait, and the other assumes two QTLs per chromosome-trait. An IUT is used to choose between the one-

Table 6.2 *Composite p-value methods.*

Method	Fisher	Tippett	Inverse Normal	Average	Pearson	Maximum
Composite p	$\prod_{i=1}^{n} p_i$	$\min_{1 \le i \le n} p_i$	$\dfrac{\sum_{i=1}^{n} \Phi^{-1}(p_i)}{\sqrt{n}}$	$\sum_{i=1}^{n} p_i/n$	$1 - \prod_{i=1}^{n}(1 - p_i)$	$\max_{1 \le i \le n} p_i$

and two-QTL models. For chromosome-traits in which two loci are deemed significant under the single-QTL model, the two single-QTL models (one for each locus) are each nested within the double-QTL model, and an F-test is used to compare the goodness of fit of each of the single-QTL models against the two-QTL model. The null composite union hypothesis is rejected in favor of the two-QTL model only if the F-test rejects the single-QTL model at *both* loci.

6.3 Assessing the significance of a composite hypothesis test

Table 6.2 summarizes six methods for computing a composite p-value from multiple hypothesis tests: Fisher's inverse chi-square method (Fisher, 1932), Tippett's minimum (Tippett, 1931), the inverse normal (Stouffer et al., 1949), the average, Pearson's (Pearson, 1933), and the maximum method. These methods are sometimes referred to as *omnibus* procedures because they combine multiple p-values from individual tests into a composite p-value without regard to the details of the hypothesis testing procedures. All but the maximum method are suited to UITs and not IUTs. Under the first five methods, p-values from false nulls influence the composite p-value, which assesses the strength of evidence against the union of all component null hypotheses.

Consider testing a two-component composite hypothesis, as described in the caloric restriction example, and determining its significance with a composite p-value based on the maximum method. In UITs, each component of the composite null hypothesis is determined to be rejected or not at the multiplicity-adjusted threshold that controls an overall experiment-wide Type I error rate. The rejection region for this UIT is the union of rejection regions that correspond to the individual tests (see Figure 6.1(a)). In contrast, IUTs based on the maximum method maintain a pre-specified Type I error rate without adjustment for multiple components. The rejection region for this IUT is the intersection of the rejection regions corresponding to the two individual tests, that is,

$$\bigcap_{i=1}^{2} R_i = \{x: \ \min(T_1(x), T_2(x)) \ge c_\alpha\},$$

where for the ith null hypothesis, R_i is the rejection region, $T_i(x)$ is an appropriate test statistic, and c_α is the threshold value associated with the Type I error rate of α (see Figure 6.1(b)). The p-value for the minimum of the test statistics, which is the

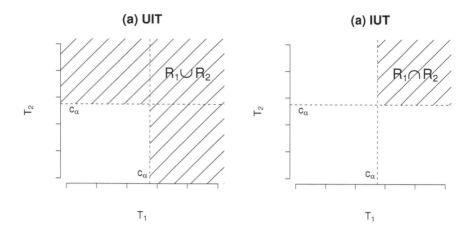

Figure 6.1 *Rejection regions of parameter space corresponding to the alternative hypothesis H_A in UITs (left panel) and IUTs (right panel).*

maximum p-value, is used to determine whether the composite null hypothesis will be rejected, regardless of the magnitudes of any other p-values.

As stated above, the maximum method is an omnibus procedure suited to IUTs, but it can be quite conservative, potentially controlling Type I error at a level much lower than the α-level selected for the IUT's constituent tests. In the most extreme case, in which n true null hypotheses are tested by n independently conducted experiments, the resulting composite significance level would be α^n. On the other hand, Berger (1982; 1997) gives conditions under which the composite significance level of the IUT is exactly α. Stated loosely, this occurs when exactly one of the null hypotheses is true, and the power of each of the $n - 1$ other tests (which pertain to untrue nulls) approaches one. We expect nearly all scientifically relevant IUTs to fall between these two extremes, but the omnibus nature of the maximum method makes it impossible to determine where, without further knowledge and analysis.

That is, despite the appealing property of omnibus procedures for combining p-values, they do not provide a precise quantitative significance measure of uncertainty against the null hypothesis. Therefore, the statistical significance of omnibus tests is insufficient to draw inferential decisions about the magnitude, direction, and consistency of variation on phenotypic responses of interest across multiple experiments. To obtain quantitative measures of significance, a Bayesian measure, expressed in terms of posterior probability that the null hypothesis is true, provides quantitative evidence for or against the composite null hypothesis. In the next section, we discuss such a Bayesian procedure, which utilizes information about the *distributions* of individual test statistics and p-values.

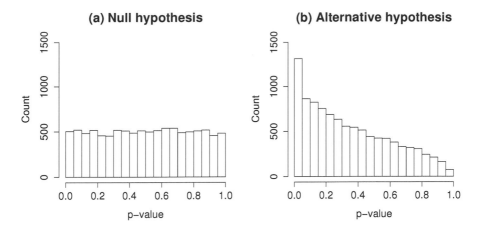

Figure 6.2 *Mixture model approach from Allison et al. (2002). (a) Under the null hypothesis, the distribution of p-values is uniform on the interval* [0, 1] *regardless of the sample size and statistical test used. (b) Under the alternative hypothesis, the distribution of p-values will have more weight near zero than near one.*

6.4 Measuring Bayesian significance evidence in composite hypothesis testing

Pratt (1965) attempts to reconcile frequentist inferential statements with Bayesian statistics and shows that in a one-sided test and under a broad set of assumptions, the p-value is an approximation of the Bayesian posterior probability that the null hypothesis is true. The quality of this approximation, however, depends on several conditions which may or may not hold and, in fact, may be impossible to evaluate. Simply using the p-value as an estimate of the posterior probability of the null hypothesis, therefore, may do no better than the omnibus strategies discussed in the previous section. One advantage to our setting, in which a large number of hypotheses (i.e. genes) are being simultaneously tested, is that we observe not merely a single p-value but a collection or sample distribution of many p-values. By modeling this distribution as a realization of a mixture of true and untrue null hypotheses, a better estimate of the posterior probability of the null can be inferred from the p-value. This is an empirical Bayes approach, because the empirical distribution of p-values informs the *a priori* probability of the null hypothesis being true.

The mixture model approach described by Allison et al. (2002) is based on the theoretical distribution of p-values under null and alternative hypotheses. Under the null, the distribution of p-values (assuming the test is valid) is uniform on the interval [0, 1], while if the null is false, the density of p-values decreases monotonically on [0, 1], as illustrated in Figure 6.2. Parker and Rothenberg (1988) and others have argued for the usefulness of the beta family of distributions for modeling the distribution of p-values under null and alternative hypotheses. If there is at least one gene

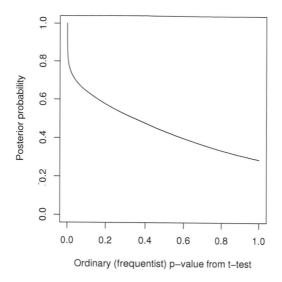

Figure 6.3 *Posterior probability of the alternative hypothesis as a function of p-value.*

differentially expressed between two groups, therefore, the density of p-values will be greater near zero than near one and can be modeled as a mixture of betas; recall that the uniform distribution on $[0, 1]$ is a special case of the beta distribution with both shape parameters set to one.

The log-likelihood function of the mixture model with $m + 1$ independent beta components can be written

$$L_{m+1} = \sum_{i=1}^{k} \ln \left[\lambda_0 \beta(1, 1)(x_i) + \sum_{j=1}^{m} \lambda_j \beta(r_j, s_j)(x_i) \right], \qquad (6.1)$$

where $\beta(r, s)(x)$ is the density function for the beta distribution with shape parameters r and s, x_i is the p-value for test i, λ_0 is the probability that a randomly chosen hypothesis test is true, and λ_j is the probability of a randomly chosen test of a false null hypothesis from component j of beta distribution. Although the expression in Equation 6.1 is only properly termed the log-likelihood of the data when all k p-values are independent, the expression remains a useful index of model fit regardless of independence. If any of the m non-uniform components of the mixture model is not zero, then there is at least one gene that is differentially expressed between the two groups. As described in Allison et al. (2002), the parameters $\lambda_0, \ldots, \lambda_m$, $r_1, \ldots, r_m$, and $s_1, \ldots, s_m$ can be estimated by maximum likelihood.

Once parameters are estimated, the posterior probability of the null being true can be computed using Bayes' rule, with $\lambda_0, \ldots, \lambda_m$ being the prior probabilities of component membership and the $\beta(r_j, s_j)(x_i)$, for $j = 0, \ldots, m$, being the component-wide likelihoods for the observed p-value x_i. The estimated mixture distribution, in other words, provides a "map" between the (frequentist) p-value and the (Bayesian) posterior probability of the null being true or false, as illustrated in Figure 6.4. As one would expect, the posterior probability of the null being false approaches one as the p-value approaches zero. Allison et al. (2002) evaluate the reliability of this mapping through a series of computer simulations and show that a relatively modest (by contemporary standards) number of simultaneous tests is required to estimate the parameters of the distribution of p-values.

6.5 Combining posterior probabilities in a Bayesian IUT

Recall that in an IUT, the composite alternative is the intersection of the component alternatives; rejecting the composite null means rejecting all the component nulls. The Bayesian probability with which to evaluate an IUT, therefore, is the probability of all nulls being false or, conversely, all alternative hypotheses being true. If the "events" (i.e., hypotheses) to which the posterior probabilities correspond are independent, this can be computed by simply multiplying the component probabilities. Note that the independence assumption is quite strong, because it requires not only that the data used to compute the component p-values be independent, but the hypotheses themselves be independent. In the caloric restriction experiments, for example, this would mean that whether a gene is differentially expressed under caloric restriction in experiment A has no relation to whether that same gene is differentially expressed under caloric restriction in experiment B. It would be difficult to imagine a scientifically relevant IUT for which this type of independence would hold. If we assume, however, that the hypotheses are not negatively correlated, but are independent or positively correlated, then the product rule yields conservative estimates of the posterior probability of the intersection alternative.

As an example of the Bayesian IUT approach, consider two studies comparing two groups, one relatively lean and the other fatter, in two different species, humans and mice. The first study is of adipocyte (fat cell) RNA from 20 lean and 19 obese Pima Indians. Biopsies were taken after an overnight fast, and none of the individuals had any manifested diseases; these data were generated at the NIDDK Phoenix by Dr. Paska Permana (Lee et al., 2005b). The second is a study of mouse adipocytes from five mice fed *ad lib* and five mice with long-term caloric restriction. The biopsies were taken after a 16-hour overnight fast; these data were generated by Dr. Kazu Higami in Dr. Richard Weindruch's Lab at University of Wisconsin-Madison (Higami et al., 2004). We wish to find homologous genes in humans and mice that are differentially expressed between the leaner and fatter groups in both species. The null hypothesis for each homologous gene-pair is that the mouse homolog is not differentially expressed in mice under caloric restriction, or the human homolog is not differentially expressed in between normal and obese, or both. The IUT of the two-component composite hypothesis can be performed by a simple extension of the mixture model approach: mixture models for the data from each species are fitted separately to obtain posterior probabilities for all genes. Then the two resulting posterior probabilities of the two species per gene are multiplied to compute the joint posterior probability for the use of IUT. One would consider a gene to have a conserved response across the two species only if the joint posterior probability is sufficiently high; consequently, one can also estimate the number of genes for which the null hypothesis is false in both mice and human by calculating the sum of all joint posterior probabilities across all genes. Such conserved genes are the best investment in further comparative genomic studies of global patterns of gene expression relevant to caloric restriction and its influence on obesity.

The application of IUTs can also be extended to many other types of genetic investigations in which multiple results (either from multiple studies or within a single study) are combined. Examples include QTL-mapping studies or association studies in which multiple traits (phenotypes) are modeled. Pleiotropy, defined here as a single gene influencing multiple traits, is of increasing interest and importance in genetic research and must be evaluated with multi-trait models. At least two kinds of pleiotropy have been distinguished in the literature. *Mosaic pleiotropy* refers to situations in which a locus independently affects two or more traits. *Relational pleiotropy* refers to situations in which a locus affects one trait which in turn affects a second trait, such that the locus is associated with both traits, but only directly affects the first. An IUT can be used to test whether a QTL (or, ultimately, a gene) affects all of the traits of interest as follows: first, for each trait, a genome-wide analysis can be conducted to estimate genetic effects and their corresponding significance scores (e.g., p-values or lod scores) at each genotyped locus across the genome, which generates a set of k genetic effect estimates (k being the total number of loci). Next, a mixture-model-based analysis for a set of the p-values for each trait is performed, yielding a set of posterior probabilities. Third, the IUT for the two-component hypothesis is performed to calculate the joint posterior probability by multiplying two resulting posterior probabilities for each locus. This joint posterior probability can be interpreted as the probability that pleiotropy exists at the locus, which is one advantage of using a Bayesian approach.

However, the above approach would merely test for some form of pleiotropy. To test for mosaic pleiotropy specifically, one would repeat the process, but in the initial linkage or association scan for the first trait, one should condition on the second, and in the initial linkage or association scan for the second trait, one should condition on the first. One is therefore testing for independent effects of the locus on the traits, and if the probability of both such effects being zero is low (obtainable via the IUT posterior probability) then mosaic pleiotropy can be tentatively concluded. When investigators subsequently prioritize candidate genes or regions for further study, they may wish to seek loci that have the highest joint posterior probabilities of independently affecting more than one trait of interest.

6.6 Issues and challenges

Although intersection-union tests enable one to test a composite hypothesis for multiple experiments without multiplicity correction, there are a few issues related to the underlying assumptions of this approach. These include cross-experiment variation due to different experimental conditions and dependency on phenotypic measurements between experiments. For example, experiments can differ in the use of different microarray platforms (Moreau et al., 2003; Parmigiani et al., 2004) and the number of genes (or genome QTLs) under study (Choi et al., 2003), especially when two different, distant species are compared with each other. This inconsistency may cause results to be biased and incomplete by mismatching or missing pairs of genes across experiments.

A second challenging issue, already mentioned, is dependence among the component hypotheses of a composite null hypothesis. Again, the issue here is not merely whether the data generating the posterior probabilities are independent. Rather, the states of nature or propositions (null hypotheses) to which these probabilities apply must also be independent for a simple product rule involving posterior probabilities to be useful. If these propositions are not independent in this sense, then the multiplication rule for computing the joint probability from marginals no longer holds, i.e. $P(A \cap B) \neq P(A)P(B)$. If we are not prepared to assume that the propositions are independent, a joint distribution of p-values from the multiple domains could be modeled to accurately reflect the dependency. However, unlike the now commonplace mixture modeling procedures for single vectors of p-values (Gadbury et al., 2008), there do not seem to be well-established methods for mixture modeling multiple correlated vectors of p-values. This is an area where further research would be useful.

6.7 Summary

Comparisons of multiple experiments are often useful for understanding the genetic pathways of complex diseases. The common strategy involves confirming that genes which influence a phenotypic change in one experiment also influence the change in other experiments. This confirmatory analysis that performs multiple tests for the multiple experiments individually requires greater efforts, both statistically and computationally, until a comprehensive conclusion from the multiple experiments is drawn. Therefore, methods for combining all tests into a single hypothesis-based test are very attractive. In this chapter, we have discussed theoretical and practical aspects of composite hypothesis testing built on intersection-union tests (IUTs) in the search for genes which affect variation of a phenotypic response of interest across multiple experiments. We have also provided a basic understanding of the traditional IUT and have shown how this method can be applied to genetic or genomic research. Moreover, we have addressed limitations of the IUT. As an alternative approach to overcome some of these limitations, we have proposed a hybrid approach which incorporates both frequentist and Bayesian methodologies and improves the flexibility and efficiency of traditional IUTs by incorporating information of data properties. As illustrated in the context of high-dimensional biology, the application of a mixture model approach to IUTs is expected to be more useful than IUTs which use only individual p-values. The continuing development of these statistical methods will be useful to assist in the discovery of genes that influence complex diseases.

6.8 Software availability

Software that includes the methods in this paper is available at the web site of the Section on Statistical Genetics of the Department of Biostatistics at the University of Alabama, Birmingham: `http://www.soph.uab.edu/ssg/software`.

CHAPTER 7

Frequentist and Bayesian error pooling methods for enhancing statistical power in small sample microarray data analysis

Jae K. Lee, HyungJun Cho, and Michael O'Connell

7.1 Introduction

Each gene's differential expression pattern in a microarray experiment is usually assessed by (typically pairwise) contrasts of mean expression values among experimental conditions. Such comparisons can be made as fold changes whereby genes with greater than two- or three-fold changes are selected for further investigation. However, it has been frequently found that a gene showing a high fold change between experimental conditions might also exhibit high variability and hence its differential expression may not be significant. Similarly, a modest change in gene expression may be significant if its differential expression pattern is highly reproducible. A number of authors have pointed out this fundamental flaw in the fold change-based approach (e.g., Jin et al., 2001). And, in order to assess differential expression in a way that controls both false positives and false negatives, a standard approach is emerging as one based on statistical significance and hypothesis testing, with careful attention paid to reliability of variance estimates and multiple comparison issues.

In early studies, the classical two-sample t-statistic was used to test for differential expression for each gene, one at a time; procedures such as the Westfall-Young step-down method (Westfall and Young, 1999) have been suggested as a correction for multiple hypothesis testing to control family-wise error rate (FWER; Dudoit et al., 2002). These t-test procedures, however, rely on reliable estimates of reproducibility or within-gene error, requiring a large number of replicated arrays. When only a small number of replicates are available per condition, e.g., duplicate or triplicate, the use of naive, within-gene estimates of variability does not provide a reliable hypothesis testing framework. For example, a gene may have very similar differential expression values in duplicate experiments by chance alone. This can lead to inflated signal-to-noise ratios for genes with low but similar expression values. Furthermore, in small sample studies, the comparison of means can be misled by outliers with dramatically smaller or bigger expression intensities than other replicates. As such, error estimates constructed solely within genes can result in underpowered tests for differential expression comparisons and also result in large numbers of false positives.

A number of approaches to improving estimates of variability and statistical tests of differential expression have thus recently emerged. Several variance function methods have been proposed, including a simple regression estimation of local variances (Kamb and Ramaswami, 2001) and a two-parameter variance function of mean expression intensity (Durbin et al., 2002). These variance function methods borrow strength across genes in order to improve reliability of variance estimates in differential expression tests. The local-pooled-error (LPE) estimation strategy has also been introduced for improving such within-gene expression error estimation (Lee and O'Connell, 2003; Jain et al., 2003). LPE variance estimates for genes are formed by pooling and smoothing the error variability of genes with similar expression intensities from replicated arrays. From this error pooling, the LPE approach effectively handles many statistical artifacts in large screening analysis, e.g., where a gene with low expression may have very low variance by chance and the resulting signal-to-noise ratio is unrealistically large, or vice versa. This, in turn, leads to a dramatically improved statistical testing framework for the discovery of biologically relevant differential expression patterns in a microarray study: an LPE test for comparing two contrasting conditions and empirical-Bayes heterogenous error model (HEM) for identifying differentially expressed genes under multiple experimental conditions (Cho and Lee, 2004, 2008).

Kooperberg et al. (Chapter 8 in this volume) summarizes a comparison study of several different analysis approaches to small-sample microarray data with practical data sets and realistic simulation settings. For real data applications, LPE performs well in the context of a small number of replicates per condition that are actually technical replicates since LPE assumes a relatively well-regulated error distribution close to Gaussian. This assumption may not be true when such replicates are different biological samples since both technical and heterogeneous biological errors will be confounded in such cases. Note also that LPE and HEM are based on the observation of systematic error in microarrays – monotonic decreasing, e.g., reverse-exponential relationship between log-transformed gene expression intensities and their error. However, since the underlying assumptions of LPE, HEM, or any alternative approaches may not be true in certain microarray data sets, we generally recommend that researchers compare the statistical results from two or more different approaches to their data analysis.

The LPE-based approaches are available as open source R software packages LPE and HEM as part of the BioConductor project (Gentleman et al., 2004; http://www.bioconductor.org); the LPE method is also available through the commercial software S+ArrayAnalyzer^{TM}, based on S-PLUS® (O'Connell, 2003).

7.2 Local pooled error test

7.2.1 Method

The Local Pooled Error (LPE) method constructs error variance estimates by pooling variance estimates for genes with similar expression intensities from replicated arrays within experimental conditions (Jain et al., 2003). LPE carefully leverages the observation that genes with similar expression intensity values often show similar variability within each experimental condition and that error variability of (log) expression is a decreasing (or non-increasing) function of intensity in practical microarray data. The latter is due to the fact that microarray instrumentation exhibits common background noise that is a bigger proportion of gene expression intensity in a low intensity region than that in a high intensity region. Figure 7.1 illustrates this phenomenon with LPE-estimated baseline error distributions under three different experimental conditions for a mouse immune response microarray study described below.

To take into account heterogeneous error variability across different intensity ranges, the LPE method is applied as follows (refer to Jain et al. (2003) for a more detailed technical description). For oligonucleotide array data (see Section 1.4), let $y_{1_{ik}}$ and $y_{2_{ik}}$ be the observed expression intensities at gene i for replicate k under two conditions. For duplicate arrays, $k = 1$ or 2. Plots of $M = \log_2(y_{1_{ik}}/y_{2_{ik}})$ vs. $A = \log_2(y_{1_{ik}}y_{2_{ik}})/2$ can facilitate the investigation of between-duplicate variability in terms of overall intensity. This MA plot provides a very raw look at the data and is useful in detecting outliers and patterns of intensity variation as a function of mean intensity (Dudoit et al., 2002). At each of the local intensity regions of the MA plot under a particular biological condition, the unbiased estimate of the local variance is obtained. A cubic spline is then fit to these local variance estimates to obtain a smooth variance function. The optimal choice of the effective degrees of freedom df_λ of the fitted smoothing spline is obtained by minimizing the expected squared prediction error or generalized cross validation error (GCV). This two-stage error estimation approach – estimation of error of M within quantiles and then non-parametric smoothing on these estimates – is used because direct non-parametric estimation often leads to unrealistic (small or large) estimates of error when only a small number of observations are available at a fixed-width intensity range.

Based on the LPE estimation above, statistical significance of the LPE-based test is evaluated as follows. First, each gene's medians under the two compared conditions are calculated to avoid artifacts from outliers. The approximate normality of medians can be assumed with a small number of replicates based on the fact that the individual log-intensity values within a local intensity range follow a normal distribution (see the supplementary data in Jain et al. (2003)). The LPE statistic for the median (log-intensity) difference z is then calculated as:

$$z = (\hat{\mu}_1 - \hat{\mu}_2)/\hat{\sigma}_{pooled}, \tag{7.1}$$

where $\hat{\mu}_1$ and $\hat{\mu}_2$ are the median intensities in two comparing array experimental conditions Y_1 and Y_2, and $\hat{\sigma}_{pooled} = [\hat{\sigma}_1^2(\hat{\mu}_1)/n_1 + \hat{\sigma}_2^2(\hat{\mu}_2)/n_2]^{1/2}$ is the pooled standard

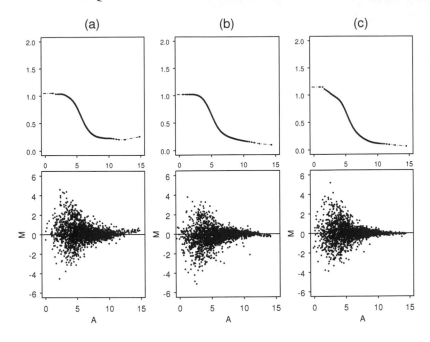

Figure 7.1 *Log intensity ratio* $M = \log_2(y_{1_{ik}}/y_{2_{ik}})$ *as a function of average gene expression* $A = \log_2 \sqrt{y_{1_{ik}} y_{2_{ik}}}$. *Top row of panels represent local pooled error (LPE) for (a) naive, (b) 48-hour activated, and (c) T cell clone D4 conditions for the mouse immune response microarray study reported by Jain et al. (2003). Variance estimates in percentile intervals are shown as points, and a smoothed curve superimposing these points is also shown. Bottom row of panels represents the corresponding MA plot. The horizontal line represents identical expression between replicates.*

error from the LPE-estimated baseline variances of $\hat{\sigma}_1^2$ and $\hat{\sigma}_2^2$. The LPE approach shows a significantly better performance than two-sample t-test, Significance Analysis of Microarrays (SAM) (Tusher et al., 2001), and the Westfall-Young permutation tests (Westfall and Young, 1999), especially when the number of replicates is smaller than ten. In a simulation study from a Gaussian distribution without extreme outliers, the LPE method showed a significant improvement of statistical power with three and five replicates, as reported in Jain et al. (2003).

7.2.2 Example: Microarray study of T cell immune responses

Cytotoxic T cells play a central role in the pathophysiology of many inflammatory lung diseases wherein they accumulate in the alveolar space and/or in the interstitium. A microarray study was performed to investigate the role of T cells among three different immune exposure populations: naive (no exposure), 48-hour activated, and CD8$^+$ T cell clone D4 (long-term mild exposure), using triplicate microarrays of Affymetrix murine chip, MG-U74A with 12,488 genes (Jain et al., 2003). Signal intensity values are MAS 5.0 values obtained from the Affymetrix MicroArray Suite software (Affymetrix, 2002). Many genes exhibiting significant differential expression patterns were identified by the LPE test. The LPE method identified genes that are well known in the literature for their mouse immune response function. Other hypothesis testing methods, such as the Westfall-Young procedure and the two-sample t-test, were not able to identify some of these genes.

In order to examine the relationship between the LPE p-values and fold change values more systematically, a scatter plot (or volcano plot) of each gene's LPE p-values vs. fold change for the combined naive and the CD8$^+$ T cell D4 clone conditions is shown in Figure 7.2. The two horizontal lines represent 2-fold changes in both directions and the vertical line the Bonferroni-adjusted LPE p-value 0.05. The numbers of genes in each sector of the left panel are also shown. Note that the two RNA samples (naive and CD8$^+$ T cell clone) are biologically quite heterogeneous, and a large number of differentially expressed genes were identified both by LPE test and fold change. In this figure a weak correlation is found between significant differential expression and fold change. This suggests differential-expression discovery based on fold change alone is misleading because a large number of insignificant genes are identified with high fold changes in the low intensity region as displayed in black in Figure 7.2(b).

7.2.3 Resampling-based FDR estimation for LPE tests

An extremely large number of genes (e.g., $> 40,000$) can be represented on a microarray; thus, comparisons between experimental conditions for all genes must take false positive error rate and multiple comparison issues into account. There are several possibilities for error rate control in multiple testing (Ge et al., 2003; Dudoit et al., 2004; van der Laan et al., 2004b,a). Traditional statistical methods often control the family-wise error rate (FWER), the probability of incorrectly accepting at least one false positive hypothesis (i.e. Type I error rate) among all hypotheses. For example, the commonly-used Bonferroni correction divides the Type I error rate α by the total number of hypotheses for each single-gene test of differential expression. This method is based on the Bonferroni inequality, which gives a bound for the FWER assuming independence. However, the independence assumption is unlikely to be true in microarray data, as functions of many genes are interrelated to varying degrees. Moreover, the methods controlling FWER are frequently found to be too conservative to identify many important genes in biological applications. There are step-down procedures that apply the severe Bonferroni correction only to the most extreme value of the test statistic, modifying subsequent corrections in a sequential manner (Šidák, 1967; Westfall and Young, 1999). However, these methods result in high false negative error rates, likely missing many genes that are truly differentially expressed.

Benjamini and Hochberg (1995) (BH) suggested that controlling false discovery rate (FDR), or the expected proportion of false positives among all positive (or rejected)

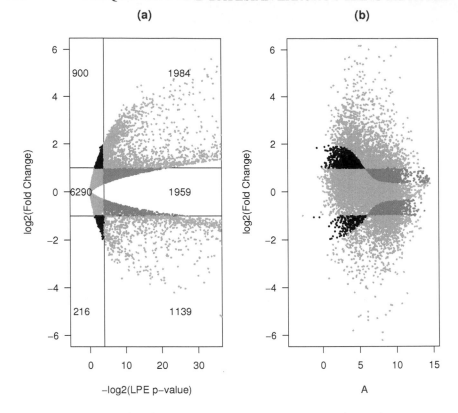

Figure 7.2 *Fold change* ($\log_2$) *of gene expression and LPE p-values* ($-\log_2$) *for naive mice and CD8$^+$ T cell D4 clone (left panel) conditions. The two horizontal lines mark the two-fold change threshold and the vertical line marks the threshold of cutoff Bonferroni-adjusted p-value = 0.05. Genes shown in dark gray undergo low fold change but changes are significant – these genes are missed by fold change method alone. Genes shown in black have high differential expression but are not significant and would be detected as false positives by a fold change method. The right panel shows the distribution of genes as an MA plot. There is no clear-cut relation between significant and high fold change genes, and hence LPE is required for such differentiation. Numbers shown in each sector of the left panel represent the number of genes in that sector.*

hypotheses, is more appropriate than FWER control for large screening problems. Benjamini and Yekutieli (2001) (BY) proposed a new FDR procedure considering a certain dependency structure among the test statistics. However, both the BH and BY procedures may still be too conservative when applied to real microarray data analysis (Dudoit et al., 2002). This is mainly due to the fact that the independence or the artificial dependency assumptions made in these approaches may not be supported in real microarray data applications. Furthermore, microarray experiments are often conducted with a small number of replicates due to limited availability of RNA samples and/or budgetary constraints as mentioned earlier.

One of the key issues in estimating FDR is the assumption regarding the underlying null distribution. The Significance Analysis of Microarrays (SAM) method (Tusher et al., 2001) uses a full permutation strategy, sampling across all genes and conditions to generate such a null distribution (mix-all). However, this strategy breaks many intrinsic correlation structures and does not generate a realistic or *biologically relevant null* distribution for microarray data. Chip-by-chip permutation strategies, which randomly shuffle all the columns (chips) and preserve gene structure, are not applicable when the sample size is small because the number of independent permutations is too small to generate a null distribution with sufficient resolution to support desired significance calculations. To provide more stable estimation of such FDR values, a method based on the spacings LOESS histogram (SPLOSH) was also proposed based on a certain assumption about the *p*-value distribution (Pounds and Cheng, 2004). In order to further improve the FDR estimation in practical microarray data analysis, a *rank-invariant resampling* (RIR) approach can be applied to microarray data with a small number of replicates as follows.

Generation of biologically relevant null distribution

It is critical to generate an underlying null distribution as close as possible to real microarray data because the statistical significance for a gene can be dramatically different under different underlying null distributions. Therefore, a resampling strategy needs to be designed to preserve the biological structure of each microarray data set as much as possible. We first define an algorithm for constructing intervals in the resampling strategy. A naive approach for construction of intervals is to partition intensity ranges so that each interval has an equal number of genes. This approach may yield overly large test statistics at high intensity levels because intensities are very sparse at high levels and condense in the middle levels. In order to obtain the local intervals of the genes with homogeneous variances, adaptive intervals are constructed by the following algorithm.

Adaptive Interval (AI) Algorithm

1. Estimate a baseline variance function for all data under consideration (within each experimental condition) by LPE.
2. Obtain medians and variance estimates for each gene.
3. Order the medians and variances by the medians.
4. Obtain the first interval with threshold values $\xi_{(1)}$ and $\xi_{(1)} + \sigma_{(1)}$.
5. Obtain the next interval with $\xi_{(2)}$ and $\xi_{(2)} + \sigma_{(2)}$, where $\xi_{(2)}$ is the smallest median such that $\xi_{(2)} \geq \xi_{(1)} + \sigma_{(1)}$.
6. Repeat Step 5 to obtain the next intervals with $\xi_{(i)}$ and $\xi_{(i)} + \sigma_{(i)}$, where i is the index of the smallest median such that $\xi_{(i)} \geq \xi_{(i-1)} + \sigma_{(i-1)}$, until all the data are assigned to certain intervals.

Note that the number of genes in each interval is forced to be between given minimum and maximum numbers, typically between 10 and (total number of genes)/100 for the minimum and maximum numbers, respectively. Note also that this AI algorithm is applied to the replicated array data under each experimental condition separately.

The RIR procedure for generating null data is then as follows.

1. Calculate medians for each gene and obtain the ranks of these medians within each experimental condition.
2. Calculate rank differences between two conditions for each gene.
3. Construct the first intensity intervals using the AI algorithm above and retain rank-invariant genes by eliminating a certain percentage of genes with largest rank differences within each interval.
4. Construct the final intensity intervals of rank-invariant genes using the AI algorithm.
5. Obtain a set of null data by resampling intensities of rank-invariant genes within each interval.
6. Repeat the above step B times, e.g., 1,000, to obtain B independent sets of resampled null data.

In Step 3 of the above procedure, a certain percentage of genes are eliminated to retain only rank-invariant expressed genes. In this current application, 50% of all genes with largest rank differences are eliminated in this step. Note that the AI algorithm is used twice in this RIR procedure. The first time is to remove rank-variant genes evenly throughout the whole intensity range. Without this step, many genes in the low intensity range would be disproportionately removed due to the larger variability in those ranges. This is a particularly important issue for Affymetrix data that have been summarized using MAS 5.0.

RIR-based FDR estimation

Suppose Z^0 is an LPE Z-statistic calculated from null data as described above. Generation of the null data is repeated many times independently. Let Z be an LPE Z-statistic computed from the real data. FDR at a threshold value Δ can be estimated as

$$\widehat{FDR}(\Delta) = \frac{\hat{\pi}_0(\lambda)\bar{R}^0(\Delta)}{R(\Delta)}, \tag{7.2}$$

where $\bar{R}^0(\Delta)$ is the average number of Z^0 equal to or greater than Δ and $R(\Delta)$ is the number of Z equal to or greater than Δ. The proportion $\pi_0(\Delta)$ of true null genes in real data can be estimated by the number of $\{Z \leq \lambda_q\}$ divided by the average number of $\{Z^0 \leq \lambda_q\}$, where λ_q is the qth quantile of Z^0 as suggested by Storey and Tibshirani (2003a), e.g., 0.9 for q. The FDR value for a gene might be estimated as zero when no gene in the resampled null data exceeds its Z; in these cases the minimum estimate of FDR is forced to be the reciprocal of the product between the numbers of genes and resampled null data sets, which is the finest resolution of this RIR method of FDR estimation. The confidence bounds for $\widehat{FDR}(\Delta)$ at each threshold value c can also be obtained from the B resampled null data sets.

Table 7.1 *Numbers of differentially expressed genes discovered by five methods.*

FDR cutoff	BY	BH	SPLOSH	Mix-all	RIR
0.0001	1397	1730	2876	2542	2074
0.001	1730	2162	3134	2958	2485
0.01	2160	2849	3467	3694	3382
0.05	2670	3661	5654	4594	4548

Comparison with other FDR estimation methods

The full permutation (or *mix-all*) strategy of SAM randomly samples all intensity values across genes and conditions to generate null data, for which FDR estimation can be similarly performed as described above for our RIR approach. Benjamini and Hochberg (1995) proposed the step-up procedure to control FDR. These approaches can be compared with our RIR approach based on the LPE statistics in the following manner. Let $z_{(1)} \geq z_{(2)} \geq \cdots \geq z_{(G)}$ be LPE z-statistics for discovery of differential expression of G genes. Denote the corresponding ordered raw p-values as $p_{(1)} \leq p_{(2)} \leq \cdots \leq p_{(G)}$. BH adjusted p-values are defined as $\tilde{p}_{(i)} = \min_{k=1,\ldots,G}\{\min(p_{(k)}G/k, 1)\}$. For control of FDR at level α, a gene i is claimed as significant if $\tilde{p}_{(i)} \leq \alpha$. Thus, the BH estimate of FDR at a given critical value c can conservatively be defined as $\tilde{p}_{(i^*)}$, where i^* is $\min\{i : z_{(i)} \geq c\}$. The BY adjusted p-values are defined as $\tilde{p}_{(i)} = \min_{k=1,\ldots,G}\{\min(p_{(k)}G\sum_{j=1}^{G}(1/j)/k, 1)\}$. Utilizing the full information in the p-value distribution, the SPLOSH FDR estimate is derived as $h_{(i)} = \min_{k \geq i}(r_{(k)})$, where $r_{(k)}$ is the FDR estimate of gene k (Pounds and Cheng, 2004). A comparison between these five FDR estimation methods and the RIR method is shown in Table 7.1 for the mouse immune response microarray data.

Table 7.1 displays the numbers of the selected differentially expressed genes at FDR 0.0001, 0.001, 0.01, or 0.05. The results show that BH and BY are more conservative than others, whereas the SPLOSH and mix-all methods are more liberal than the others. Table 7.2 shows the FDR estimates for the five well-known genes that were reported and confirmed in the original study (Jain et al., 2003). The FDR estimates of several genes among them were greater than 0.01 by conservative BH and BY. FDR estimates for one or more genes were greater than 0.01 by SPLOSH and mix-all, whereas RIR identified all of these genes with FDR < 0.01.

7.3 Empirical Bayes heterogeneous error model (HEM)

7.3.1 Background

Microarray experiments are often performed under multiple experimental conditions. The statistical testing methods discussed above are inefficient and restrictive for analyzing such data sets because they have to be applied to each pairwise comparison

Table 7.2 *FDR estimates of well-known genes found to be differentially regulated.*

Gene Title	BY	BH	SPLOSH	Mix-all	RIR
CD97 antigen	.0230	.0023	.0489	< .0001	.0006
GATA-binding protein-3	.0208	.0021	.0489	< .0001	.0006
CD40 ligand transcript	.1005	.0103	< .0001	.0007	.0034
Granzyme K	.2768	.0277	.0524	.0037	.0091
Fas-associated factor-1	1	.1100	< .0001	.0335	.0038

among many different combinations of the multiple conditions. Analysis of variance (ANOVA) approaches have been suggested to examine and evaluate the statistical significance of differential expression one gene at a time, controlling for the random chance of false positives among all candidate genes in microarray data (Kerr and Churchill, 2001; Wolfinger et al., 2001).

Under a Bayesian testing framework, several approaches have been developed for analyzing microarray data, including Bayesian parametric modeling (Newton et al., 2001), Bayesian regularized t-test (Baldi and Long, 2001), Bayesian hierarchical modeling with a multivariate normal prior (Ibrahim et al., 2002), and Bayesian hierarchical error model with two error components (Cho and Lee, 2004). In order to improve the error estimation accuracy of large-screening microarray data, empirical Bayes (EB) techniques have also been applied (Efron et al., 2001; Newton and Kendziorski, 2003). In these cases, empirical Bayes priors are used for mixture distributions of equivalently and differentially expressed genes. Certain Bayesian approaches, including the Bayesian model presented in Ibrahim et al. (2002), have considered heterogeneous error variability in microarray data.

However, the error estimation in these classical and Bayesian approaches is not accurate when the number of replicated arrays is small. Furthermore, these modeling approaches are limited in that they are not able to capture heterogeneous error components accurately in microarray data due to the unidentifiability and computational restrictions of numerous error components. Consequently, these approaches do not provide a reliable statistical inference framework when the number of array replicates is very small, as is typically the case in investigating complex biological and biomedical mechanisms (Lee, 2002).

To remedy these restrictions, a heterogenous error model (HEM) approach is suggested to estimate heterogeneous technical and biological errors in microarray data separately and accurately (Cho and Lee, 2004). In particular, using LPE-estimated empirical Bayes prior specifications (Cho and Lee, 2008), HEM takes into account the fact that these two heterogeneous error components can often be observed separately, the former at different intensity ranges and the latter for different genes and conditions. Similar to the RIR-based FDR evaluation for LPE, a resampling-based evaluation of the FDR is also used for HEM, fully utilizing the distributional information of the original raw data.

7.3.2 Heterogeneous error modeling

Suppose that y_{ijkl} is the lth technically replicated gene expression value of the ith gene for a particular k^{th} individual sample under the jth biological condition, where $i = 1, \ldots, G; j = 1, \ldots, C; k = 1, \ldots, m_{ij}; l = 1, \ldots, n_{ijk}$. Assume that data are properly normalized and log-transformed (typically base 2). The heterogeneous error model (HEM) with two layers of error is considered as follows. HEM first separates the technical error e_{ijkl} from the observed expression value y_{ijkl} to obtain the expression value x_{ijk} free of the technical error. The first layer, thus, is defined as

$$y_{ijkl} | \{x_{ijk}, \sigma^2_{e_{ijk}}\} = x_{ijk} + e_{ijkl}. \tag{7.3}$$

The technical error e_{ijkl} is assumed to be i.i.d. $N(0, \sigma^2_{e_{ijk}})$, where its heterogeneous variance is defined to be a function of x_{ijk}, i.e., $\sigma^2_{e_{ijk}} = \sigma^2_e(x_{ijk})$. This assumption is based on the fact that such technical error variances vary on different intensity levels in microarray data.

In the subsequent layer, expression intensity x_{ijk} is decomposed into additive effects of gene, condition, and interaction:

$$x_{ijk} | \{\mu_{ij}, \sigma^2_{b_{ij}}\} = \mu_{ij} + b_{ijk} = \mu + g_i + c_j + r_{ij} + b_{ijk}, \tag{7.4}$$

where μ is the parameter for the grand mean; g_i and c_j are the parameters for the gene and condition effects, respectively; r_{ij} is the parameter for the interaction effect of each gene-condition combination; and b_{ijk} is the error term for the biological variation, assuming i.i.d. $N(0, \sigma^2_{b_{ij}})$. The biological variance parameter $\sigma^2_{b_{ij}}$ is allowed to be heterogeneous for each combination of gene i and condition j because each gene can have its inherent, distinctive biological variation under a specific biological condition. Note that this two-layer HEM is conceptually similar to the two-consecutive regression fitting suggested by Wolfinger et al. (2001); HEM inference is based on the complete likelihood of the two layers, whereas the two-stage ANOVA models are separately fit in the latter. Note also that the above two-layer HEM is suitable for analyzing microarray data with both biological and experimental replicates. This HEM method is slightly modified in the later section when only one of the biological and technical replicates is available. The following section describes the LPE-derived empirical Bayes prior specifications for HEM (non-mathematical readers may skip this section and refer to Cho and Lee (2008) for a more detailed technical description).

7.3.3 LPE-based empirical Bayes prior specifications

The two-layer HEM contains unobserved data as well as a large number of parameters. Most of these parameters can be efficiently estimated in a Bayesian framework, using conjugate priors such as a uniform distribution for μ and normal distributions for g_i, c_j, and r_{ij} with mean zero and variances σ^2_g, σ^2_c, and σ^2_r, respectively. Prior information is negligible or posterior distributions consistently converge to their target distribution when there are a large number of replicates. However, estimation of variance parameters, $\sigma^2_{b_{ij}}$ and $\sigma^2_{e_{ijk}}$, with a small number of replicates heavily depends upon the choice of priors; hence, constant gamma priors such as in Newton and Kendziorski (2003) are not enough to precisely estimate heterogeneous variance parameters in this case. In order to correctly estimate heterogeneous variances in microarray data with limited replication, strong informative priors are needed. Thus, informative LPE-estimated EB priors are used with a non-constant gamma prior Gamma($\alpha_b, \beta_{b_{ij}}$) for $\sigma^{-2}_{b_{ij}}$ with varying hyperparameters and a non-parametric prior for $\sigma^{-2}_{e_{ijk}}$. These LPE-based EB priors are constructed as follows.

Suppose that there are two biological replicates (1 and 2) and two technical replicates (a and b) in a condition, i.e., Y_{1a}, Y_{1b}, Y_{2a}, and Y_{2b}. For the technical error distribution, the variances of $Y_{1a} - Y_{1b}, Y_{1b} - Y_{1a}, Y_{2a} - Y_{2b}$, and $Y_{2b} - Y_{2a}$ are pooled to derive its baseline distribution because no biological variability between replicates a and b is involved. (Both differences, e.g., $Y_{1a} - Y_{1b}$ and $Y_{1b} - Y_{1a}$ are used to guarantee that the local mean is zero.) Similarly, for the biological error distribution, the variances of $Y_{1a} - Y_{2a}, Y_{2a} - Y_{1a}, Y_{2a} - Y_{1b}$, and $Y_{1b} - Y_{2a}$ are pooled to obtain its error distribution. Note, however, that the latter error distribution is for the *total* variance containing both technical and biological variances since both technical and biological errors are involved between replicates 1 and 2. Therefore, the biological variance estimate is obtained by LPE, subtracting the technical variance estimate from the corresponding total variance estimate at each local intensity region under their (orthogonal) independence assumption.

Based on the above modified-LPE estimates of the two error variances, the hyperparameters of the EB priors can be defined. The inverse of the biological variance, $\sigma^{-2}_{b_{ij}}$, is assumed to have a Gamma($\alpha_b, \beta_{b_{ij}}$) prior satisfying $E(\sigma^{-2}_{b_{ij}}) = \alpha_b/\beta_{b_{ij}}$. It follows that $E(\sigma^2_{b_{ij}}) = \beta_{b_{ij}}/(\alpha_b - 1)$ and $Var(\sigma^2_{b_{ij}}) = \beta^2_{b_{ij}}/(\alpha_b - 1)(\alpha_b - 2)$. For positive expectations and variances, a value α_b is chosen such that $\alpha_b > 2$ (e.g., $\alpha_b = 3$). Given α_b and modified-LPE estimates $\tilde{\sigma}^2_{b_{ij}}$, $\beta_{b_{ij}}$ is obtained by the method of moments, i.e., $\beta_{b_{ij}} = (\alpha_b - 1)\tilde{\sigma}^2_{b_{ij}}$. This provides Gamma priors that are dependent on each combination of gene and condition, so that a gene under a different condition has its specific error distribution for biological error. In contrast, technical error varies on different intensity levels, so that its baseline distribution can be estimated precisely. Therefore, the inverse of the technical error variance, $\sigma^{-2}_{e_{ijk}}$, is assumed to have a nonparametric prior rather than a Gamma prior. In order to fully utilize the LPE-estimated baseline distribution of technical error, a nonparametric EB prior specification is used for $\sigma^{-2}_{e_{ijk}}$ based on the following resampling algorithm:

1. Given a probability p, find sample quantiles $0 = \xi_0 < \xi_1 < \xi_2 < \ldots < \xi_Q$ of median intensities corresponding to probabilities $0 < p < 2p < \ldots < (Q-1)p < Qp = 1$.

2. Randomly sample gene vectors with replacement at each quantile range proportionally, and obtain a dataset $D^{(b)}$ with size G.

3. Apply LPE to $D^{(b)}$ with the above quantiles, and so obtain quantile experimental variance estimates $\tilde{\sigma}_{e_{qj}}^{2(b)}$, where $q = 1, \ldots, Q$.

4. Repeat steps 2 and 3 with B times, and obtain $\tilde{\sigma}_{e_{qj}}^{2(b)}$ where $q = 1, \ldots, Q; b = 1, \ldots, B$.

The above algorithm is performed for each condition for a Markov Chain Monte Carlo (MCMC) update described below.

7.3.4 HEM inference

The joint probability of the observed and unobserved variables for the two-layer HEM is $\Pr(\boldsymbol{y}, \boldsymbol{x}; \boldsymbol{\theta}) = \prod_{ijkl} \phi\left(\frac{y_{ijkl} - x_{ijk}}{\sigma_{e_{ijk}}}\right) \times \prod_{ijk} \phi\left(\frac{x_{ijk} - \mu - g_i - c_j - r_{ij}}{\sigma_{b_{ij}}}\right)$, where $\boldsymbol{\theta} = (\mu, \boldsymbol{g}, \boldsymbol{c}, \boldsymbol{r}, \sigma_{\boldsymbol{b}}^2, \sigma_{\boldsymbol{e}}^2)$ and ϕ is the density function of the standard normal distribution. With the above prior specification, the posterior distribution $\pi(\boldsymbol{x}, \boldsymbol{\theta}|\boldsymbol{y})$ of the unobserved data $\boldsymbol{x}$ and the parameters $\boldsymbol{\theta} = (\mu, \boldsymbol{g}, \boldsymbol{c}, \boldsymbol{r}, \sigma_{\boldsymbol{b}}^2, \sigma_{\boldsymbol{e}}^2)$, given the observed data $\boldsymbol{y}$, is proportional to

$$\Pr(\boldsymbol{y}, \boldsymbol{x}; \boldsymbol{\theta}) \prod_i \phi(g_i/\sigma_g) \prod_j \phi(c_j/\sigma_c) \prod_{ij} \phi(r_{ij}/\sigma_r) \prod_{ij} \Gamma(\sigma_{b_{ij}}^{-2}; \alpha_b, \beta_{b_{ij}}) \prod_{ijk} h(\sigma_{e_{ijk}}^{-2}),$$

where $\Pr(\boldsymbol{y}, \boldsymbol{x}; \boldsymbol{\theta})$ is the joint probability, $\Gamma(*; \alpha, \beta)$ is the density function of a Gamma distribution with mean α/β and variance α/β^2, and h is an unknown distribution.

In order to estimate such a large number of parameters and unobserved data, the MCMC technique is used to sample the parameters or unobserved data from their posterior conditional distributions (Gilks et al., 1995). Unobserved data and parameters except for technical error $\sigma_{e_{ijk}}^{-2}$ can be estimated by Gibbs sampling. The conditional posterior distribution of technical error $\sigma_{e_{ijk}}^{-2}$ cannot be obtained explicitly, so the Metropolis-Hastings algorithm is applied.

HEM summary statistic

Suppose that posterior estimates of parameters are obtained. Denote posterior estimates by $\bar{\mu}, \bar{g}_i, \bar{c}_j, \bar{r}_{ij}, \bar{\sigma}_{b_{ij}}^2$ and $\bar{\sigma}_{e_{ijk}}^2$, and let $\bar{\mu}_{ij} = \bar{\mu} + \bar{g}_i + \bar{c}_j + \bar{r}_{ij}$. Based on these posterior estimates, one still needs a guiding statistic to evaluate the significance of overall differential expression patterns. Therefore, the HEM summary statistic, H-score is defined by utilizing the posterior estimates:

$$H_i = \sum_{j=1}^{C} \frac{w_{ij}(\bar{\mu}_{ij} - \bar{\bar{\mu}}_i)^2}{(\bar{\sigma}_{b_{ij}}^2 + \sum_{k=1}^{m_{ij}} \bar{\sigma}_{e_{ijk}}^2/m_{ij})}, \tag{7.5}$$

where $w_{ij} = m_{ij}/\sum_{j=1}^{C} m_{ij}$ and $\bar{\bar{\mu}}_i = \sum_{j=1}^{C} \bar{\mu}_{ij}/C$. The form of H-score is similar to the ANOVA F-statistic; however, H-score utilizes variance estimates that are non-constant over conditions as well as genes, and separately account for change of each condition divided by its own variance. The HEM H-statistic does not follow any parametric distribution. Differentially expressed genes have large H-scores, so gene selection can be performed based on the magnitude of its score. A rigorous selection criterion of differentially expressed genes is detailed by Bayesian FDR evaluation below.

7.3.5 Resampling-based Bayesian FDR evaluation for HEM

As it was for the RIR-based FDR evaluation for LPE, it is again important to generate *biologically relevant* null distributions of small-sample microarray data for the HEM application. In order to obtain null data simulating biological microarray data, gene and condition identities are preserved in our resampling. That is, all of $y_{ij,1,1}$, $\ldots$, $y_{ij,m_{ij},n_{ijk}}$ for gene i and condition j are sampled simultaneously for a simulated gene under each condition. For example, consider a microarray study with two conditions. Suppose gene i is selected for condition 1. For condition 2, gene i' is then selected with the normal probability for $(\mu_{i,1} - \mu_{i',2})$, so that genes with means closer to the mean of gene i are more likely to be sampled. Gene vectors i and i' are then combined as a gene vector in the (simulated) null data. Similarly, gene vectors are selected for all conditions simultaneously if there are multiple conditions. This strategy maintains gene and chip identities so that their corresponding variance structure can be retained.

Suppose H-statistics and and H^0-statistics are computed from raw data and generated null data, respectively. Generation of the null data is repeated B times independently. Given a critical value Δ, the estimate of Bayesian FDR is calculated by Equation 7.2. In the equation,

$$\bar{R}^0(\Delta) = \#\{H_{ib}^0|H_{ib}^0 \geq \Delta, \; i = 1,\ldots,G, b = 1,\ldots,B\}/B$$

is the average number of significant genes in the null data, and

$$R(\Delta) = \#\{H_i|H_i \geq \Delta, \; i = 1,\ldots,G\}$$

is the number of significant genes in the raw data. The estimate of a correction factor with the λ-quantile m_λ of H_{ib}^0 is $\hat{\pi}_0(\lambda) = \#\{H_i|H_i \leq m_\lambda\}/\#\{H_{ib}^0|H_{ib}^0 \leq m_\lambda\}$, which is required because of the different numbers of true null genes in the raw data and the null data.

7.3.6 HEM with only one type of replication

The two-layer HEM described above is for microarray data when both technical and biological replicates are available. If a microarray study does not have technical replicates but has some biological replicates, the technical error distribution cannot be separately observed and two error distributions are therefore confounded. In contrast, if only technical replicates are available, the biological variability cannot be observed from the data; this kind of array experiment may be performed for examining the effects of specific biological treatments and conditions on a single subject. In these cases, HEM is reduced into a model with one layer as follows:

$$y_{ijk}|\{\mu, g_i, c_j, r_{ij}, \sigma^2_{\epsilon_{ij}}\} = \mu + g_i + c_j + r_{ij} + \epsilon_{ijk}, \tag{7.6}$$

where $i = 1, \ldots, G, j = 1, \ldots, C, k = 1, \ldots, m_{ij}$ and ϵ_{ijk} is the error term for the biological and experimental error variation, assuming i.i.d. $N(0, \sigma^2_{\epsilon_{ij}})$. Note that m_{ij} is the number of technical (or biological) replicates in this model. The other parameters are the same as those in the two-layer model and the l-subscript is suppressed in this model.

For the one-layer HEM, the joint probability of the observed variables is $\mathbb{P}r(y; \theta) = \prod_{ijk} \phi(y_{ijk} - \mu - g_i - c_j - r_{ij})/\sigma_{\epsilon_{ij}}$, where $\theta = (\mu, g, c, r, \sigma^2_b)$. In this case the H-score summary statistic is defined slightly differently as

$$H_i = \sum_{j=1}^{C} w_{ij}(\bar{\mu}_{ij} - \bar{\bar{\mu}}_i)^2/\bar{\sigma}^2_{\epsilon_{ij}}.$$

The LPE-based nonparametric prior distribution is used for variance parameter $\sigma^2_{\epsilon_{ij}}$, similarly as before.

7.3.7 Examples

Example 1. Ionizing radiation response data.   The two-layer HEM with microarray data is applied to the transcriptional response microarray data of lymphoblastoid cells to ionizing radiation of which details can be found in Tusher et al. (2001). In brief, two wild-type human lymphoblastoid cell lines (1, 2) were grown in an unirradiated state (U) or in an irradiated state (I) 4 hours after exposure to a modest dose of 5 Gy of ionizing radiation. RNA samples from each combination of the two cell lines and two states were labeled and divided into two identical aliquots (A, B) that were hybridized independently to the Affymetrix HuGeneFL GeneChip, generating eight hybridized microarrays (U1A, U1B, U2A, U2B, I1A, I1B, I2A, I2B). Signal intensity values were obtained using MAS 5.0 and $\log_2$-transformed.

The scatter plots of log-expression values between the two aliquots, two cell lines, and two conditions demonstrate that larger variability exists between the two cell lines than between the two aliquots (data not shown). This implies that the biological variability is distinguishable from technical variability. Accordingly, these data are fit to the two-layer HEM with $m_{ij} = 2$ biological replicates and $n_{ijk} = 2$ technical replicates for $G = 7129$ genes under $C = 2$ biological conditions. For the prior distributions $g_i \sim N(0, \sigma^2_g)$, $c_j \sim N(0, \sigma^2_c)$, $r_{ij} \sim N(0, \sigma^2_r)$, we assume $\sigma^2_g = \sigma^2_c = \sigma^2_r = 1$. The LPE-based EB prior specifications is used for a Gamma prior on biological error and a nonparametric prior for technical error. For the MCMC runs, 2,000 burn-ins and 10,000 main iterations are executed and their updated parameters values collected.

Figure 7.3 displays the top 100, 500, and 1,000 genes on the MA plots of $(U - I)$ against $(U + I)/2$. With HEM the selected genes are well distinguished from the distribution of random genes (first column, Figure 7.3), demonstrating that HEM successfully identifies differentially expressed genes by capturing heterogeneous error variability in microarray data. In contrast, with SAM (Tusher et al., 2001) the boundary between the selected and unselected genes becomes obscure as the number of the selected genes increases and many random genes seem to be identified significantly (second column, Figure 7.3). The results of HEM and SAM differ because HEM estimates heterogeneous technical and biological variances separately, whereas SAM includes a variance stabilizing factor that is common to all genes.

In these data, the variability between two conditions is not as large as it is within conditions; this implies that it is difficult to identify differential expression or that there are a small number of differentially expressed genes. When $\widehat{FDR} = 0.01$ and 0.05, the thresholds of HEM H-scores are 2.98 and 1.69, respectively. In each of the cases, 11 and 17 genes are claimed as significant while no genes are claimed as significant under these levels of FDR by the corresponding SAM analysis. This confirms that the between-condition variation is not large in this case (the opposite case is described next).

Example 2. T cell immune response data. A one-layer HEM is applied to the mouse immune response microarray data (Jain et al., 2003) since this experiment comprises three experimental conditions and only technical replicates. Many genes are found to be differentially expressed with a small FDR value, e.g., 2,464 genes are claimed as significant with an FDR less than 0.001 because variations between conditions are much larger than those within each condition. Many important genes also have large HEM H-scores. Table 7.3 displays the top ten genes selected by HEM, including their scores from HEM and SAM. The genes with the largest HEM H-scores have large SAM d-scores as well; five of them are in the top ten of SAM. We note that unlike the above ionizing radiation data example, which has both technical and biological replicates, this T cell immune response microarray study has only technical replicates. Therefore, both HEM and SAM provide error estimates based on a single error term, so that the difference between their error estimates is relatively small in this study. However, when there exists unusual (very low or high) variability by chance among a small number of replicates for a gene (compared to other genes in the same local intensity range), its HEM error estimate can be significantly different from its SAM error estimate. Recall that the latter still is heavily weighted by the within-gene error estimate from the small number of replicates.

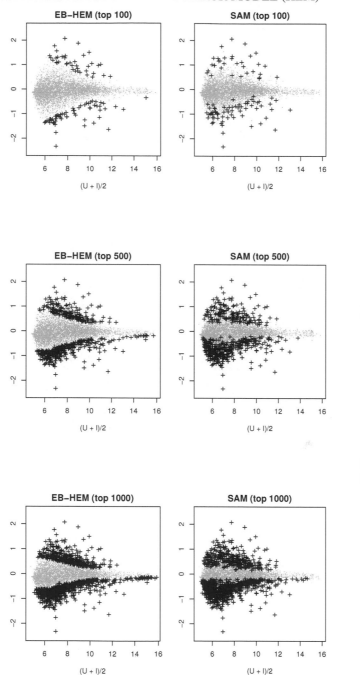

Figure 7.3 *Identification of differentially expressed genes by HEM or SAM for the ionizing radiation response data; dark (+) points represent top genes with large HEM H-scores (left panels) or SAM d-scores (right panels).*

Table 7.3 *Top 10 genes selected by HEM for the T cell immune response data.*

	HEM		SAM	
Gene name	H-score	Rank	d-score	Rank
Granzyme A	1362.7	1	2.02	3
Ubiquitin specific protease 18	1163.3	2	1.84	8
Chemokine (C-C) receptor 2	1148.7	3	1.95	7
S100 calcium binding protein A6 (calcyclin)	1142.4	4	1.56	22
Disintegrin and metalloprotease domain 8	1033.8	5	1.96	5
Granzyme K	984.0	6	2.12	2
Cytotoxic T lymphocyte-assoc. protein2 α	969.7	7	1.46	34
Chemokine (C-C motif) receptor 5	831.9	8	1.42	38
Annexin A1	817.2	9	1.70	14
Chemokine (C-X-C motif) receptor 3	686.6	10	1.22	72

7.4 Conclusion

In this chapter we introduced LPE and LPE-based empirical Bayes HEM methods to accurately capture varying technical and/or biological error variances among different genes, experimental conditions, and intensity ranges in microarray data. It is difficult to estimate such numerous, heterogenous error components using classical statistical estimation and standard Bayesian approaches in small-sample microarray data analysis. To overcome this limitation, we used advanced error-pooling techniques, such as Local Pooled Error estimation and LPE-based empirical Bayes specifications. These information-pooling approaches not only enabled us to precisely capture heterogeneous error components with limited replication, but also dramatically improved the statistical power for identification of differential expression genes, compared to widely-used SAM and ANOVA approaches in small-sample microarray data analysis.

However, as both parametric and nonparametric tests are often used simultaneously in traditional statistical analysis, researchers may want to compare the statistical results from two or more different methods in this kind of small-sample microarray analysis, especially since error distribution structures can vary substantially between different microarray data sets. We also found in applying these tests that permutation or resampling-based evaluation of significance is more accurate than analytic derivation based on certain distributional assumptions.

Acknowledgments

This research was supported by US NIH grant 1R01HL081690 to J.K.L.

CHAPTER 8

Significance testing for small microarray experiments

Charles Kooperberg, Aaron Aragaki, Charles C. Carey, and Suzannah Rutherford

8.1 Introduction

When a study has many degrees of freedom it is sometimes less critical which significance test is carried out, as most analyses will give approximately the same result. However, when there are few degrees of freedom the choice of which significance test to use can have a strong effect on the results of an analysis. Unfortunately, this small degrees of freedom situation is often the case for microarray experiments, as many research laboratories perform such experiments with only a few repeats. Reasons for the small number of repeats include specimen availability and economics. Kooperberg et al. (2005) compare several approaches to significance testing for experiments with a small number of oligonucleotide arrays (a one-color technology; see Section 1.4). This chapter summarizes results from that analysis and describes a similar comparison for methods of carrying out significance testing for two-color arrays (e.g., cDNA arrays).

The large variability that even the most precise microarray platforms have makes small-sample comparisons unattractive. A standard t-test for an experiment with six two-color arrays has, depending on whether other variables are controlled for, at most five degrees of freedom. The resulting two-sided test, with $\alpha = 0.05$ and a Bonferroni correction for 10,000 genes requires a t-statistic value of at least 20.6 for significance. The lack of degrees of freedom drives the extremely large significance threshold for t-statistics: the same α and Bonferroni correction for 20 arrays requires a t-statistic of 6.3 while a normal distribution only requires a Z-statistic of 4.6. On the other hand, reducing the number of genes of interest on the original array from 10,000 to 500 only reduces the required t-statistic to 11.3.

Nonparametric (Wilcoxon) or permutation tests do not provide a simple solution to the significance problem. For example, for an experiment with n two-color arrays, a p-value for a permutation test can be no smaller than 2^{-n}; a two-sided test with $\alpha = 0.05$ and a Bonferroni correction for 10,000 genes requires n to be at least 19.

Reducing the number of genes to 500 reduces the minimum n to only 15. Similarly, for a one-color array the p-value for a permutation test with n cases and n controls cannot be smaller than $\binom{2n}{n}n$; so for a two-sided test with $\alpha = 0.05$ and a Bonferroni correction for 10,000 genes, at least $2n = 22$ arrays are needed. Reducing the number of genes to 500 reduces the minimum number of arrays to 18.

There is thus a need for a better estimate of the residual variance to overcome the lack of repeats. Combining information can be helpful in this regards. There are two obvious choices available: combine different genes in the same experiment or combine different experiments, if similar experiments were carried out. For combining genes, we can choose either to combine those genes for which the general expression level is similar (see e.g., Huang and Pan (2002) and Jain et al. (2003)) or to combine all genes. Alternative approaches to obtain more power with small experiments are to add a stabilizing constant to the estimate of the variance for each gene or to use a (Bayesian) model for the expression levels. Significance Analysis of Microarrays (SAM; Tusher et al., 2001) is a methodology that adds a constant to the estimate of the SD. The approaches by Baldi and Long (2001), Lönnstedt and Speed (2002), Smyth (2004) and Cui et al. (2005) are four related (empirical) Bayesian approaches. Wright and Simon (2003) discuss a closely related frequentist approach.

In practice, when carrying out tests for many thousands of genes simultaneously, a multiple testing correction is essential (Section 7.2.3; see Dudoit et al. (2003) for an extensive overview). However, the focus here is on obtaining a well-calibrated marginal p-value, so we do not control for multiple comparisons.

8.2 Methods

Most of the methods that we compare here can be used either for one-color arrays or for two-color (spotted) arrays. We assume that the arrays have been properly normalized; see Section 8.6 for preprocessing details for the experiments we analyze here.

8.2.1 Notation

Two-color arrays.   For each gene and each two-color array, the value x_{ijl}^M summarizes the ($\log_2$-)expression ratio (M-value; see Section 1.4.1) between experimental conditions $k = 1$ and $k = 2$ (these may be different between experiments) for gene $i = 1, \ldots, g$ in experiment $j = 1, \ldots J$ on replicate array $l = 1, \ldots, L_j$. For each gene on each array there is also an estimate of the overall expression level x_{ijl}^A, typically this will be the average of the normalized $\log_2$ expression for both channels of the array. Unless there is confusion we write x_{ijl} instead of x_{ijl}^M for the $\log_2$-expression ratios.

Let μ_{ij} be the "true" (mean) $\log_2$-expression ratio of gene i in experiment j for condition 1 relative to condition 2. Set $\widehat{\mu}_{ij} = \sum_l x_{ijl}/L_j$, $s_{ij}^2 = \sum_l (x_{ijl} - \widehat{\mu}_{ij})^2$, and $x_{ij}^A = \sum_l x_{ijl}^A/L_j$.

One-color arrays. Similarly, for each gene and each one-color array let x_{ijkl} be the ($\log_2$-)expression value for experimental condition $k = 1$ or $k = 2$, for gene $i = 1, \ldots, g$ in experiment $j = 1, \ldots J$ on replicate array $l = 1, \ldots, L_{jk}$.

Let μ_{ijk} be the "true" mean ($\log_2$-)expression level of gene i in experiment j under condition k. Set $\widehat{\mu}_{ijk} = \sum_l x_{ijkl}/L_{jk}$ and $s_{ijk}^2 = \sum_l (x_{ijkl} - \widehat{\mu}_{ijk})^2$.

8.2.2 Significance tests

All significance tests that we consider can be written in the form

$$\frac{\widehat{\mu}_{ij}}{\widetilde{\sigma}_{ij}/\sqrt{L_j}}$$

for two-color arrays and

$$\frac{\widehat{\mu}_{ij1} - \widehat{\mu}_{ij2}}{\widetilde{\sigma}_{ij}\sqrt{\frac{1}{L_{j1}} + \frac{1}{L_{j2}}}}$$

for one-color arrays; $\widetilde{\sigma}_{ij}^2$ is an estimate of the variance of x_{ijl}, so $\widetilde{\sigma}_{ij}$ estimates the standard deviation (SD). The methods discussed here differ primarily in how the estimate $\widetilde{\sigma}_{ij}$ is obtained. The traditional test statistics estimate $\widetilde{\sigma}_{ij}$ based only on the data on gene i in experiment j. Approaches that inflate the variance or that combine genes also use data on genes i^*, $i^* \neq i$, either implicitly, to estimate hyperparameters for the empirical Bayes approach that inflates the variance, or explicitly, to smooth the estimates for $\widetilde{\sigma}_{ij}^2$. Finally, the approaches that combine experiments use data on experiments j^*, $j^* \neq j$. Most of the methods below have a defined reference (null) distribution, but alternatively significance levels can be obtained using permutations (see Section 8.2.3); in fact, some authors recommend permutations as the method to obtain p-values.

Below we describe the test statistics included in the comparison. We provide details for the two-color arrays; modifications for one-color arrays are indicated. All these approaches are either already implemented in R packages available from Bio-Conductor (http://www.bioconductor.org) or CRAN (http://cran.r-project.org), or are easily programmed in R code.

Traditional single gene within-experiment method

t-statistic. The traditional t-statistic is

$$t_{ij} = \frac{\widehat{\mu}_{ij}}{\widehat{\sigma}_{ij}/\sqrt{L_j}},$$

where $\widehat{\sigma}_{ij}^2 = s_{ij}^2/(L_j - 1)$, provided $L_j > 1$. The reference distribution is the t-distribution with $L_j - 1$ degrees of freedom, and the main assumption is that for each gene i and experiment j the x_{ijkl} are independent having a normal distribution with variance σ_{ij}, although the t-test is generally considered to be robust against departures from normality.

The two-sample t-statistic is the equivalent test for one-color arrays. Use of this statistic assumes that the variance for both experimental conditions is the same. An alternative is the Welch (1938) two-sample t-statistic that does not make that assumption. This approach has almost no power for small sample sizes (Kooperberg et al., 2005), and should probably be avoided for small microarray experiments.

Methods combining genes: smoothing the variance

There have been several proposals in the literature to combine the estimates of the variance for several genes to obtain better estimates, so that the resulting test has more degrees of freedom. Typically the assumption that is made is that genes with the same expression level have approximately the same variance. Under this assumption estimates for the variance can be obtained by smoothing the variance as a function of the expression level. For one-color arrays there are methods which smooth the variances jointly and methods which smooth variances separately for both experimental conditions.

LPE. Jain et al. (2003) describe the Local Pooled Error test method (LPE), applicable to one-color arrays where both experimental conditions are measured separately. This method is outlined here and described in detail by Lee, Cho, and O'Connell (Chapter 7 of this volume). In this approach, let $\widehat{\sigma}^2_{ijk}$ be the the sample variance of the x_{ijkl}, for $l = 1, \ldots, L_{jk}$. LPE regularizes these estimates for each j and k separately by smoothing the $\widehat{\sigma}^2_{ijk}$ versus $\widehat{\mu}_{ijk}$. The assumption being made here is that genes with the same expression level for the same experiment and the same condition have (approximately) the same variance. Since the smoothing spline that is used effectively involves averaging a large number of genes, the authors use a normal reference distribution.

Loess. Huang and Pan (2002) make several related proposals. The main difference between their approach and *LPE* is that they first compute $\widehat{\sigma}^2_{ij}$ and smooth these estimates against $\widehat{\mu}_{ij} = \widehat{\mu}_{ij1} + \widehat{\mu}_{ij2}$ for one-color experiments and against x^A_{ij} for two-color experiments. Their simulation results show that, not unexpectedly, for the null-model a normal reference distribution is appropriate.

Methods combining genes: (empirical) Bayesian model for σ Rather than smoothing the variance explicitly as a function of the expression level, we can include information from other genes for the analysis of a particular gene by making assumptions about the distribution of the variance for all genes. The information about the other genes then allows us to estimate (hyper)parameters that can be used to stabilize the variance estimate. There are several such methods, based on different motivations: *ad hoc* (Tusher et al., 2001), (empirical) Bayes argument (Baldi and Long, 2001; Lönnstedt and Speed, 2002; Smyth, 2004), James-Stein-type estimation (Cui et al., 2005), or a frequentist approach (Wright and Simon, 2003).

Some approaches combine the sample variance $\widehat{\sigma}^2_{ij}$ with another estimate σ^2_{0ij} that has d_{ij} degrees of freedom, yielding a variance estimate

$$\widetilde{\sigma}_{ij}^2 = \frac{d_{ij}\sigma_{0ij}^2 + (L_j - 1)\widehat{\sigma}_{ij}^2}{d_{ij} + L_j - 1}, \tag{8.1}$$

that can be used in a t-test with $d_{ij} + L_j - 1$ degrees of freedom. The methods *Cyber-T* and *Limma* use this approach; they differ primarily in the methods to obtain σ_{0ij}^2 and d_{ij}.

Cyber-T. The *Cyber-T* approach of Baldi and Long (2001) is motivated as a fully Bayesian procedure. However as implemented in practice (Baldi and Long, 2001, Section 5) the test is carried out using a t-test on (for two-color arrays) $\nu_0 + L_j - 1$ degrees of freedom, and an variance estimate (compare Equation 8.1)

$$\widetilde{\sigma}_{ij}^2 = \frac{\nu_0\sigma_{0ij}^2 + (L_j - 1)\widehat{\sigma}_{ij}^2}{\nu_0 + L_j - 1}, \tag{8.2}$$

where σ_{0ij}^2 is an estimate of the "prior variance" that is obtained as a running average of the variance estimates of the genes in a "window" of size w of similar x_{ij}^A. Thus, the *Cyber-T* approach uses the average of a smoothed variance (like *LPE* and *Loess*, just using a different smoother) with the regular variance of the t-statistic. A non-Bayesian interpretation of *Cyber-T* is thus that it combines a smoothed estimate (as in *Loess* and *LPE*) with a traditional estimate from the t-test.

We use the default values $\nu_0 = 10$ and window width $w = 101$ in the R software available at http://cybert.microarray.ics.uci.edu. (Note that in Baldi and Long (2001) a different default value of $\nu_0 = 10 - L_j$ is mentioned.)

Limma. Smyth (2004) generalizes the approach of Lönnstedt and Speed (2002). The main assumption in Smyth's model is a prior distribution on the variances σ_{ij}^2:

$$\frac{1}{\sigma_{ij}^2} \sim \frac{1}{d_{0j}s_{0j}^2}\chi_{d_{0j}}^2.$$

The model also includes priors on the coefficients for each gene in a linear regression model, which in the two-sample case reduces to the difference between the mean expression for the two groups. By the method of moments, estimates of d_{0j}, s_{0j}^2, and other parameters are obtained. An inflated variance

$$\widetilde{\sigma}_{ij}^2 = \frac{d_{0j}s_{0j}^2 + (L_j - 1)\widehat{\sigma}_{ij}^2}{L_j + d_{0j} - 1} \tag{8.3}$$

(compare Equation 8.2) is used for a "moderated t-test" with $d_{0j} + L_j - 1$ degrees of freedom. Thus, a main difference between the *Limma* approach of Smyth (2004) and the *Cyber-T* approach of Baldi and Long (2001) is that *Limma* uses one single estimate for the prior variance (s_{0j}^2) for all genes and estimates the prior degrees of freedom d_{0j} based on the data, while *Cyber-T* uses a smooth estimate for the prior variance σ_{0ij}^2, but it uses a fixed number of prior degrees of freedom ν_0. The approach of Smyth (2004) is implemented in the BioConductor package limma (Smyth, 2005).

Shrinking. Cui and Churchill (2003) and Cui et al. (2005) develop a James-Stein shrinkage estimate $\widetilde{\sigma}_{ij}^2$. After appropriate transformations this estimator "shrinks" the t-test estimate $\widehat{\sigma}_{ij}^2$ toward the mean variance $\sum_{i=1}^{n} \sigma_{ij}^2/In$, where the exact amount of shrinkage differs from gene to gene, and depends on the variability for that gene. Easy to implement formulas are given in Cui et al. (2005). The authors of this method recommend a permutation approach (see Section 8.2.3) to obtaining p-values. We include this approach without permutations using a normal reference distribution, as well as with the permutation p-values.

Methods combining experiments

Instead of simply combining different genes *within* one experiment, we can also combine expression levels of the same gene *between* experiments carried out using the same microarray platform. This would potentially be useful if there are several smaller experiments for which it is reasonable to assume that for each gene the variance in each experiment is approximately the same.

Pooled-t. We define the pooled t-test statistic, combining experiments, as

$$c_{ij} = \frac{\widehat{\mu}_{ij}}{\widehat{\sigma}_i\sqrt{\frac{1}{L_j}}},$$

where $\widehat{\sigma}_i^2 = \sum_j s_{ij}^2/L$ and $L = \sum_j(L_j - 1)$, provided $L > 0$. The reference distribution is the t-distribution with L degrees of freedom, and the main assumption is that the x_{ijl}^M are independent for each j and l, having a normal distribution with mean μ_{ij} and variance σ_i^2.

It is in principle also possible for the other methods discussed above to pool different experiments in obtaining a single variance estimates. Since these methods already regularize the estimates for σ^2 in some way, pooling typically has little or no effect, and the corresponding combined method behaves similarly to the "parent" method (Kooperberg et al., 2005).

8.2.3 Permutation p-values

Permutation of the arrays in an experiment can be an alternative to using a parametric reference distribution for a test statistic. Assume that we have a two-color experiment with L arrays, and that the test statistic for the ith gene is T_i. To compute the significance of T_i we also compute the test statistics for all genes for each of the $m = 1, \ldots, 2^L$ experiments that are obtained by "flipping" the signs of the x_{il}^m for some of the l. (We omit the index of experiment j.) Note that one of these permutations will be the original design. Let T_i^m be the test statistic for the ith gene for the mth permutation. We estimate the p-value corresponding to T_i as

$$\sum_{i^*=1}^{n} \sum_{m=1}^{2^L} I(T_i < T_{i*}^m)/n2^L,$$

where $I(\cdot)$ is the indicator function. If L is larger than, say, eight, it may be preferred to sample permutations (rather than computing all possible permutations) to save computing time.

These estimates will be unbiased if (i) each T_i has the same distribution under the null-hypothesis, and (ii) no genes are differentially expressed. The first assumption is not as severe as it might appear, since no particular parametric form for the common distribution is assumed. The second assumption is much more severe, and it will lead to conservative p-values when in fact a substantial number of genes are differentially expressed (Storey and Tibshirani, 2003b).

For one-color arrays, we randomly rearrange the L_1 arrays with the first experimental condition and the L_2 arrays with the second experimental condition, and proceed in a similar manner.

8.3 Data

We analyze two sets of data. One comes from an unpublished study of Drosophila, and the other comes from a one-color experiment that is analyzed in Kooperberg et al. (2005).

Table 8.1 *Organization of the two-color data. Experiments whose code starts with a D (different) are expected to have differences between both groups, while those starting with an S (same) are repeats; the digit "2" refers to a two-color array. The arrays for experiments D2.3 and D2.4 and those for D2.5 and D2.6 are different; experiment S2.1 are arrays from a cell-line not used for the other experiments.*

Exp.	Sample 1	Sample 2	L_j	Different
S2.1	KC cell	KC cell	4	no
S2.2	SAM	SAM	2	no
S2.3	SAM	SAM	2	no
S2.4	SAM	SAM	4	no
D2.1	SAM	D-recomb 304	2	yes
D2.2	SAM	D-recomb 220	2	yes
D2.3	SAM	D-pure	2	yes
D2.4	SAM	D-pure	4	yes
D2.5	SAM	E-pure	4	yes
D2.6	SAM	E-pure	4	yes
D2.7	SAM	F-pure	6	yes

The two-color experimental data come from a series of spotted microarrays (13,440 spots) of *Drosophila melanogaster* that were grown in Suzannah Rutherford's lab at the Fred Hutchinson Cancer Research Center. All experiments are "dye-swapped": i.e., half of the arrays have sample one on the red channel (and therefore sample two in green), the other half have sample two on the red channel (with sample one in green). The arrays that we compare here include some experiments that are self-self hybridizations, and some experiments where both samples are genetically different (see Table 8.1). In a self-self hybridization the two labeled samples are from the same source, so no genes are in fact differentially expressed and those identified as such are false positives. Thus, the experiments S2.1, S2.2, S2.3, and S2.4 are intended to establish that the tests have the right size Type I error, and the experiments D2.1, D2.2, D2.3, D2.4, D2.5, D2.6, and D2.7 are intended to establish power properties of the tests.

One-color experimental data was obtained using Affymetrix Mu 11K-A microarrays (6,595 probe sets) generated for a series of experiments on Huntington's disease (HD) mouse models. The results of these experiments are reported in a series of related papers (Chan et al., 2002; Luthi-Carter et al., 2002a,b). For this analysis we compare cerebellar gene expression in similarly aged mice carrying either a wild type or mutant form of the HD gene. Every comparison reported in Chan et al. (2002), Luthi-Carter et al. (2002a) and Luthi-Carter et al. (2002b) shows some differentially expressed genes, although the amounts of differential expression differ considerably between the experiments. For each of the experiments both groups had between two and five mice. Thus, all the repeats use different samples (sometimes referred to as "biological replicates") and are not repeat arrays using the same samples (sometimes referred to as "technical replicates"). The one-color experiments are listed in Table 8.2. Again, test size is examined with the S experiments and power with the D experiments.

Table 8.2 *Organization of the one-color (Affymetrix) data. HD: Huntington's disease mouse, WT: wild type mouse. Experiments whose code starts with a D are expected to have differences between both groups, while those starting with an S are repeats; the digit "1" refers to a one-color (Affymetrix) array.*

Exp.	Tissue	Mouse	Group 1	Group 2	L_{j1}	L_{j2}	Different
S1.1	cerebellum	DRPLA 26Q	HD	HD	2	2	no
S1.2	cerebellum	DRPLA 26Q	WT	WT	2	2	no
S1.3	cerebellum	YAC	HD	HD	3	2	no
S1.4	cerebellum	YAC	WT	WT	3	2	no
D1.1	cerebellum	DRPLA 65Q	HD	WT	4	4	yes
D1.2	cerebellum	R6/2 12 weeks	HD	WT	2	2	yes
D1.3	cerebellum	N171	HD	WT	4	4	yes

8.4 Results

We analyze the experiments listed in Section 8.3 using the methods described in Section 8.2.2. For the experiments where both groups are different (D2.x and D1.x) we prefer methods with the largest percentage of significant genes (the largest power), provided that the method does have the correct percentage of significant genes in the experiments where both groups are the same (S2.x and S1.x): i.e., at most $\alpha\%$ significant genes when tested at significance level $\alpha\%$. This power comparison is fair when the Type I error rate is controlled at the same level α.

We show results for $\alpha = 1\%$ and $\alpha = 0.01\%$. For the two-color arrays there are approximately 11,000 genes after removal of spots (genes) whose intensities are too close to the background level (see Section 8.6). Assuming independence of genes, a 95% confidence interval for the percentage of significance genes based upon the binomial distribution is between 0.8 and 1.2% at $\alpha = 1\%$ and between 0 and 0.03% at $\alpha = 0.01\%$. For the one-color arrays there are 6,595 genes, thus these confidence intervals are slightly larger (0.75 through 1.25% at $\alpha = 1\%$ and 0 and 0.045% at $\alpha = 0.01\%$). When we average four experiments and (incorrectly) assume independence, we expect between about 0.9 and 1.1% significant genes at $\alpha = 1\%$ and between 0 and 0.025% at $\alpha = 0.01\%$ for both array types.

8.4.1 Bandwidth selection for smoothers

The methods *Cyber-T*, *LPE*, and *Loess* require the choice of a bandwidth or smoothing parameter. For *LPE* and *Loess* this determines over how many genes the variance is "averaged". For *Cyber-T* the averaged variance is combined with the variance for the individual genes.

Table 8.3 summarizes the results for the two-color experiment for the *Loess* approach. The parameter span for the loess function in R (Ihaka and Gentleman, 1996) is approximately linear in the bandwidth for a local linear smoother. Table 8.3 shows that the bandwidth has very little influence on the results. The explanation for this is that even for the smallest bandwidth the variances of several dozen genes are effectively averaged. Smaller values of span are not useful, as they lead to numerical problems in regions with little data.

For all four choices of span and for all S2.x experiments at $\alpha = 0.01\%$ and for two of the four of these experiments at $\alpha = 1\%$, the percentage of genes called significant is much too large. This was concluded by Kooperberg et al. (2005) for the one-color arrays.

For the remainder of the comparisons we use a span of 0.1, which yields the lowest average number of significant results for both $\alpha = 1\%$ and $\alpha = 0.01\%$ for the four S2.x experiments. As the influence of the bandwidth appears minimal, we use *Cyber-T* and *LPE* with their default values.

Table 8.3 *Performance of the Loess approach with varying bandwidth* (span) *for the two-color experiments. We report the percentage of genes called differentially expressed at levels* $\alpha = 1\%$ *and* $\alpha = 0.01\%$. *Ideally the four S2.x experiments would have* α *differentially expressed genes, while the seven D2.x would have many such genes.*

span	$\alpha = 1\%$				$\alpha = 0.01\%$			
	10	1	0.1	0.01	10	1	0.1	0.01
S2.1	1.1	1.1	0.7	0.7	0.340	0.306	0.198	0.159
S2.2	7.8	7.0	5.8	6.6	2.884	2.507	1.528	1.915
S2.3	2.2	2.1	2.0	2.0	0.984	0.922	0.982	0.942
S2.4	0.7	0.6	0.6	0.6	0.262	0.262	0.230	0.212
S2-ave	3.0	2.7	2.3	2.5	1.118	0.999	0.735	0.807
D2.1	25.8	25.9	26.8	27.1	11.941	11.994	12.698	12.827
D2.2	31.7	31.8	32.3	32.9	16.817	17.000	17.682	18.300
D2.3	53.5	53.6	53.8	53.8	38.170	38.354	38.368	38.457
D2.4	54.3	54.4	54.4	54.7	37.709	37.858	37.774	38.043
D2.5	43.3	43.5	43.5	44.2	28.006	28.190	28.225	28.574
D2.6	73.0	73.2	76.5	76.6	62.230	62.431	66.313	66.501
D2.7	62.1	62.3	64.3	64.3	47.863	48.003	50.124	50.471
D2-ave	49.1	49.2	50.2	50.5	34.677	34.833	35.883	36.168

8.4.2 Comparison of methods

Tables 8.4 and 8.5 show the results of the methods described in Section 8.2.2 when applied to the data described in Section 8.3 (results for the *LPE* method are not available for the two-color data). Cui et al. (2005) recommend permutations to obtain p-values for the *Shrinking* approach. In Tables 8.4 and 8.5 and Figures 8.1 and 8.2 we use a normal reference distribution. Tables 8.6 and 8.7 and Figures 8.3 and 8.4 use the permutation approach. The choice of distribution has a substantial impact on the results.

Figure 8.1 gives a graphical display of how well these methods adhere to the nominal significance levels, Figure 8.2 displays power results. These figures are probability-probability plots on a logit-scale. For a given method and a particular experiment let p_i be the two-sided (sometimes called signed) p-values; that is, if p_i is close to 0 there is evidence of under-expression and if p_i is close to 1 there is evidence of over-expression of group one relative to group two. We now combine all p_i for a group of experiments and sort them. Assume that there are N p-values. The sorted p-values are plotted on the horizontal axis, with $(1, \ldots, n)/(N+1)$ on the vertical axis. For a self-self experiment, these plots should ideally follow the identity line, as that implies that the significance levels are "unbiased." Curves that flatten out are particularly worrisome, as they suggest significantly differentially expressed genes that are in fact false positives. Curves that are more vertical than the identity line suggest statistics that are too conservative: something that is not a concern when there is in fact no difference, but would likely be harmful when using the same method to analyze data where some genes are in fact differentially expressed. For the D experiments, where there is a difference between the two sample types, the ideal curve is more horizontal, as long as the method does not generate a substantial number of false positives in the S experiments.

Figure 8.1 shows that the *Loess* and *LPE* approaches identify substantially more differentially expressed genes than the nominal levels for the S experiments. The *Cyber-T* approach shows a mild number of increases, and none of the other approaches shows serious bias. For both groups of experiments, the *Shrinking* approach with a normal reference distribution appears too conservative.

Table 8.4 elaborates these observations. Although most methods appear to be rather conservative, at a significance level of $\alpha = 1\%$ the *Loess* method shows a substantial anticonservative bias, in five out of eight data sets. For microarray experiments, the more stringent level $\alpha = 0.01\%$ is very relevant, as multiple testing corrections generally imply selecting genes at low significance levels. Again, the *Loess* shows substantial bias. The *LPE* approach also indicates ten times more significant genes than the nominal value; this bias is present for three of the four data sets. At this significance level, the *Cyber-T* method shows a modest bias overall, being substantial for only one dataset (two-color experiment S2.2). The excess percentage of significant genes for the *Pooled-t* approach is minimal, and could be due to chance.

In Figure 8.2 it is seen that for all methods far more genes are identified as differentially expressed by the two-color experiments than by the one-color experiments, as the curves for the two-color experiments are much more horizontal than those for the one-color experiments. This is largely an effect of the particular data used, as the two-color Drosophila experiments involved substantially altered flies, while the differences between the mice involved in the one-color Huntington's disease experiments are much more subtle. This figure does indicate though that the ordering of the methods is largely unchanged, suggesting that since the conclusions remain the same for two dramatically different experiments (different technologies, different amounts of differential genes) they appear to be fairly robust and may well generalize to many other situations.

In both the two-color and the one-color experiments the *Loess* approach produces the most genes identified as differentially expressed. This is not a surprise, since the method does not maintain the correct significance levels in the self-self (S) experiments. Similarly, it is not surprising that the *LPE* method identifies more differential expression for the one-color experiments, since it also does not adequately control test size here. Among the remaining methods, which tend to maintain significant levels rather conservatively, the *Pooled-t* approach performs best for the two-color experiments, followed by *Cyber-T* and *Limma*, while for the one-color experiments *Cyber-T* and *Limma* approach seem slightly more powerful than *Pooled-t* (Table 8.5). Interestingly for the D2.x (two-color) experiments, *Pooled-t* seems more powerful in

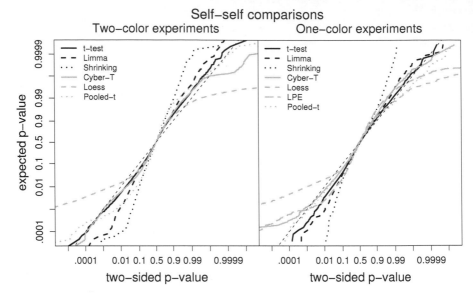

Figure 8.1 *Method performance using a defined null reference distribution in self-self (S) experiments. For unbiased methods the curves should follow the identity line.*

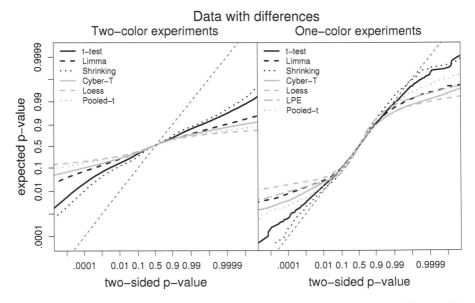

Figure 8.2 *Method performance using a defined null reference distribution in difference (D) experiments. If there is appropriate Type I error control, curves that are more horizontal correspond to more powerful methods.*

those experiments with two arrays (D2.1, D2.2, and D2.3). Maybe this is not surprising: borrowing degrees of freedom between experiments, as *Pooled-t* does, is particularly useful when the number of degrees of freedom is small.

8.4.3 Permutation p-values

As detailed in Section 8.2.3, an alternative to obtaining p-values is a permutation approach in which the test statistics for all genes are combined. Figure 8.3 gives a graphical display of how well each method adheres to the significance levels when p-values are determined using such an approach. Figure 8.4 displays curves related to power for these situations. We do not show permutation results for *Pooled-t*: since this procedure combines arrays from different experiments a permutation procedure is less standard, and in any case the results using a t-distribution are already satisfactory.

The permutation approach for computing p-values yields approximately unbiased, if somewhat conservative, results for all approaches since all curves in Figure 8.3 follow the diagonal. However, as expected, the permutation approach is associated with a reduction in the number of genes called differentially expressed. Figure 8.4 shows that the procedures based on permutation produce considerably fewer differentially expressed genes than the procedures that do not use permutation (Figure 8.2). In fact, the curves in Figure 8.4 all stay within a "band" of the diagonal. This result is a consequence of using the permutation approach with a small number of repeats: irrespective of the actual number of differentially expressed genes, there is a maximum number of genes that can be identified as differentially expressed at any particular significance level due to the experimental design. A detailed explanation is given below in the discussion of Table 8.7.

Tables 8.6 and 8.7 summarize results for the permutation-based procedures. Although the permutation approach does control the significance level α appropriately, there is correspondingly less differential expression identified for these data and methods. The part of Table 8.7 for the D2.x experiments clearly illustrates an artifact of the permutation approach. As already seen above, the D2.x experiments have very many genes identified as differentially expressed (see Table 8.5). But in Table 8.7 there seems to be a cap: at a significance level of $\alpha = 1\%$ for experiments D2.1, D2.2, and D2.3 all methods suggest at most 2% differentially expressed genes, for experiments D2.4, D2.5, and D2.6 all methods suggest at most 8% differentially expressed genes, and for experiments D2.7 all methods suggest at most 32% differentially expressed genes. We focus on experiment D2.4, which uses 4 arrays. There thus result at most $2^4 = 16$ permutations from "flipping" the arrays. Since each permutation arises twice (when all arrays are flipped relative to the first analysis), only 8 of these permutations are unique. Assume that for this experiment 40% of the genes are differentially expressed (as Table 8.5 suggests), and therefore that these 40% of the genes have very large test statistics. With about 10,000 genes on these arrays, there are thus about 4,000 large test statistics, say larger than a value A. Now

Table 8.4 *Percentage of differentially expressed genes in self-self (S) experiments identified using a defined null reference distribution at significance levels $\alpha = 1\%$ and $\alpha = 0.01\%$. For unbiased methods the percentage of differentially expressed genes should be close to α.*

$\alpha = 1\%$	t-test	Limma	Shrinking	Cyber-T	Loess	LPE	Pooled-t
S2.1	0.2	0.1	0.0	0.1	0.7	NA	0.3
S2.2	1.1	0.1	0.0	2.3	5.8	NA	0.3
S2.3	0.6	0.2	0.0	0.3	2.0	NA	0.4
S2.4	0.2	0.1	0.0	0.0	0.6	NA	0.1
S2-ave	0.5	0.1	0.0	0.7	2.3	NA	0.3
S1.1	0.4	0.2	0.0	0.4	0.7	0.4	0.0
S1.2	0.6	0.3	0.0	1.4	2.7	1.1	0.2
S1.3	0.8	0.1	0.0	0.3	3.9	0.3	3.2
S1.4	0.3	0.0	0.0	0.1	2.6	0.1	1.3
S1-ave	0.5	0.2	0.0	0.6	2.5	0.5	1.2

$\alpha = 0.01\%$	t-test	Limma	Shrinking	Cyber-T	Loess	LPE	Pooled-t
S2.1	0.000	0.000	0.000	0.000	0.198	NA	0.017
S2.2	0.009	0.000	0.000	0.277	1.528	NA	0.061
S2.3	0.018	0.000	0.000	0.000	0.982	NA	0.009
S2.4	0.000	0.000	0.000	0.000	0.230	NA	0.009
S2-ave	0.007	0.000	0.000	0.069	0.735	NA	0.024
S1.1	0.015	0.030	0.000	0.061	0.197	0.106	0.000
S1.2	0.000	0.000	0.000	0.045	0.697	0.243	0.000
S1.3	0.000	0.000	0.000	0.015	0.500	0.061	0.091
S1.4	0.000	0.000	0.000	0.000	0.728	0.000	0.000
S1-ave	0.004	0.008	0.000	0.030	0.531	0.102	0.023

Table 8.5 *Percentage of differentially expressed genes in difference (D) experiments identified using a defined null reference distribution at significance levels $\alpha = 1\%$ and $\alpha = 0.01\%$. If there is appropriate Type I error control, a larger percentage of differentially expressed genes corresponds to a more powerful method.*

$\alpha = 1\%$	t-test	Limma	Shrinking	Cyber-T	Loess	LPE	Pooled-t
D2.1	1.9	12.1	0.0	15.8	26.8	NA	30.9
D2.2	2.3	16.0	0.0	21.9	32.3	NA	28.9
D2.3	4.0	34.8	0.0	43.6	53.8	NA	48.2
D2.4	31.0	44.8	22.6	45.5	54.4	NA	62.7
D2.5	20.9	31.6	13.1	35.1	43.5	NA	52.4
D2.6	53.6	66.5	46.3	66.9	76.5	NA	58.6
D2.7	51.8	57.6	46.9	55.9	64.3	NA	56.3
D2-ave	23.7	37.6	18.4	40.7	50.2	NA	48.3
D1.1	2.6	3.4	2.0	4.0	6.4	2.7	3.3
D1.2	1.2	5.3	0.1	5.6	6.7	5.0	1.5
D1.3	1.6	1.6	1.0	1.6	3.0	0.9	0.8
D1-ave	1.8	3.4	1.1	3.7	5.4	2.9	1.9

$\alpha = 0.01\%$	t-test	Limma	Shrinking	Cyber-T	Loess	LPE	Pooled-t
D2.1	0.009	0.864	0.000	2.148	12.698	NA	10.835
D2.2	0.026	1.219	0.000	5.051	17.682	NA	11.928
D2.3	0.027	7.699	0.000	19.441	38.368	NA	26.722
D2.4	1.994	15.378	0.296	21.732	37.774	NA	44.632
D2.5	1.083	4.752	0.201	10.856	28.225	NA	31.806
D2.6	7.729	39.769	2.858	47.705	66.313	NA	40.295
D2.7	17.023	29.986	11.971	34.357	50.124	NA	38.347
D2-ave	3.984	14.238	2.189	20.184	35.883	NA	29.224
D1.1	0.121	0.349	0.030	1.046	2.593	0.788	0.516
D1.2	0.000	2.153	0.000	1.668	2.835	2.092	0.243
D1.3	0.106	0.243	0.061	0.379	1.410	0.288	0.182
D1-ave	0.076	0.915	0.030	1.031	2.280	1.056	0.313

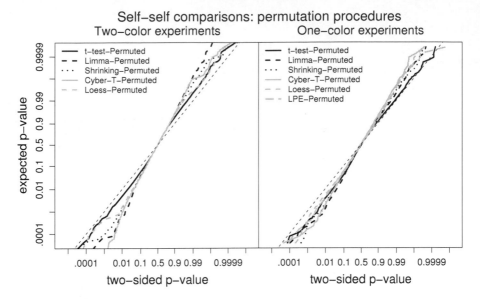

Figure 8.3 *Method performance using a permutation reference distribution in self-self experiments. For unbiased methods the curves should follow the identity line.*

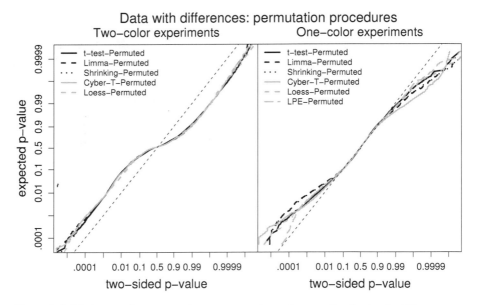

Figure 8.4 *Method performance using a permutation reference distribution in difference experiments. Curves that are more horizontal correspond to more powerful methods.*

Table 8.6 *Percentage of differentially expressed genes in self-self (S) experiments identified using a permutation distribution at significance levels $\alpha = 1\%$ and $\alpha = 0.01\%$. For unbiased methods the percentage of differentially expressed genes should be close to α.*

$\alpha = 1\%$	t-test permuted	Limma permuted	Shrinking permuted	Cyber-T permuted	Loess permuted	LPE permuted
S2.1	0.1	0.0	0.0	0.0	0.0	NA
S2.2	1.0	0.0	0.2	0.4	0.6	NA
S2.3	0.6	0.1	0.1	0.0	0.4	NA
S2.4	0.2	0.1	0.1	0.0	0.2	NA
S2-ave	0.5	0.1	0.1	0.1	0.3	NA
S1.1	0.3	0.1	0.1	0.1	0.1	0.1
S1.2	0.6	0.4	0.4	0.3	0.4	0.4
S1.3	1.1	0.5	0.4	0.2	0.5	0.5
S1.4	0.3	0.1	0.1	0.1	0.4	0.2
S1-ave	0.6	0.2	0.2	0.1	0.4	0.3

$\alpha = 0.01\%$	t-test permuted	Limma permuted	Shrinking permuted	Cyber-T permuted	Loess permuted	LPE permuted
S2.1	0.000	0.000	0.000	0.000	0.000	NA
S2.2	0.000	0.000	0.000	0.000	0.000	NA
S2.3	0.017	0.000	0.000	0.000	0.000	NA
S2.4	0.000	0.000	0.008	0.000	0.000	NA
S2-ave	0.004	0.000	0.002	0.000	0.000	NA
S1.1	0.000	0.000	0.000	0.000	0.000	0.000
S1.2	0.000	0.000	0.000	0.000	0.000	0.000
S1.3	0.000	0.000	0.000	0.015	0.000	0.015
S1.4	0.000	0.000	0.000	0.000	0.000	0.000
S1-ave	0.000	0.000	0.000	0.004	0.000	0.004

assume that among the 7 other permutations none of the test statistics is larger than A. Then out of $8 \times 10,000 = 80,000$ test statistics, there are 4,000 larger than A. However, at the $\alpha = 1\%$ level at most $0.01 \times 80,000 = 800$ can be called signif-

Table 8.7 *Percentage of differentially expressed genes in difference (D) experiments identified using a permutation distribution at significance levels $\alpha = 1\%$ and $\alpha = 0.01\%$. If there is appropriate Type I error control, a larger percentage of differentially expressed genes corresponds to a more powerful method.*

$\alpha = 1\%$	t-test permuted	Limma permuted	Shrinking permuted	Cyber-T permuted	Loess permuted	LPE permuted
D2.1	1.6	2.0	1.8	2.0	2.0	NA
D2.2	1.5	2.0	2.0	2.0	2.0	NA
D2.3	1.9	2.0	2.0	2.0	2.0	NA
D2.4	7.7	8.0	8.0	8.0	8.0	NA
D2.5	7.4	8.0	8.0	7.9	7.5	NA
D2.6	8.0	8.0	8.0	8.0	0.0	NA
D2.7	30.5	31.8	30.5	31.8	24.8	NA
D2-ave	8.4	8.8	8.6	8.8	7.8	
D1.1	2.8	3.8	3.8	3.6	2.8	2.8
D1.2	1.2	3.0	2.6	2.7	2.7	2.7
D1.3	1.9	1.8	1.8	1.4	1.3	1.0
D1-ave	2.0	2.9	2.7	2.6	2.3	2.1
$\alpha = 0.01\%$	t-test permuted	Limma permuted	Shrinking permuted	Cyber-T permuted	Loess permuted	LPE permuted
D2.1	0.008	0.008	0.008	0.008	0.017	NA
D2.2	0.017	0.017	0.017	0.017	0.026	NA
D2.3	0.009	0.008	0.000	0.009	0.018	NA
D2.4	0.068	0.076	0.076	0.068	0.079	NA
D2.5	0.075	0.083	0.059	0.084	0.079	NA
D2.6	0.075	0.075	0.075	0.025	0.068	NA
D2.7	0.308	0.315	0.283	0.308	0.314	NA
D2-ave	0.080	0.083	0.074	0.074	0.086	NA
D1.1	0.121	0.258	0.212	0.243	0.106	0.030
D1.2	0.000	0.000	0.015	0.015	0.015	0.015
D1.3	0.136	0.243	0.258	0.212	0.121	0.045
D1-ave	0.086	0.167	0.162	0.157	0.081	0.030

icant (at $\alpha = 1\%$) from the permutation distribution. This makes 8%, rather than the 40% that are differentially expressed, of all the genes on the array. (In fact the percentage called differentially expressed is slightly lower as a few rare permuted genes also have large statistics.) We cannot ignore the original permutation to obtain percentiles of the permutation distribution, since doing so violates the assumption of exchangeability under the null hypothesis of no differential expression. This artifact disappears when the number of differentially expressed genes is much smaller or when the number of arrays increases, since then more permutations can be created.

8.4.4 Relation between average signal and variance

The local smoothing approaches generally assume that genes with the same expression level have approximately the same variance, then estimate the variance by smoothing as a function of the expression level. We examine this relationship here. Figure 8.5(a) contains an MA plot for an individual two-color array, showing the relation between the difference between the logs of the two signals (i.e., the log-ratio, or M-value) and the average of the logs of the signals. For one of the two-color experiments (Figure 8.5) and one of the one-color experiments (Figure 8.6), the relation between the variance and the average signal is shown. As can be seen, the relation between average signal and variance is minimal. In fact, the correlation between the variance from one experiment to the next experiment for the same gene is much larger than the correlations in these figures (data not shown).

8.5 Discussion

We have seen here that the choice of significance test in microarray experiments with low replication can dramatically influence the results. We focus on p-values, rather than for example the false discovery rate (FDR), as we believe that an appropriately obtained p-value will yield a more reliable multiple testing correction, and that the multiple testing adjustment cannot by itself save a procedure that yields badly calibrated p-values.

The two groups of experiments analyzed here differ in another important aspect besides technology: the one-color experiments have a modest number of differentially expressed genes, while the two-color experiments have many such genes. Given this difference between the experiments, the similarity in results is striking.

The main conclusions are:

1. The t-test has almost no power when the sample size is small. When there are fewer than, say, six to eight repeat arrays some of the alternative solutions are much more powerful. Kooperberg et al. (2005) conclude that the lack of power is even more extreme for the Welch statistic, which suffers at least in part because the variance estimate is not pooled.

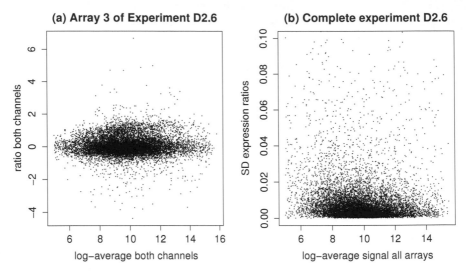

Figure 8.5 *(a) Relation between log expression ratio and average log expression for one nor-malized two-color array, and (b) SD of log expression ratios vs. average log expression ratio for all arrays from the experment (D2.6).*

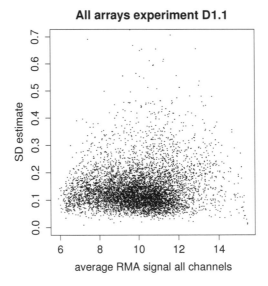

Figure 8.6 *Residual SD vs. average RMA value for one of the one-color experiments (D1.1).*

2. A permutation approach to obtaining p-values also severely reduces the number of genes that are identified as differentially expressed for small experiments with a lot of differential expression. This limits our conclusions about the *Shrinking* approach (Cui et al., 2005), since for this approach it is the only suggested method to obtain p-values.

3. Combining an estimate of the overall variance with an estimate of the individual variance, such as is done for *Limma* (Smyth, 2004) and *Cyber-T* (Baldi and Long, 2001), appears to be very effective. Apparently such a regularization reduces the noise in the variance estimates in an effective manner. Because of the similarity of the results for these two approaches, and the much worse results for the smoothing approaches, it appears that for the *Cyber-T* approach the running average estimate of σ_{0ij}^2 is in effect estimating an overall variance, rather than a local variance. In the analyses here, *Limma* performs slightly better than *Cyber-T*.

4. The *Pooled-t* approach proposed by Kooperberg et al. (2005), which borrows degrees of freedom from other experiments, performs equally well as *Limma* and *Cyber-T*. In fact, when the sample size is minimal ($n = 2$) it seems to perform slightly better. An obvious question concerns which experiments to combine. A small simulation study carried out by Kooperberg et al. (2005) suggests that there can be a fair degree of experiment-to-experiment variation without seriously inflating the Type I error. The fact that we were able to combine here information on experiments carried out on such diverse material as cell-lines and RNA harvested from fruit flies lends support to this conclusion.

5. The approaches to combining information here do not all perform equivalently. Methods which use only a (locally) smoothed estimate of the variance, such as *LPE* (Jain et al., 2003) and *Loess* (Huang and Pan, 2002), can give severely biased results by inflating the percentage of significant genes well beyond a pre-specified level α when in fact there are no differences between the two samples. For *Loess* this is evident at $\alpha = 1\%$ and $\alpha = 0.01\%$, for *LPE* it is only evident at $\alpha = 0.01\%$. However, due to multiple testing in microarray experiments very small significance levels are generally used, so it would seem better to avoid methods relying solely on smoothing. One reason for this bias might be that with the improved normalization methods now available, the relation between variance and expression level has been considerably reduced (see Section 8.4.4). Thus, locally averaging the variances will sometimes yield variances that are too large and sometimes yield variances that are too small. When the variance is too small there is a substantial chance of incorrectly identifying a gene as differentially expressed. Another, more fundamental reason is due to the experimental design itself. The *LPE* approach is more appropriate for technical replicates, for which the error distributions are closer to Gaussian. The error distribution for biological replicates, such as those we analyze here, will confound technical variability with heterogeneous biological variability, leading to the observed

bias. Lee, Cho, and O'Connell (Chapter 7 in this volume) provide additional detail along with methods to address these issues.

8.6 Appendix: Array preprocessing

For all arrays we carried out a graphical quality assessment, which indicated that all arrays were of good quality.

Two-color arrays. For the two-color arrays we first exclude all spots with a $\log_2$-expression ratio of less than 5 and spots whose background level was higher than the foreground level for either channel. This excludes about 11.5% of the spots, primarily those that do not hybridize well. In particular, of the 13,440 spots on the arrays, 1,296 are excluded on all 36 arrays: of the remaining spots only about 2% are excluded. We then subtract the background and use a print-tip loess correction (Yang et al., 2002), carried out using the `limma` function `normalizeWithinArrays()` with defaults. Any spot that had at least two estimates for a particular experiment was included in the analysis.

One-color arrays. Gene expression is quantified using RMA (Irizarry et al., 2003b) on all arrays simultaneously. We also carried out the analyses using $\log_2$ of the MAS 5.0 summary (Affymetrix, 2002; see also Section 1.4) and again using RMA separately within each experiment. In both cases, the results are very similar to those reported in this chapter.

Acknowledgments

Charles Kooperberg was supported in part by NIH grants CA 74841, CA 53996, and HL 74745. Aaron Aragaki was supported in part by NIH grant CA 74841. Charles C. Carey was supported in part by NIH interdiciplinary training grant 2T32CA80416-06. Suzannah Rutherford was supported in part by NIH grant GM 06673. The authors thank Andy Strand, Jim Olson, and the HDAG group for allowing us to use the one-color data.

CHAPTER 9

Comparison of meta-analysis to combined analysis of a replicated microarray study

Darlene R. Goldstein, Mauro Delorenzi, Ruth Luthi-Carter, and Thierry Sengstag

9.1 Introduction

The widespread use of microarrays has resulted in a large-scale, rapid expansion of data. Many research groups throughout the world are engaged in gene expression studies of the same or similar conditions – specific cancers, for example. Data from many microarray studies are deposited in publicly available databases such as Gene Expression Omnibus (GEO; Edgar et al., 2002; Barrett et al., 2005). It is hoped that ready access to primary data will facilitate the integration of information across studies.

Each microarray study gives rise to its own list of "interesting" genes. The lists from different studies, however, may not exhibit substantial concordance. Discordant results may produce scientific confusion or disagreement regarding the underlying biology, as well as lost time and misused resources. Consequently, the ability to synthesize information across these studies is essential.

Given the limited size of many microarray studies, meta-analysis seems a natural approach to the problem of integrating conclusions across studies. Indeed, there is a recent and increasing literature for meta-analysis of microarray studies (Rhodes et al., 2002, 2004a; Ghosh et al., 2003; Choi et al., 2003; Stevens and Doerge, 2005; Wirapati et al., 2008; Borozan et al., 2008; Ma and Huang, 2009).

In this chapter, we examine properties of different methods for combining information from a replicated experiment carried out with Affymetrix GeneChips (Lockhart et al., 1996; see also Section 1.4). Our aim is to demonstrate that even in this almost ideal situation, several issues concerning appropriate data normalization and combination still arise. We first give some background on the study, then describe the statistical analyses and present results, and conclude with a discussion.

9.2 Study description

The data are obtained from two experiments on the R6/2 mouse. The R6/2 mouse line is transgenic for exon 1 of the human Huntington's disease (HD) gene, thus serving as an experimental model for the disease (Mangiarini et al., 1996). These mice exhibit mRNA changes weeks in advance of neuronal death or gliosis phenotypes (Luthi-Carter et al., 2000, 2002a).

Two separate studies were carried out to investigate the effects on gene expression of different drugs on HD and normal (or wild type (WT)) mice in order to identify genes differentially expressed between HD and WT mice. Each experiment was designed as a 2×2 factorial layout, where one factor is drug/placebo treatment and the other is HD/WT mouse.

We consider only the control groups for the two studies, which received the placebo (injected with normal saline 30-60 minutes prior to sacrifice). In Study I there are 8 control mice, while in Study II there are 6 control mice. In each study, half of the mice are HD and half WT.

The two experiments were carried out by the same laboratory a few months apart. In each experiment, the same protocols were used throughout with regard to mouse breeding, care and sacrifice, mRNA extraction, and hybridization to the microarray. Thus, these data are those of a completely replicated study.

The studies were carried out with the Affymetrix MOE 430A (Mouse Expression Array). These chips contain in total 22,690 probe sets, to which, with a slight abuse of terminology, we refer henceforth as "genes."

9.3 Statistical analyses

There are several components of the data analyses to be carried out. First, quality of the hybridizations should be assessed so that low quality chips are removed from further analysis. We quantify expression of each gene for each individual, then compute for each gene a statistic for assessing genes for differential expression between the HD and WT mice. A determination of significance is also be made for these statistics, taking into account the multiplicity of hypotheses tested.

We compare analyses carried out under two scenarios: one is a mega-analysis, where the data are combined and analyzed as a single set; for the second, the two data sets are analyzed separately and their results are combined via meta-analysis. The steps are described in detail below. All analyses reported here were coded in the R statistical programming environment (Ihaka and Gentleman, 1996; R Development Core Team, 2005), using the following packages from R and BioConductor citepBIOC: affy (Irizarry et al., 2008), affyPLM (Bolstad, 2008), car (Fox, 2008), limma (Smyth, 2005), qvalue (Dabney and Storey, 2008), and rmeta (Lumley, 2008).

9.3.1 Chip quality assessment and expression quantification

We assessed all chips for quality with the RMA-QC approach described in Collin (2004) and Brettschneider et al. (2008), and implemented in the package `affyPLM` (Bolstad, 2008). In this method, gene expression is modeled as the sum of chip and probe effects, with the model fit by robust regression (i.e. outliers are downweighted; see Equation 1.5, p. 16 of Section 1.4.2). Pseudoimages of the robust regression weights or residuals for each probe provide a graphical means to assess chip quality; numerical measures indicative of quality were also computed.

By these criteria, all 14 chips were of similar and suitably high quality that none required exclusion.

For measuring expression, we use RMA, indexrobust multi-array average (RMA) due to its demonstrated favorable properties (Irizarry et al., 2003a,b; Bolstad et al., 2003). RMA values were computed with the `affy` package.

9.3.2 Identifying differential expression

A commonly addressed problem in microarray experiments is detection of genes differentially expressed under two or more conditions. A substantial number of statistical papers propose methods for this purpose, with new ones still being introduced (for an overview see Goldstein and Delorenzi (2004)). The high dimensionality of microarray data has also brought to the fore multiple hypothesis testing issues. The approach we adopt is described here.

Moderated t-statistic

Perhaps the most readily interpretable measure of differential expression is given by the fold change (ratio) in expression of a given gene between two types of samples (HD and WT here). It is more convenient to consider fold change on the logarithmic scale, M = (average) $\log_2$(fold change).

The measure M has the shortcoming of not taking into account differing variability of different genes. The variability of M, though, is not the same across the range of signal intensities. In particular, genes with larger variance across arrays are likely to produce large values of M even when they are not truly differentially expressed between the two sample types.

An obvious way to deal with differing variability is by standardization. Here M is divided by its standard error, which is estimated based on expression measures of the corresponding gene. Thus, the difference in average expression between sample types is quantified with a t-statistic. However, a problem here is that the t-statistic performs very poorly at identifying true differential expression with the small sample sizes found in typical microarray studies (see e.g., Kooperberg et al., Chapter 8 in this volume).

Bayesian and empirical Bayes methods have been proposed as a compromise between single gene estimates of variability and no estimate of variability at all. These use data from all genes to improve estimation of differential expression for single genes (Lönnstedt and Speed, 2002; Smyth, 2004). These methods have been shown to perform well, in terms of true and false positive and negative rates, at identifying differential expression. In addition, the methods have been extended to be applied to a large variety of experimental designs through a linear modeling approach (Smyth, 2004; Lönnstedt et al., 2001).

We follow the linear modeling approach here. For each gene g in a given study, the measured gene expression vector Y_g across samples is modeled as

$$Y_g = X\beta_g + \epsilon_g,$$

where X is the design matrix, β_g is a vector of coefficients, and ϵ_g is a vector of error terms. The design matrix X reflects the experimental conditions (here, HD or WT) for each sample, and is the same for all genes within a study.

The moderated t-statistic for coefficient j and gene g is given by

$$\text{mod } t_{gj} = \frac{\hat{\beta}_{gj}}{\tilde{s}_g \sqrt{v_{gj}}},$$

where $\hat{\beta}_{gj}$ is the estimate of coefficient j for gene g, $\tilde{s}_g$ is the square root of the empirical Bayes estimated variance, and v_{gj} is the scaling for the variance, reflecting sample size. That is, mod t is the ratio of M to its standard error, which has now been estimated taking into account expression levels not only of gene g but of all genes. It is similar to the ordinary t-statistic, but with a moderated standard error estimate and correspondingly an increased number of degrees of freedom. For a detailed explanation, refer to Smyth (2004). We base inference about effects on mod t.

Multiple hypothesis testing

The biological question of differential expression can be restated as a problem in multiple hypothesis testing: the simultaneous test for each gene of the null hypothesis of identical mean expression in the two sample types.

A multiplicity problem arises when attempting to assess the statistical significance of the results on tests carried out on several thousands of genes simultaneously. With whole genome coverage arrays consisting of probes for many thousands of genes, most genes will not be differentially expressed between the conditions under investigation. Thus, even a nominal p-value of say 0.01 cannot be characterized as "significant," since many such small p-values will occur by chance when such a large number of tests are made.

The (nominal, unadjusted) p-value of a test reflects significance only for a single gene considered in isolation. When controlling the false discovery rate (FDR; Benjamini and Hochberg, 1995; Reiner et al., 2003; see also Section 7.2.3), the q-value of a test measures the proportion of false positives (FDR) incurred among rejected nulls when that test is called significant. It has been described as the expected proportion of false positives among all test results as or more extreme than the one obtained (Storey, 2002; Storey and Tibshirani, 2003b), and is analogous to a p-value that accounts for multiplicity.

We make use of q-values to take into account the large number of individual hypotheses tested. The mod t p-values from a set of single gene tests are transformed to q-values with the `qvalue` package (Dabney and Storey, 2008). We call a test result significant by fixing a q-value (or FDR) threshold, usually at 0.05. When the different conditions studied are more similar than the extremely varied ones considered here (HD and WT), a higher threshold may be more relevant. For example, an FDR of 0.25 still suggests that three of four significant findings are real.

9.3.3 Combined data analysis

In the combined data analysis, we consider all 14 chips as a single data set from the same experiment. This is not a completely artificial treatment, as most large experiments take place over a period of time and include hybridizing groups of chips at different times. RMA measures are obtained by processing all chips together.

The linear modeling approach is used to identify genes differentially expressed between HD and WT mice. We consider a series of models, each of which includes an effect on gene expression of HD over WT. There might also be additional variability due to study, so we also allow for a study ("batch") effect as well as entertain the possibility of an HD by study interaction.

The design matrices are set up using treatment contrasts so that the effects of interest are included as coefficients in the models. Thus, for each gene g, the three combined data linear models are given by (the subscript g is suppressed; I represents an indicator random variable):

Model A: $y = \beta_0 + \beta_{HD} I_{\{HD=1\}} + \epsilon$

Model B: $y = \beta_0 + \beta_{HD} I_{\{HD=1\}} + \beta_{batch} I_{\{batch=1\}} + \epsilon$

Model C: $y = \beta_0 + \beta_{HD} I_{\{HD=1\}} + \beta_{batch} I_{\{batch=1\}} + \beta_{HD \times batch} I_{\{HD \times batch=1\}} + \epsilon.$

The coefficients are estimated by ordinary least squares. The `limma` package is used to compute for each gene under each model the statistic mod t and corresponding p-values as a prelude to obtaining q-values.

9.3.4 Meta-analysis

In the meta-analyses, each experiment is first analyzed as a separate study. After heterogeneity analysis, results from the two studies are combined under three meta-analytic techniques: fixed effects meta-analysis, random effects meta-analysis, and

Fisher p-value combination. Computations were done with the R package `rmeta` (Lumley, 2008).

In the separate study analyses, gene expression is again quantified with RMA, but values are computed using only chips from the same study (8 chips for Study I or 6 chips for Study II). Linear modeling is carried out as above, but because each study is analyzed individually the model includes only the HD effect (Model A). Fitting the model produces effect estimates (coefficients) for each gene, while the empirical Bayes procedure produces estimates of variance, moderated t-statistics and p-values, which in turn yield q-values and gene rankings based on evidence in favor of differential expression between HD and WT mice.

Heterogeneity analysis

In addition to those mentioned in Section 1.2.2, heterogeneity issues more specific to the microarray context include: differences in the technology used for the study, heterogeneity of measured expression from the same probe occurring multiple times on the array, multiple (different) probes for the same gene, variability in probes used by different platforms, and differences in quantification of gene expression, even when the same technology is used.

Graphical methods for assessing between-study heterogeneity, such as forest plots of individual study confidence intervals, seem of limited usefulness in the microarray setting, as one such plot would be required for each individual gene. We thus depend on numerical assessments to screen genes for heterogeneous treatment effects across studies. For this we use the Q-statistic of Cochran (1954a) (Equation 1.1). It should also be kept in mind that there is one homogeneity test per gene, so the usual caveats regarding multiple hypothesis testing apply.

For the meta-analyses, we apply both fixed and random effects models as well as Fisher's method (see Sections 1.2.3, 1.2.3, and 1.2.4). We compute p-values for the Fisher combined p-value statistic X^2 (Equation 1.3) in two ways: first, with the χ_4^2 approximation and second, by resampling as proposed by Rhodes et al. (2002). In the resampling procedure, rather than choosing a p-value at random from $U(0, 1)$ a p-value is instead chosen at random from each of the sets of p-values from the two studies. These are then combined as in Equation 1.3 (p. 7 of Section 1.2.4) into a randomized summary statistic X_R^2. We obtain an empirical distribution of X_R^2 by repeating the resampling procedure 100,000 times. The p-value for the Fisher X^2 statistic is estimated as the proportion of the resampling-generated statistics X_R^2 greater than or equal to the original observed value X^2. This method yields a more conservative estimate for the p-value of X^2 because the distribution of actual study p-values is not uniform.

9.4 Results

Here we present detailed results of the analyses outlined above for the combined data set and for meta-analyses of separate experimental study outcomes.

9.4.1 Combined data

Combined vs. separate gene expression quantification

Because the first step of analysis requires a measure of gene expression, we compared quantification of expression with the combined data (RMA values based on all 14 chips) to individual study quantification (we separately compute RMA values based on the 8 chips from Study I and RMA values from the 6 chips from Study II). Figure 9.1 contains plots to explore this comparison.

Figures 9.1(a) and (c) show separate versus combined RMA values for one chip from each study (chip 1 from Study I and chip 1 from Study II, or Chips I-1 and II-1). These chips are representative of all chips in the respective studies – all plots were quite similar within study – so our remarks on the plots apply to all chips. Each gene on the chip is represented by a point in the plot, with the diagonal line representing equal expression by each method.

As it is difficult to detect differences from the line of equality in these scatter plots, we have also plotted the corresponding rotated and rescaled version as an MA plot (Tukey, 1977) (Figures 9.1(b) and (d)). In this representation, the difference between RMA values computed separately and combined is plotted against the average of the two values. If both RMA values were identical, all points would fall on the horizontal line at 0. Differences are more readily detected in this version of the plot.

It is easily seen that Study I chips tends to have higher RMA values when all chips are combined, while Study II chips have lower RMA values in the combined data set. The tendency persists throughout the range of $(\log_2)$ signal intensities.

Many investigators have assumed that normalization of a set of chips together would remove artifacts of this nature. In fact, this does not appear to be the case at all. The persistence of the study batch artifact can be seen, for instance, using cluster analysis (Everitt et al., 2001; Kaufman and Rousseeuw, 1990). When samples (chips) are clustered based on gene expression, the ones from the same study cluster together. The clustering details (algorithm, dissimilarity measure, number of genes) do not seem to affect the cluster results to any great degree.

Figure 9.2 shows an example of a dendrogram obtained clustering samples. The major cluster split occurs between Study I chips and Study II chips. There is a minor, and less clean, split on HD status: Study I samples 1 – 4 and Study II samples 1 – 3 are from the HD mice, the rest are WT mice. Thus, we see that even in what we might expect to be quite homogeneous studies, the most striking difference is in fact purely artifactual and is exactly of the sort that normalization is meant to remove. It

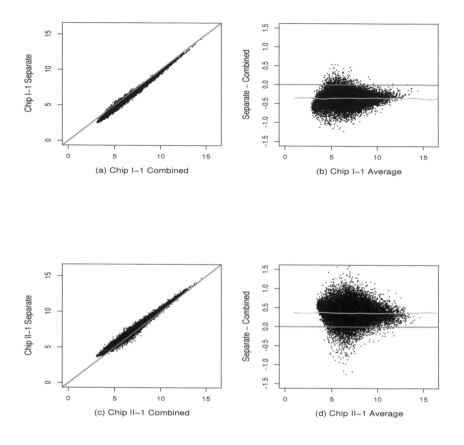

Figure 9.1 *Comparison of RMA values when studies combined and separate. (a) RMA values for chip I-1 computed within Study I vs. values computed from all chips combined; (b) Difference (Separate – Combined) vs. Average RMA values for chip I-1; (c) RMA values for chip II-1 computed within Study I vs. values computed from all chips combined; (d) Difference vs. Average RMA values for chip II-1. Diagonal and horizontal lines indicate equal values under both methods, curves through points are loess fits.*

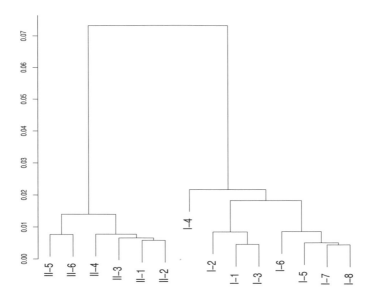

Figure 9.2 *Cluster dendrogram of combined data RMA values. Samples are clustered using all genes, Ward's method and a dissimilarity measure of 1–correlation.*

is not quite clear why the effects persist, they do not appear to vary systematically with intensity (data not shown). However, quantile normalization of the entire set together does not remove the study batch effect. The batch effect must be removed in another way before reliable inference relating to differential expression can take place.

HD-study batch interaction

We next turn attention to linear modeling of gene expression in terms of the effects of interest. Here, gene expression is obtained by computing RMA values for the combined set of 14 chips. Although the primary focus is on the HD effect, we must also consider the ramifications of other potential terms for the model. We have just seen the need to include study batch in the model. We now consider Model C to assess the need to include the HD by batch interaction term.

Histograms of p-values and q-values for the estimated interaction effects are shown in Figures 9.3(a) and (b). There are 2,242 genes with unadjusted p-values less than 0.05, but only three q-values less than 0.10 and thus indication of interaction between HD status and batch for only a few genes. In the face of this mild evidence, we discard Model C and ignore the possibility of interaction in the rest of the analyses.

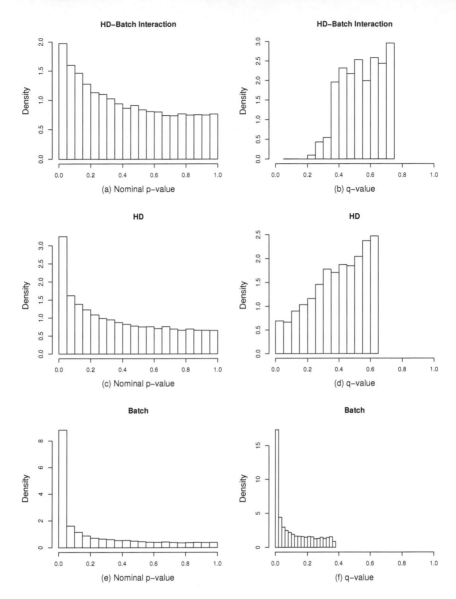

Figure 9.3 *Histograms of p-values and q-values for Model C interaction term (a, b), and Model B HD (c, d) and study batch (e, f) effects.*

Detection of differentially expressed genes

On the other hand, we have seen that there is strong evidence of batch effects for many genes. Using Model B to estimate HD and batch effects, we find evidence of significant HD effects for several genes along with a staggering number of genes with strong batch effects (Figure 9.3(b) – (d)). While there are 785 genes with HD q-values < 0.05, nearly one half of the genes (10,571 out of 22,690) have batch effects with q-values < 0.05. It is not necessary to believe in the exactness of the p- and q-value estimation to conclude that there are many genes with strong batch effects.

We consider Model A, which contains only the HD term, in order to compare genes identified as differentially expressed between HD and WT mice with and without batch effects. Not surprisingly, the significance of the HD effect is always higher (q-value lower) for Model B, where a study batch effect is included in the model. This is because we have controlled for an important source of variability here by introducing an effective stratification factor (batch). There is within stratum homogeneity but heterogeneity between strata, resulting in increased power to detect HD differences.

Figure 9.4 displays the q-values for individual gene HD effects both with and without batch effects in the model. This plot shows the importance of the batch effect in uncovering differential expression due to HD status. Use of Model B produces an additional 681 significant genes at the same FDR of 0.05 (points in black to the left of the vertical line and above the horizontal line at 0.05).

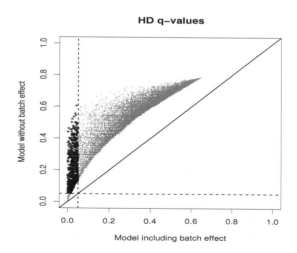

Figure 9.4 *q-values for HD effects without (Model A) vs. with (Model B) batch effect. Solid diagonal line indicates equal values; dashed vertical and horizontal lines indicate a FDR of 0.05. Highlighted points are genes with significant HD effects for Model B but not Model A.*

A list of genes identified by the analysis as differentially expressed between HD and WT mice can be produced upon selection of a significance threshold for the HD effect q-value, such as a FDR of 0.05. For comparison with results of meta-analyses of Studies I and II (below), we retain not only the gene list but all of the mod t p-values and corresponding q-values obtained using Model B.

9.4.2 Meta-analysis of Study I and Study II

Another strategy for dealing with study batch effects is to quantify gene expression separately for each study and then combine the studies by meta-analytic techniques. This is how the problem would necessarily be handled in the case of unrelated studies carried out by different research groups. Here, we are able to examine how meta-analysis would compare with a combined data analysis.

Heterogeneity analysis

We investigate heterogeneity of gene-specific HD estimates from Study I and Study II by computing the statistic Q (Equation 1.1, p. 5 of Section 1.2.2) as well as estimating the between-study standard deviation (SD) σ_B for each gene. Characteristics of these are plotted in Figure 9.5.

There is evidence of HD effect heterogeneity for some genes. Several genes have small nominal, unadjusted p-values: 3273 for $p < 0.05$; 5230 for $p < 0.10$ (Figure 9.5(a)). If we choose a more stringent criterion of significance, say a q-value of 0.10, there are still 802 genes with significant heterogeneity. This is substantially larger than the number of genes for which an interaction effect was detected, but also very much smaller than the number with a significant batch effect.

The quantile-quantile plot (Figure 9.5(c)) shows some deviation from the assumed χ_1^2 null distribution. This could be due to inadequacy of the χ_1^2 approximation, but as it is unlikely that all the nulls are in fact true we instead interpret this as indicative of the presence of genes for which the alternative holds, i.e. there is some true heterogeneity. Since the χ^2 test has low power to detect heterogeneity for the small study number and sample sizes that we have, there is likely to be a greater degree of heterogeneity, and for more genes, than suggested here.

The distribution of estimated between-study SD $\hat{\sigma}_B$ is highly skewed. Over 70% (16,428) of genes have $\hat{\sigma}_B < 0.01$ (Figure 9.5(d)). The value of $\hat{\sigma}_B$ corresponding to a FDR of 0.10 is about 0.066. This gives some idea of what a "large" value of σ_B is in this context. An example of intensities for a gene displaying heterogeneity is provided in Table 9.1, which gives the individual study RMA values of each chip for a gene with $\hat{\sigma}_B \approx 0.1$.

One purpose of testing homogeneity is for deciding between the FE and RE model for combining effect estimates. There will not be a large difference between FE and RE meta-analysis for genes with small σ_B. A general recommendation which has

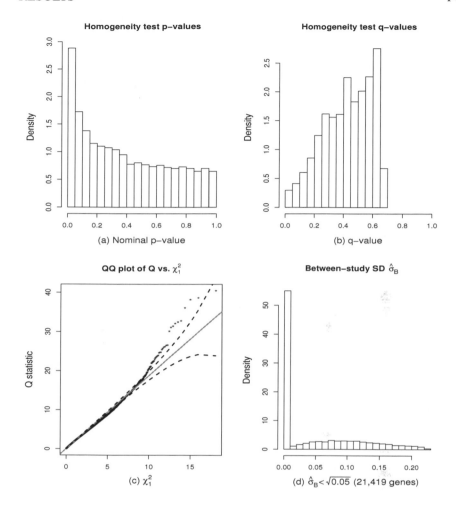

Figure 9.5 *Plots for heterogeneity analysis: histograms of (a) p-values and (b) q-values for the homogeneity statistic Q; (c) quantile-quantile plot of Q compared to the theoretical χ_1^2 distribution with 95% confidence region (dashed lines); (d) histogram of gene-specific estimated between-study SD for the 21,419 genes for which $\hat{\sigma}_B < \sqrt{.05}$.*

Table 9.1 *Individual chip RMA values for a gene with $\hat{\sigma}_B \approx 0.1$.*

| | Chip number | | | | | | | | Summary | |
	1	2	3	4	5	6	7	8	Mean	SD
Study I	8.67	8.88	8.91	8.64	9.08	9.02	9.02	9.27	8.94	0.21
Study II	9.86	9.75	9.96	9.83	9.57	9.73			9.78	0.13

been made is to carry out both, then compare similarity of results. If the results are similar then there is unlikely to be important heterogeneity and the FE model would typically be reported. If results are different, it is usually considered preferable to use RE meta-analysis to estimate the mean and SD of the effect size distribution. In the case of extreme, unexplained heterogeneity, it is probably more suitable to avoid combining the study results at all.

Fixed effects and random effects meta-analysis

Based on the results of the heterogeneity analyses, we would choose to adopt the RE model for combining HD effect size estimates. Nevertheless, we investigate both approaches here in order to compare them.

Figure 9.6(a) compares the combined HD (mean) effect estimated by the RE model versus the FE model combined estimate. With the exception of a few genes, these combined estimates tend to be remarkably similar.

Due to the additional variability included by the RE model, however, there is a great deal of difference between the standardized estimates (Figure 9.6(b)). Here, we can see that the FE standardized estimates are stochastically larger than the RE ones: about half of the genes have identical results for FE and RE, but for only 949 genes (or 4%) is the RE standardized estimate larger than that of FE. This phenomenon is also clearly reflected by the distributions of the corresponding q-values (Figures 9.6(c) and (d)) – at any FDR, many more genes are called differentially expressed between HD and WT by the FE model.

Figure 9.7 shows how the methods compare for different degrees of heterogeneity. In Figures 9.7(a), (b) and (c), q-values are transformed by $-\log_{10}$ so that larger values are more significant. We see that significant effects in the RE meta-analysis tend to be for the genes with more homogeneous effects across studies (Figures 9.7(a), (b)); that is, genes for which the between-study variability does not overwhelm the size of the estimated effect. The q-values from FE and RE are compared directly in Figure 9.7(c), where it is seen that the FE combined HD effect estimate is more significant than that of RE for virtually all genes. Finally, the location (centered at 0) and flatness of the loess curve in Figure 9.7(d) show that heterogeneity is not more frequently found for larger estimated mean HD effect sizes.

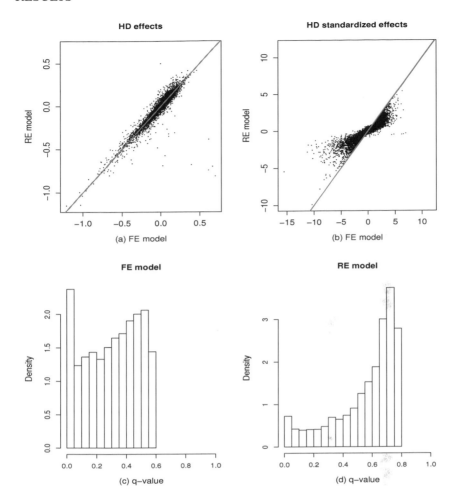

Figure 9.6 *Comparison of HD effects for FE and RE meta-analysis. (a) HD effects estimated by RE vs. FE; (b) HD standardized effects estimated by RE vs. FE; q-values for HD (standardized) effects estimated by FE (c) and RE (d). Diagonal lines indicate equal values under both methods.*

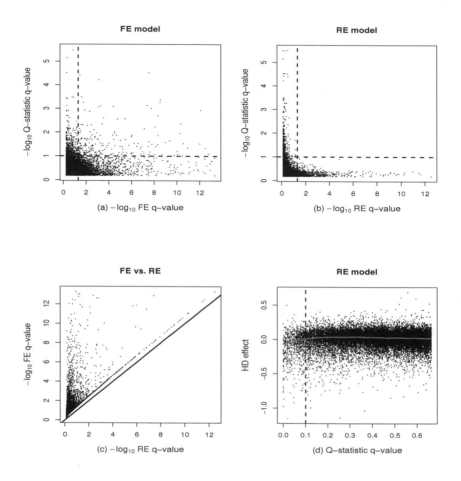

Figure 9.7 *Characteristics of FE and RE inference for varying heterogeneity.* $-\log_{10}$ *q-value of homogeneity Q-statistic vs.* $-\log_{10}$ *q-value of combined HD effect estimate from FE (a) and RE (b); vertical line indicates HD effect FDR of .05, horizontal line indicates Q-statistic FDR of .10; (c)* $-\log_{10}$ *q-values for FE vs. RE, diagonal line indicates equal values; (d) RE model estimated mean HD effect size vs. q-value of Q-statistic; vertical line indicates Q-statistic FDR of .1, horizontal smooth line is a loess curve.*

Fisher p-value meta-analysis

Figure 9.8 displays results obtained by combining for each gene the HD mod t p-values from Study I and Study II. The distribution of q-values obtained from p-values derived from the χ^2_4 distribution is compressed downward toward significance (a). Resampling p-values are exceedingly conservative compared to χ^2_4-derived p-values (b). Compared to RE model p-values the χ^2_4 p-values are liberal (c), while the resampling p-values are again conservative, although somewhat less than in comparison to the χ^2_4 p-values (d).

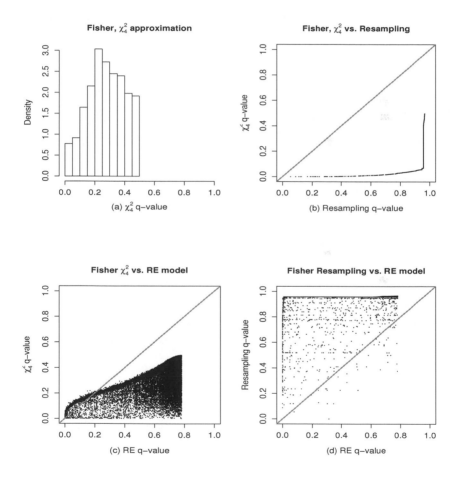

Figure 9.8 *Comparison of Fisher X^2 statistic q-values. (a) Histogram of χ^2_4 q-values; scatter plot of q-values obtained by χ^2_4 vs. resampling (b), χ^2_4 vs. RE (c), resampling vs. RE (d). Diagonal lines indicate equal values under both methods.*

The lack of agreement indicates the degree to which inference depends on the specific method chosen. Even where the same statistic is used, there is a real problem determining its p-value – the χ_4^2 assumption results in very different p-values from the resampling-based ones. Caution must therefore be exercised in choosing a method and interpreting results.

Comparison of results stratified by heterogeneity status

The findings presented thus far consider the entire set of genes in aggregate. However, the set of all genes can be viewed as a mixture of two types: genes for which the HD effects are homogeneous and genes for which the effects are heterogeneous across studies. It is therefore worth looking at characteristics of the analyses when genes are stratified by heterogeneity status.

Defining heterogeneity status requires a criterion for significance. In the microarray context, one must consider its impact on the subsequent identification of differential expression. To be more conservative in calling a gene differentially expressed, a fairly liberal heterogeneity criterion would seem in order. Taking into consideration the outcome of the heterogeneity analysis above, we decided on a FDR cut-off of 0.10 for Q. For this threshold, the number of genes for which studies are heterogeneous is 802; there are thus 21,888 homogeneous ones.

Table 9.2 gives the proportions of genes with significant HD effects for the four meta-analysis methods, as well as the combined data, for all genes together and also stratified by heterogeneity status (Hom. or Het.) at varying FDR for the HD effect. The methods are: C = combined data, FE = fixed effects model, RE = random effects model, FX = Fisher p-value combination method, χ^2 p-values, and FR = Fisher p-value combination method, resampling method. The Fisher resampling method proportions are extremely low, so the numbers of genes are also reported. For RE, proportions given as zero are actual zeros.

Table 9.2 *Significance proportions for meta-analysis methods.*

Method	Sig. at FDR = .10			Sig. at FDR = .05			Sig. at FDR = .01		
	All	Hom.	Het.	All	Hom.	Het.	All	Hom.	Het.
C	0.07	0.06	0.19	0.03	0.03	0.12	0.01	0.01	0.05
FE	0.18	0.17	0.38	0.12	0.11	0.30	0.06	0.05	0.21
RE	0.06	0.06	0.01	0.04	0.04	0.00	0.02	0.02	0.00
FX	0.08	0.06	0.70	0.04	0.03	0.29	0.01	0.01	0.10
FR	0.00	0.00	0.00	0.00	0.00	0.00	0.00	0.00	0.00
FR Number	(5)	(3)	(2)	(3)	(2)	(1)	(3)	(2)	(1)

FE finds the most significant effects, followed by Fisher χ^2 (FX). These two methods as well as the combined data method also find drastically higher rates of significant effects for studies which are heterogeneous, pointing to the need for caution when combining information. In contrast, RE finds lower rates of significant effects under heterogeneity.

Pairwise agreement of meta-analysis results

Lastly, we look at agreement for pairs of methods stratified by heterogeneity status, varying the FDR for calling a gene differentially expressed between HD and WT (Figure 9.9). The simple agreement rate is just the proportion of genes for which both methods agree on whether or not the HD effect is significant. The correspondence between plotting symbol and comparison pair is given in Table 9.3.

Table 9.3 *Correspondence of Plotting Symbol and Pair.*

Symbol	0	1	2	3	4	5	6	7	8	9
Pair	FE	FE	FE	FE	FX	FX	FX	C	C	RE
	FR	RE	C	FX	RE	FR	C	FR	RE	FR

The pairs appear to fall roughly into three groups when homogeneity and heterogeneity rates are considered jointly. Pairs 0 – 3 form one group. This group consists of all pairwise comparisons with FE. These pairs have the lowest agreement under homogeneity. Pairs 4, 5 and 6 have high homogeneity agreement and lower, but increasing with decreasing FDR, heterogeneity agreement. FX appears in each of these pairs. Finally, Pairs 7, 8 and 9 have highest agreement under both homogeneity and heterogeneity. These are all pairs are formed from C, RE, FR.

9.5 Discussion

Pooling raw data from different studies for analysis is not always possible; even when it is possible it might not be recommended (see Section 1.2.1). However, in the simple setup we have described here, where a single lab has carried out the same experiment twice, one would think that combining the raw data should be a fundamentally sound approach. In particular, carrying out the normalization step on the aggregated data would seem not only desirable but also necessary.

We have illustrated, however, that even in such an uncomplicated scenario, without issues of different platforms or experimental designs and protocols, integrating the available information might not be completely straightforward. There are persistent batch effects that must be taken into account. We would recommend that new methods developed for more complex situations also be tested in simpler cases so that the properties of the methods may be better understood.

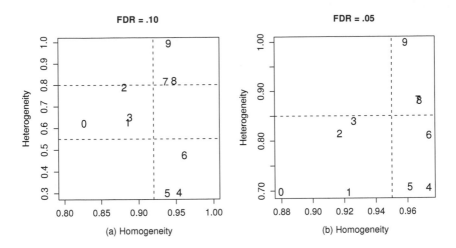

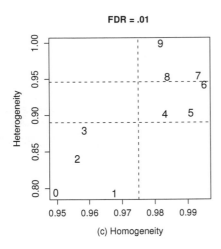

Figure 9.9 *Simple agreement rates of pairs of methods for varying FDR, see Table 9.3 for symbol legend. Agreement rate under heterogeneity vs. agreement rate under homogeneity for (a) FDR = 0.10, (b) FDR =0.05 and (c) FDR = 0.01. Dashed vertical and horizontal lines separate groups of pairs.*

The importance of accounting for variability across different labs has been noted in the literature (Irizarry et al., 2005); our work here suggests that within lab variability may also need to be considered.

Our results also have substantial implications for large single studies, where patients are recruited over time and arrays are not all hybridized at the same time. Avoidance of problems before they arise calls for careful study design in advance. In addition, comprehensive exploratory data analyses are required once data are collected, to identify and adjust for sources of variability which could obscure the underlying biology. The presence of strong batch effects may be indicative of aspects of laboratory practice in need of improvement. Analysis of data for batch effects can reveal such problems and could therefore help in their rectification.

In this work we can compare results from different methods of analysis, but we are unable to rigorously assess method performance or robustness because it is not feasible at present to determine the truth of the findings. To date we can identify as true positives only a subset of genes that are likely to be differentially expressed in R6/2 mice (Luthi-Carter et al., 2000, 2002a), and we also do not yet know which identified genes are not (false positives), or which genes missed are in fact differentially expressed (false negatives). It is advisable to build some truth into the experiment where feasible, for example by using spike-in controls (specific RNAs added to the sample in known quantities). Nevertheless, we hope that this survey of single methods and method agreement can provide some guidance to investigators in selecting appropriate procedures.

In a similar investigation, Stevens and Doerge (2005) use a model to generate a known truth and simulate data from their model to examine properties of meta-analysis of microarray studies. Although this approach may provide some useful broad guidelines, further empirical evidence is required to gain more refined insight into the sources and magnitudes of variability and their effects on properties of meta-analysis of microarray studies.

In most studies, researchers have the resources for further investigation into only a few of the findings. A typical validation study consists of following up on a few genes, often in the range of 5 – 20. The research community would benefit from larger scale follow-up studies to enable the properties of different methodologies for synthesis to be judged more critically.

Although a number of intriguing methods have been introduced for meta-analysis of microarray data, the literature in this field is not yet fully developed. Clearly there is a need for further empirical and theoretical research in this challenging area. Large-scale validation studies would provide a welcome opportunity to advance both biological and methodological knowledge.

Acknowledgments

This work was supported by funds from the Swiss National Science Foundation to the National Centers for Competence in Research (NCCR) in Plant Survival (DRG) and

Molecular Oncology (MD, TS), and also by the US National Institutes of Health (RL-C). The authors also acknowledge R. J. Ferrante and J. K. Kubilus for the samples used to generate the HD data set for this analysis.

CHAPTER 10

Alternative probe set definitions for combining microarray data across studies using different versions of Affymetrix oligonucleotide arrays

Jeffrey S. Morris, Chunlei Wu, Kevin R. Coombes, Keith A. Baggerly, Jing Wang,

and Li Zhang

10.1 Introduction

Pooling data across multiple microarray studies should provide increased power to detect new relationships between gene expression levels and outcomes of interest. In the chapter of Goldstein et al. (Chapter 9), the data from two different microarray studies are derived from the same type, making pooling relatively straightforward. In general, though, pooling is complicated by the fact that gene expression measurements from different microarray platforms are not directly comparable. In this chapter, we discuss two methods for combining information across different versions of Affymetrix oligonucleotide arrays (Lockhart et al., 1996; see also Section 1.4). Each involves a new approach for combining probes on the array into probe sets.

In the first (Partial Probe Sets) approach, "matching probes" present on both chips are assembled into new probe sets based on UniGene clusters. We demonstrate that this method yields comparable expression level quantifications across chips without sacrificing much precision or substantially altering the relative ordering of the samples. We apply this method to combine information across two lung cancer studies performed using the HuGeneFL and HG-U95Av2 chips, revealing some genes related to patient survival. It appears that the gain in statistical power from the pooling is key to identifying many of these genes, since most were not found by equivalent analyses performed separately on the two data sets.

We have found that the Partial Probe Sets approach is not feasible for combining information across the HG-U95Av2 and HG-U133A chips, which share fewer probes

157

in common. We thus introduce a second method using full-length transcript-based probe sets. This approach redefines a probe set as a group of probes matching the same full-length mRNA transcripts in current genomic databases. This method yields comparable expression levels across HG-U95Av2 and HG-U133A chip types, and has better correlation across chip versions than the Affymetrix corresponding probe set definitions.

10.2 Combining microarray data across studies and platforms

In recent years, microarrays have been used extensively in biomedical research. This is evident from the fact that since 2000 there are over 28,000 articles indexed in PubMed (`http://www.ncbi.nlm.nih.gov/sites/entrez`) that involve microarrays. Generally, these studies involve the identification of individual genes or sets of genes whose expression profiles are related to clinical or biological factors of interest, including tissue type, disease status, disease subtype, patient prognosis, and biological pathway, to list a few. Due to cost limitations, most studies are performed using only a small number of samples. As a result, individual studies often have limited power to detect relevant biological relationships.

More recently, there has been a movement within the scientific community to make data from microarray studies publicly available. This movement has been propelled by the establishment of standards for minimal information to provide when reporting data (MIAME; Brazma et al., 2001) and the requirement of many major journals to make such data publicly available. There are currently several public repositories in which microarray data are posted, including ArrayExpress `http://www.ebi.ac.uk/arrayexpress/` and Gene Expression Omnibus (GEO; `http://www.ncbi.nlm.nih.gov/geo/`). The resulting vast quantity of publicly available data makes it possible to consider meta-analyses that combine information across multiple studies. If done properly, this pooling of information across studies can provide increased power to detect small consistent relationships that may have gone undetected in the individual analyses, and can provide results that are more likely to prove reproducible.

A growing number of studies are aimed at combining information across multiple data sets. Methods of combination vary across the information spectrum (Section 1.2). Approaches that have been used in the microarray context fall roughly into the following classes:

1. Identify an intersection of genes that are significant across multiple studies.
2. Combine results across studies to find genes whose results are consistent across studies.
3. Validate results from a single individual study using data from other studies.
4. Perform a single analysis after combining data across multiple studies.

We now briefly discuss the merits and drawbacks of each approach.

Intersecting gene lists. The idea behind this approach is that if a gene is truly differentially expressed, then this differential expression should be manifest across multiple data sets. This procedure is simple to carry out and gene lists are widely available for most microarray studies so it is no problem to collect the necessary input. However, this Venn diagram-based approach often reveals a shockingly small number of genes found to be differentially expressed in multiple data sets. For example, in a study comparing normal and CLL B-cells, Wang et al. (2004) found that only 9 genes were found to be differentially expressed in all three studies conducted on three different microarray platforms, out of 1,172 that were differentially expressed in at least one study. Similarly, in a study involving pancreatic cells, Tan et al. (2003) found only 4 genes differentially expressed across 3 different platforms, among the 185 deemed differentially expressed on at least one platform. Although perhaps identifying the most reliably differentially expressed genes, this approach results in reduced sensitivity for detecting biological relationships, since each (perhaps underpowered) study must find the gene significant before it is declared so. This leads to the paradoxical situation that including more studies results in fewer identified genes.

Combining results. Less conservative approaches focus on identifying genes that are consistent across studies include combining p-values across studies (Rhodes et al., 2002, 2004a) and the integrative correlation method of Parmigiani et al. (2004), which involves computing gene-gene pairwise correlations on the expression levels and/or tests statistics for each individual study, then computing a "correlation of correlations" across studies. This approach results in a list of reproducible genes whose absolute or relative expression levels are correlated across studies and platforms. For detecting biological relationships, combining such summaries does not provide additional power over that obtained by an analysis of combined data with a similar sample size.

Validation. Another approach is to identify biological relationships using the data from a single study, then using data from other studies for "validation" of these relationships (Beer et al., 2002; Sørlie et al., 2003; Stec et al., 2005; Wright et al., 2003). Since the studies may differ with respect to patient population, microarray platform, and sample handling and processing, results surviving this stringent form of validation are likely to be real. However, this use of multiple data sets does not yield any additional power for detecting biological relationships since only a single data set is used in the discovery process.

Combined analysis. In this approach, the data themselves are combined across studies and a single analysis is performed on the pooled data set. This is our primary interest in this chapter. The clear advantage of this approach is the possibility of increased power for detecting biological relationships, since power increases with the number of independent samples and the pooled data set will be significantly larger than any of the individual data sets. The difficulty is that there are important differences between the studies that must be taken into account before it is possible to successfully pool the data. These differences can be manifest in both the clinical outcomes and the microarray data, and may affect the genes in a differential manner (see Section 1.2.2 for clinical heterogeneity and Section 9.3.4 for sources of hetero-geneity more specific to microarray experiments). It has been shown that it is possi-ble to obtain comparable microarray data from different laboratories on a common platform if rigorous experimental protocols are established and followed across the different sites (Dobbin et al., 2005). However, data from different studies are likely to have been generated using different protocols, so these factors come into play in the meta-analysis context. These problems are further exacerbated if the studies are conducted on different microarray platforms, which have technical differences that make their gene expression levels fundamentally incomparable (Kuo et al., 2002; Tan et al., 2003; Mah et al., 2004; Marshall, 2004; Mecham et al., 2004a; Irizarry et al., 2005).

Some of this heterogeneity can be handled using standard meta-analysis approaches, modeling study effects for each gene using fixed or random effects (see Section 1.2.3) or Bayesian hierarchical models (Ghosh, 2004; Wang et al., 2004). These approaches appropriately account for between-study variability when performing inference in the meta-analysis, and provide a simple first-order correction for each gene that aligns the mean expression levels for the different studies. Other approaches also aim to make a first-order correction, but use more mathematically sophisticated methods. One method based on the singular value decomposition (Alter et al., 2000; Nielsen et al., 2002) normalizes the raw expression levels within studies using the first eigen-vector of the data matrix of expression values. This approach assumes that the first eigenvector represents between-study variability, which is assumed to dominate all other factors. Another approach (Benito et al., 2004) normalizes "distance weighted discrimination" (DWD), which performs supervised discrimination to identify linear combinations of genes associated with the study effect.

These approaches, however, do not appear to be sufficient to make data comparable across different microarray platforms. They adjust only the means of the distribu-tions for the two studies, without adjusting for higher order distributional properties like the variances or quantiles. In a study comparing data from spotted cDNA glass arrays and Affymetrix oligonucleotide arrays, Kuo et al. (2002) conclude that "data from spotted cDNA microarrays could not be directly combined with data from syn-thesized oligonucleotide arrays," and further, that it is unlikely that the data could be normalized using a common standardizing index.

Because of the nonexistence of a transform to a common measure, many studies do not attempt to combine the raw expression profiles across platforms, but instead com-bine unitless summary measures derived from the raw data. The assumption is that although the raw expression levels for the different studies may not be comparable, these unitless statistics should be. For example, Wang et al. (2004) and Choi et al. (2003) first compute the standardized log fold changes between two experimental conditions, then combine these across studies using hierarchical models. Similarly, Ghosh et al. (2003) and Tan et al. (2003) first compute t-statistics comparing two experimental conditions, then combine these t-statistics across studies. Shen et al. (2004) combine the posterior probabilities of being over-expressed, under-expressed, or similarly expressed between two experimental conditions across data sets. These approaches to combination do have increased power to detect biological relationships from the data, and can in principle be used across different microarray platforms. However, it should be even more powerful to work with the raw expression levels, if it were possible to make them comparable. In this case, we would not be limited to dichotomous comparisons, but could relate gene expression levels with any type of outcome (e.g., survival or time to progression).

Some studies explicitly use sequence information to try to obtain comparable expression levels across platforms (Morris et al., 2005; Mecham et al., 2004a; Mah et al., 2004; Wu et al., 2005; Ji et al., 2005). This idea is natural, since much of the systematic variability between expression level measurements between (and even within) platforms is attributable to sequence-related factors, such as cross-hybridization, alternative splicing, inaccurate gene sequence annotation, and RNA degradation. Cross-hybridization occurs when a gene hybridizes to "near matches" on the array, which can attenuate estimates of gene expression. Certain sequences are more likely to cross-hybridize (Zhang et al., 2003), so may result in less reliable measurements of gene expression. Also, single genes may be transcribed into multiple different mRNA variants, a phenomenon known as alternative splicing. These alternatively spliced variants may cause some sequences corresponding to different exons from the same gene to be discordant. Additionally, not all probes on microarrays map to annotated sequences in public databases. These probes tend to be less reliable (Mecham et al., 2004b), which may explain some of the lack of concordance across platforms. In a study involving matched samples run on Affymetrix and nylon cDNA arrays, Ji et al. (2005) show that the correlation of expression levels for these platforms is greater for sequences with matches in the RefSeq database. Finally, RNA degradation can have a differential effect on probe hybridization, since degradation is more likely to occur at the end of the gene, giving probe sequences closer to the end less opportunity for hybridization than those in the middle. These factors are relevant when comparing completely different technologies, e.g., spotted glass cDNA arrays and Affymetrix oligonucleotide arrays, as well as when comparing different versions of the same technologies, e.g., different versions of Affymetrix arrays or glass cDNA arrays constructed using different clones. We expect that methods that explicitly take into account these known biological and technological factors will be most successful at combining information across platforms.

10.3 Meta-analysis with Affymetrix oligonucleotide arrays

The probes are constructed based on sequence information contained in GenBank (http://www.ncbi.nlm.nih.gov/Genbank), a public archive of DNA sequence information, UniGene (http://www.ncbi.nlm.nih.gov/unigene), which partitions sequences into non-redundant clusters presumably corresponding to genes, and the Reference Sequence database RefSeq (http://www.ncbi.nlm.nih.gov/RefSeq), which is constructed by the National Center for Biotechnology Information (NCBI) to represent the state of the art in terms of the sequences of known genes. As this information has evolved over time, Affymetrix has produced different versions of its GeneChip. Chips used in human studies include the HuGeneFL, the HG-U95Av2, and the HG-U133A (and more recently the U133A Plus 2.0).

The HuGeneFL was introduced in November 1998; its sequence clusters are based upon UniGene build 18. It contains information on roughly 5600 genes, with each gene represented by roughly 20 probe pairs (PM and MM probes, see Section 1.4). The probes corresponding to the same probe set are placed together in the same region of the array. The HG-U95Av2 was introduced in April 2000, and is based upon UniGene build 95. It contains information on roughly 10,000 genes, each of which is represented by 16 probe pairs. The probes are randomly distributed across the array. The U133A was first introduced in January 2002, and is based upon UniGene build 133. It contains information on 14,500 genes, with 11 probes per gene. The probes are arranged on the array in such a way as to optimize probe synthesis efficiency.

Frequently, researchers wish to combine information across experiments that have used different versions of Affymetrix GeneChips. As new studies are conducted using more recent versions of the chips, researchers still want to use information from previous studies performed using older generation chips. Researchers may also want to carry out meta-analyses on data collected from multiple studies performed at different institutions. Merging information across chip types is not straightforward, since some genes represented on newer chips were not on previous versions. Even the genes which are in common may be represented by different sets of probes on the different chip versions, so their expression levels are not generally comparable.

Below, we describe in detail two methods that we have developed (Morris et al., 2005; Wu et al., 2005) to combine information across studies using different versions of Affymetrix chips. These methods use sequence information to define new probe sets that yield comparable expression levels across different chip types, thereby allowing data from different chip versions to be combined. For each method, we also provide an example to demonstrate the concordance of expression levels across the different chip versions.

10.4 Partial probe sets method

The incompatibility of expression levels across chip versions is largely due to the fact that different sets of probes are used to represent the same genes on different chips. We expect, however, that individual probes present on multiple chips should yield comparable expression levels. Thus, one approach for obtaining comparable expression levels across studies using two different chip versions is to use only "matching probes" that are present on both.

Consider the case of microarray data from two studies, one performed with the HuGeneFL chip and the other with the HG-U95Av2. The HuGeneFL contains a total of roughly 130,000 probes partitioned into 6,633 probe sets, each containing 20 probe pairs, while the HG-U95Av2 contains a total of roughly 200,000 probes partitioned into 12,625 probe sets, each containing 16 probe pairs. A total of 34,428 "matching probes" are present on both chips.

After identifying these matching probes, we recombine them into new probe sets based on the most current build of UniGene. We refer to these new probe sets as "partial probe sets." Because their definition is explicitly based on UniGene clusters, these partial probe sets will not correspond precisely to Affymetrix-determined probe sets. Frequently, multiple Affymetrix probe sets map to the same UniGene cluster. We eliminate any probe sets containing just one or two probes, since we expect gene expression quantification based on so few probes to be less reliable. Using UniGene build 160 results in 4,101 partial probe sets. In general, we expect these partial probe sets to contain fewer probes than the Affymetrix-defined probe sets, since only those probes which match across multiple chip versions are used in a partial probe set. Figure 10.1 contains a plot of the number of probes within each of these partial probe sets. Most partial probe sets (84%) contain 10 or fewer probes, with a median probe set size of 7. Several partial probe sets do contain more than 20 probes.

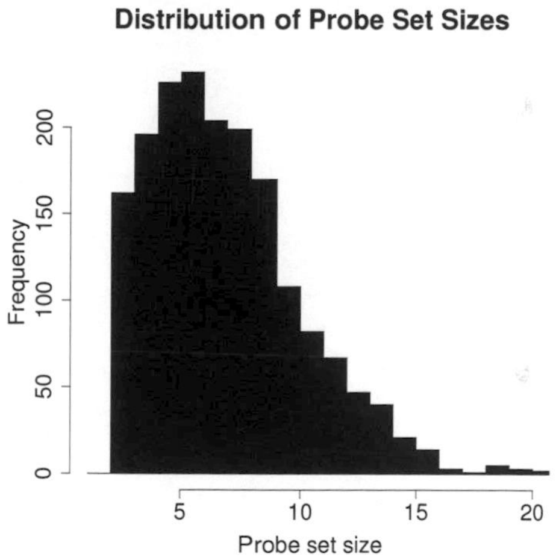

Distribution of Probe Set Sizes

Figure 10.1 *Histogram of number of probes in each partial probe set.*

10.5 Example 1: CAMDA 2003 lung cancer data

Two independent studies were performed at Harvard University (Bhattacharjee et al., 2001) and University of Michigan (Beer et al., 2002), both focusing on the same question of relating gene expression to survival in lung cancer patients. These data were part of the 2003 Critical Assessment of Microarray Data Analysis (CAMDA) competition (http://www.camda.duke.edu/camda2003). Both studies use Affymetrix GeneChips, but the Michigan study used the HuGeneFL while the Harvard study used the HG-U95Av2. Our goal here is to combine information across both data sets to identify genes whose expression levels provide prognostic information on patient survival beyond what is already provided by known clinical factors. We use partial probe sets as a basis for quantifying gene expression levels, and demonstrate that this results in comparable expression levels across the two chip versions, without any loss of precision that might be expected due to using only a subset of the probes. We are able to identify a number of prognostic genes in our pooled analysis that were not discovered in analyses of the individual studies, highlighting the benefit of pooling data across studies. We first summarize these data sets, then describe our analyses to validate the partial probe set method and identify prognostic genes. More details of this analysis can be found in Morris et al. (2005).

10.5.1 Overview of lung cancer data sets

The Harvard study used the HG-U95Av2 to analyze 186 lung tumor samples. Of these, 125 were adenocarcinomas for which there was clinical information, including gender, age, stage of disease, and survival time. Applying hierarchical clustering to these data, Bhattacharjee et al. (2001) identified four subtypes of adenocarcinoma with different molecular profiles, and further demonstrated that these subtypes had different survival prognoses.

The Michigan study used the HuGeneFL chip to analyze 86 lung adenocarcinoma samples, all of which had clinical information as above. Using univariate Cox regressions with one gene at a time, Beer et al. (2002) identified a number of genes whose expression levels were associated with patient survival. They subsequently constructed a "risk index" using the top 50 genes, and demonstrated that this risk index helped predict patient survival both in their own data and in the independent Harvard study (Bhattacharjee et al., 2001).

10.5.2 Data preprocessing

For our analysis, we first carry out quality control checks, after which we removed 10 arrays from the Michigan study and one from the Harvard study that demonstrated poor quality. This leaves a total of 200 arrays, 124 from the Harvard study and 76 from the Michigan study. Using the partial probe set definitions described above, we quantify the gene expression levels using the Positional Dependent Nearest Neighbor (PDNN) model (Zhang et al., 2003). In the PDNN model, probe signals are decomposed into three components representing gene-specific binding, nonspecific binding (cross-hybridization), and background. Other quantification methods could have been used, but we use this one here because we believe its use of probe sequence information to predict patterns of specific and nonspecific hybridization intensities can lead to more reliable and accurate quantification.

We also perform other preprocessing steps. We remove the partial probe sets whose mean expression levels across all samples are in the lower half, then normalize the log expression values using a linear transformation to force each chip to have a common mean and standard deviation (SD) across genes. We next remove those partial probe sets with the smallest variability across chips (SD < 0.20), since we consider them unlikely to be discriminatory and more likely to be spuriously flagged as prognostic. Finally, we remove the probe sets with poor relative agreement (Spearman correlation < 0.90) between the partial probe set and full probe set quantifications (see Section 10.5.3). After this preprocessing, we are left with 1,036 partial probe sets.

10.5.3 Validation of partial probe sets

Before applying our Partial Probe Set method to the data to identify prognostic genes, we assess whether it performs acceptably. First, we assess comparability of expression levels across chip versions. Specifically, we compute the median and median absolute deviation (MAD) of log expression level for each partial probe set for each data set. Since the patient populations in the two studies appear similar, we expect to see high concordance in these quantities between the two chips if the expression levels were comparable. Figure 10.2 contains plots of these quantities, demonstrating high concordance for the two studies (0.961 for the median and 0.820 for the MAD). Thus, it appears that the Partial Probe Set method yields comparable expression levels across the two chips.

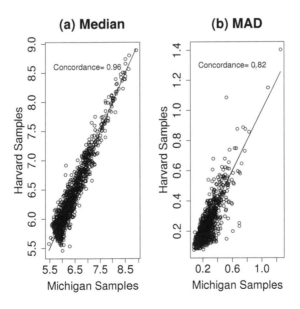

Figure 10.2 *Median (a) and median absolute deviation (MAD) (b) expression levels for each partial probe set based on the Harvard samples (HG-U95Av2) vs. the Michigan samples (HuGeneFL).*

Due to use of only the matching probes, partial probe sets are generally smaller than the Affymetrix-defined probe sets. The median size of the partial probe sets is 7, while the Affymetrix-defined probe sets for the HuGeneFL and HG-U95Av2 chips have 20 and 16 probes, respectively. Since additional probes can increase the precision of the measured expression level of the corresponding gene, one might expect a loss of precision when using the partial probe sets to quantify expression levels. To investigate this possibility, we also quantify the expression levels for the full probe sets of the Harvard samples using the PDNN model. These full probe sets consist of *all* probes on the array mapping to the UniGene cluster, not just the matching ones. Figure 10.3 shows the SD for each gene using the full probe set versus the SD for the corresponding partial probe set. There is no evidence in this plot of large precision loss, as there is strong agreement between the SDs for each gene using the two methods (concordance = 0.932). Although initially surprising this result makes sense, since we expect that the probes Affymetrix retains in formulating the new chips are likely to be in some sense the "best" ones.

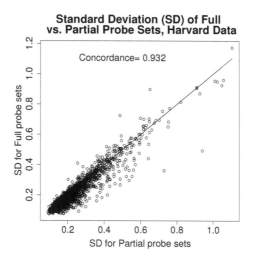

Figure 10.3 *Standard deviation across Harvard samples for each gene based on full and partial probe sets.*

We compute for each probe set the Spearman correlations between the partial and full probe set quantifications in the samples to assess whether our method preserves the relative ordering of the samples, i.e., the ranks. For example, we expect that a sample with the highest expression level for a given gene using the full set of probes should also demonstrate the highest expression level for that gene when using only the matched probes. The median Spearman correlation across all probe sets is 0.95, suggesting that the method does a good job of preserving the sample ranks. Most of the lower Spearman correlations occur for probe sets with less heterogeneous expression levels across samples and/or probe sets containing smaller numbers of probes. It appears that the Partial Probe Set method works quite well.

10.5.4 Pooling across studies to identify prognostic genes

We pool the data across the Harvard and Michigan studies to identify prognostic genes offering predictive information on patient survival. We are not primarily interested in finding genes that are simply surrogates for known clinical prognostic factors like stage, since these factors are easily available without collecting microarray data. Rather, we are looking for genes that explain the variability in patient survival remaining after modeling the clinical predictors. Thus, we fit multivariate survival models, including clinical covariates in all models used to identify prognostic genes.

We screened the 1,036 genes one at a time to find potentially prognostic ones by fitting a series of multivariate Cox models containing age, stage (dichotomized as low (stages I–II) or high (stages III–IV)), institution, and the log-expression of one gene as predictors. The institution effect is included in the model to account for differences in survival between the two studies that persist even after accounting for known clinical covariates. For each gene, we obtain a p-value for the coefficient using a permutation approach. In this approach, we first generate 100,000 data sets by randomly permuting the gene expression values across samples while keeping the clinical covariates fixed. We subsequently obtain the permutation p-value by counting the proportion of fitted Cox model coefficients for the gene more extreme than the coefficient for the original data set. A small p-value for a given gene indicates potential for that gene to provide prognostic information on survival beyond the clinical covariates. For comparison, we also obtain p-values using an asymptotic likelihood ratio test (LRT) and a bootstrapping approach.

If there were no prognostic genes, the gene coefficient p-values should follow a uniform distribution. An overabundance of small p-values would indicate the presence of prognostic genes. We fit a Beta-Uniform mixture model (BUM) to the p-values (Allison et al., 2002; Pounds and Morris, 2003; see also Section 6.4). The BUM partitions the p-value histogram into two components, a Beta distributed component containing the prognostic genes and a Uniform distributed component containing the non-significant ones. We use this model to identify a p-value cutoff that controls the corresponding false discovery rate (FDR; see Benjamini and Hochberg 1995, Section 7.2.3) to be no more than 0.20. This means that of the genes flagged as prognostic, we expect at most 1 in 5 to be false positives.

Figure 10.4 contains the histogram of permutation test p-values; the corresponding histogram for the LRT is nearly identical and is therefore not shown. The overabundance of very small p-values indicates the presence of some genes providing information on patient prognosis beyond what is offered by the modeled clinical factors. Table 10.1 contains the set of 26 genes that are flagged by the BUM method using FDR< 0.20, corresponding to those genes with p-values less than 0.0025. Many of these genes appear to be biologically interesting and worthy of further consideration. Ten of the 26 prognostic genes could be linked to lung cancer based on the existing literature. Four others could be linked to cancer in general or other lung disease in the literature. These genes are discussed in more detail in Morris et al. (2005).

None of the genes we identify appear in the list of top 100 genes from the Michi-

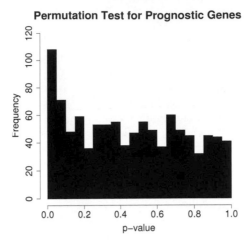

Figure 10.4 *Histogram of p-values from permutation test on single gene coefficients in Cox models containing clinical covariates for each of the 1,036 candidate genes.*

gan analysis (Beer et al., 2002), and only one (CPE) is mentioned in the Harvard study (Bhattacharjee et al., 2001). CPE is one of the genes defining a neuroendocrine cluster associated with poor prognosis. In repeating our analysis separately for two data sets (i.e. without pooling), only eight (Harvard) and one (Michigan) of the 26 genes have *p*-values less than 0.0025. The 17 remaining genes include the top gene in the combined list (FCGRT; see Table 10.1). Thus, it appears that the pooled analysis reveals new biological insights that were not identified when analyzing the data separately.

10.6 Full-length transcript-based probe sets method

The analysis presented above shows that by using partial probe sets, we are able to obtain comparable expression levels across studies conducted at different institutions using different chips (HuGeneFL and HG-U95Av2), allowing a pooled analysis that reveals new biological insights into lung cancer. Unfortunately, this approach is not feasible for combining information across the HG-U95Av2 and HG-U133A chips, since these chips share fewer probes in common than the HuGeneFL and HG-U95Av2. There are 34,428 probes (14%) on the HG-U95Av2 that are also present on the HuGeneFL, while there are only 11,582 probes (6%) that are also present on the U133A. If we form partial probe sets and eliminate those with fewer than three probes, we are left with only 628 probe sets. Thus, we explore less stringent alternatives to use for combining information across these chip versions.

A primary reason that probes yield discordant measurements is that they may be responding (hybridizing) to different transcripts alternatively spliced from the same

Table 10.1 *Genes identified as prognostic (FDR< 0.20) by applying BUM on the permutation p-values. Also included are the LRT and bootstrap p-values and estimates of the Cox model coefficient. "*" indicates the p-value reaches the BUM significance threshold. A negative coefficient indicates that larger expression levels of that gene correspond to a better survival outcome.*

Gene Symbol	Coef	Prognostic p-values		
		Permutation	LRT	Bootstrap
FCGRT	-2.07	< 0.00001*	0.00014*	0.0006*
ENO2	1.46	0.00001*	0.00002*	< 0.0001*
NFRKB	-2.81	0.00001*	0.00435	0.00404*
RRM1	1.81	0.00002*	0.00008*	< 0.0001*
TBCE	-2.35	0.00004*	0.00069*	0.0006*
Phosph. mutase 1	1.92	0.00008*	0.00020*	0.0004*
ATIC	1.81	0.00009*	0.00153*	0.0004*
CHKL	-1.43	0.00010*	0.02305	0.0260
DDX3	-2.37	0.00017*	0.00012*	0.0002*
OST	-1.64	0.00020*	0.00010*	0.0010*
CPE	0.72	0.00031*	0.00053*	0.0010*
ADRBK1	-2.20	0.00044*	0.00678	0.0030*
BCL9	-1.64	0.00067*	0.03602	0.0460
BZW1	1.33	0.00068*	0.00279*	0.0006*
TPS1	-0.64	0.00106*	0.00217*	< 0.0001*
CLU	-0.52	0.00109*	0.00239*	0.0024*
OGDH	-2.19	0.00118*	0.00405	0.0020*
STK25	2.29	0.00122*	0.00152*	0.0080
KCC2	-1.70	0.00143*	0.00988	0.0220
SEPW1	-1.29	0.00145*	0.01026	0.0160
FSCN1	0.66	0.00150*	0.00241*	0.0103
MRPL19	1.12	0.00211*	0.03213	0.0340
ALDH9	-1.18	0.00223*	0.00378*	0.0020*
PFN2	0.63	0.00248*	0.00351*	0.0020*
BTG2	-0.75	0.00232*	0.00580	0.0140

gene. When the transcripts are differentially regulated, the corresponding probes can yield conflicting signals. The current design of these arrays ignores the effects of alternative splicing. Thus, if we differentiate the probes that match sets of alternatively spliced transcripts, we may be able to resolve the discordant measurements. Based on this idea, we develop another method to regroup the probes into probe sets. In this new definition, all probes in the probe set must match the same set of full-length gene sequences. We refer to such a probe set as a "Full-Length Transcript Based Probeset" (FLTBP; Wu et al., 2005). Assuming complete inclusion of alternatively spliced transcripts, we can in principle ensure concordant behavior of the probes within these probe sets.

We now describe how to obtain these transcript-based probe sets. First, we construct a comprehensive library of full-length mRNA transcript sequences in the human genome by combining records in RefSeq and the H-Invitational Database (H-InvDB; http://hinvdb.ddbj.nig.ac.jp), an integrated database of human genes and transcripts. We estimate that collectively the two databases represent approximately 29,000 genes with 50,000 non-redundant transcripts.

We use this library as the basis for defining our probe sets. For each probe sequence on the HG-U133A and HG-U95Av2 arrays, we use the BLAST program (Altschul et al., 1997; http://www.ncbi.nlm.nih.gov/blast) to identify all matching full-length transcripts. We construct a matched target list by aggregating the IDs of the transcripts with exact matches to construct a matched target list. For 15% of the probes on the HG-U95Av2 and 13% of the probes on the U133A there is no exact match in our library. In addition, 38% of the probes on the U133A and 33% of the probes on the HG-U95Av2 match more than two targets in our library, demonstrating that it is very common for a single probe to match multiple targets.

By grouping the probes within the same matched target lists, we form 23,972 and 14,148 FLTBPs on the HG-U133A and HG-U95Av2, respectively. Because multiple probes in a probe set are essential to reduce noise and bias, we discard all small probe sets containing fewer than three probes, leaving 18,011 and 11,228 FLTBPs on the HG-U133A and HG-UAv2, respectively. Collectively, these FLTBPs contain 82% of the probes on the arrays.

These new probe sets are very different from the original ones. Only 9,893 of the original probe sets on HG-U133A and 5,257 original probe sets on HG-U95Av2 are the same after regrouping. Figure 10.5 shows the distributions of the number of probes in each FLTBP for the two chips. The probe sets outside of the major peaks reflect division and fusion of the original probe sets. Detailed information on the FLTBP probe sets is stored on our web site (http://odin.mdacc.tmc.edu/~zhangli/FLTBP), which also contains chip definition files (CDF) using FLTBPs, following the format designed by Affymetrix (http://www.affymetrix.com/index.affx). These CDF files can be used to run expression quantification algorithms in BioConductor (Gentleman et al., 2004; http://www.bioconductor.org).

By aligning the matched target lists of FLTBPs on the two arrays, we find 9,642 pairs of FLTBPs that can be mapped between the HG-U133A and HG-U95Av2. The Affymetrix probe set mapping across chip versions (http://www.affymetrix.com/Auth/support/downloads/comparisons/best_match.zip) produces 9,480 pairs of probe sets in common for the HG-U95Av2 and HG-U133A chips. There are numerous differences between the Affymetrix mappings and our FLTBPs. Only 52% of the probe sets on the HG-U133A and 48% of the probe sets on the HG-U95Av2 are mapped the same as our FLTBPs.

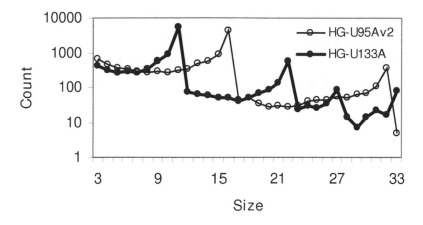

Figure 10.5 *Distributions of number of probes per FLTBP for the HG-U95Av2 and HG-U133A chips.*

10.7 Example 2: Lung cell line data

To compare our mapping method with that of Affymetrix, we use a data set consisting of 28 paired measurements obtained by hybridizing identical samples on both the HG-U133A and HG-U95Av2 arrays. Because of this paired design, we expect very little biological variability between corresponding measurements on the two arrays, so any observed differences should be attributable to technical sources. We now describe this data set and use it to demonstrate that the FLTBPs result in quantifications that are more comparable across chip types than Affymetrix-based probe sets.

10.7.1 Overview of data set

Thirty RNA samples from variant lung cancer or normal lung cell lines and one human reference sample were hybridized on both HG-U133A and HG-U95Av2 arrays. Our quality control procedures revealed that three array images had obvious defects These are discarded, leaving 28 pairs of samples for this study.

We preprocess and quantify gene expression with PDNN (Zhang et al. 2003) using the PerfectMatch software (`http://odin.mdacc.tmc.edu/~zhangli/PerfectMatch`). For comparison, we also apply a few other widely-used quantification methods (see also Section 1.4): RMA (Irizarry et al., 2003a), MAS 5.0 (Affymetrix, 2002), and dChip (Li and Wong, 2001), each with the default settings in the BioConductor `affy` package (Irizarry et al., 2008).

10.7.2 Validation of transcript-based probe sets

In order to assess comparability across chip types, we compute for each gene the correlation between the paired HG-U95Av2 and HG-U133A measurements across samples. To enhance the contrast between two different mapping methods, we focus on the probe sets that differ between the two methods. Approximately 1/3 of the probe sets are mapped differently, resulting in 3,309 and 3,527 paired probe sets for FLTBP method and Affymetrix method, respectively.

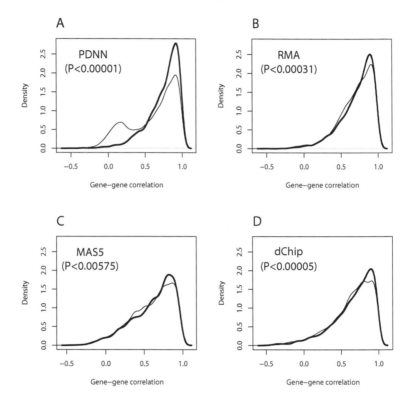

Figure 10.6 *Distributions of gene-gene correlation between probe sets on two HG-U95Av2 and HG-U133A arrays, combining information over all samples, for Affymetrix-defined probe sets (thin line) and FLTBPs (thick line). Correlations are computed using four different quantification methods: (A) PDNN, (B) RMA, (C) MAS5.0, and (D) dChip.*

Figure 10.6 contains smoothed histograms of these correlations across probe sets for the two mapping methods and four quantification methods. These histograms summarize the observed distributions of the paired correlations across probe sets. Figure 10.6(a) clearly demonstrates that, when using the PDNN quantification method, the FLTBP mapping tends to yield significantly higher correlations than the Affymetrix mapping (Kolmogorov-Smirnov [KS] test $p < 0.00001$). There are two peaks evident in the distribution of correlations for the Affymetrix mapping. The minor peak contains a large group of probe sets with poor correlation across chip types. With other quantification methods, there is also evidence that the FLTBP method tends to result in better correlation across chip types than the Affymetrix method, although this evidence is not as strong (Figures 10.6(b)-(d), KS test p-values $= 0.00031, 0.00575$, and 0.00005, respectively). This improvement from using the FLTBPs is likely due to the fact that the FLTBP adjusts for some of the heterogeneity that is due to alternative splicing.

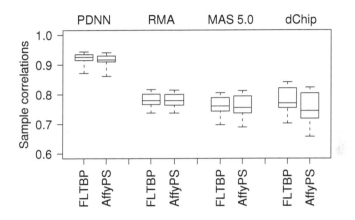

Figure 10.7 *Distributions of sample-to-sample correlation between probe sets on two HG-U95Av2 and HG-U133A arrays, combining information over all genes, for Affymetrix-defined probe sets and FLTBPs. Correlations are computed using four different quantification methods: PDNN, RMA, MAS5.0, and dChip.*

Note also that, when compared with Figure 10.6(a), the distributions in Figure 10.6(b)-(d) are shifted more toward low correlations. This suggests that, for these data, the PDNN quantification tends to yield generally higher correlations than the RMA, MAS5, or dChip quantifications. This is even more evident in the sample-by-sample correlations between the chip types computed across genes, as shown in Figure 10.7. This increased correlation with the PDNN method may reflect the manner in which the PDNN model estimates and adjusts for the effects of non-specific binding.

In Figure 10.6(a) we can see that even when using the FLTBPs, not all genes displayed high correlations across chip versions. Many of these low correlations are observed for genes that appear to have low biological variability in these data. With low biological variability, the noise component of the measurements dominate, resulting in low correlations. There are, however, some probe sets with low correlations that do not have small variances. It is possible that some of the sequences corresponding to these probe sets were strongly affected by RNA degradation, or the currently available collection of transcripts may not include certain alternatively spliced variants that were differentially expressed across the sample tests, thereby attenuating the correlations. Additional work is required to further reduce the effects of cross-hybridization and RNA degradation, which should lead to even more comparable expression levels across platforms.

10.8 Discussion

Here, a pooled analysis over two lung cancer microarray studies identified new candidate prognostic genes that were not detected by separate analyses performed on the individual data sets. Our pooled analyses are based on two new probe set definitions that result in more comparable expression levels across different versions of Affymetrix oligonucleotide chips.

The Partial Probe Set approach appears to work well but is of limited applicability, since it is only feasible to apply this method across chip versions that share many probes in common. The FLTBP method does not have this restriction – it works by recombining probes based on the set of full-length mRNA transcripts to which they map. Thus, the probe sets themselves may not match, but they do map to the same set of alternatively spliced transcripts. Combined with the PDNN quantification method, which accounts for non-specific binding, the FLTBP approach appears to result in expression levels that are more comparable across chip versions than the matched probe sets provided by Affymetrix. A major benefit of the FLTBP approach is that by not restricting attention to matched probes, it can be widely applied to combine data across any chip versions. It may even be possible to extend the method by using the underlying principle to match oligonucleotide array data with cDNA data, which could create new opportunities for cross-platform pooling of microarray data.

CHAPTER 11

Gene ontology-based meta-analysis of genome-scale experiments

Chad A. Shaw

11.1 Introduction

Genome-scale experiments and microarray expression studies have become increasingly common in biological and medical science. The proliferation of data makes the integration of results both a pressing challenge and an opportunity for discovery. The identification of commonalities between studies can add confidence to the interpretation of results, while the identification of differences gives insight to the distinctions between treatments and experimental conditions. In this chapter, we explore the utility of gene annotations – specifically Gene Ontology (GO) annotations – as a resource to facilitate meta-analysis across a family of experiments.

We consider GO-based analysis at the level of gene lists. Gene lists are the most commonly accessible summary of results from expression array or library sequencing experiments. Experimental results for microarray studies are sometimes available as numerical summaries, and there is an increasing trend to make the raw data available in publicly accessible databases such as Gene Expression Omnibus (GEO; Edgar et al., 2002; Barrett et al., 2005). However, in this chapter we focus only on gene lists.

The chapter is organized to first give an overview of the Gene Ontology system and its implementation. Statistical methods used to analyze single lists are then described. We then extend these methods to consider groups of lists arising from multiple studies. Finally, we carry out an analysis using a family of lists obtained from reports in the stem cell community.

11.2 Ontologies

Ontologies are a growing and important formalism in the world of computing. The term ontology originates in the philosophical literature, where it means system of knowledge. More recently, computational authors have defined an ontology: "An *ontology* is an explicit specification of a conceptualization" (Gruber, 1993). In the current context, we intend the term ontology to mean a generalized taxonomy for information in some field of study. Ontologies present a way to organize information to facilitate storage, retrieval, and general computing against the knowledge structure. Ontologies currently exist within a large number of different domains: engineering, medical science, linguistics, and of course molecular biology and genetics.

In practice, an ontology is usually implemented as a controlled vocabulary with well-defined relationships connecting the terms. Although the number and variety of terms as well as the types of relationships will vary across fields, most ontologies can be represented mathematically as graph structures where nodes are terms and edges connect the related terms. Ontologies are most often represented as Directed Acyclic Graphs (DAGs). A DAG is a graph with directed edges and for which a child can have more than one parent. The acyclic character of the graph comes from the directionality of edges. The graph has a direction or flow from a root node to terminal nodes. Although there may be more than one path from the root to descendant nodes, there are no cycles or loops starting from ancestral nodes that lead down through the graph and then back up to ancestor nodes.

The flood of information available online through the internet has made ontologies increasingly important. Ontologies provide a rational and flexible way to unify information within a field of activity. Ontologies are increasingly seen as a way to provide more intelligent data retrieval through smart internet searches (Berners-Lee et al., 2001). Ontologies enhance information retrieval because query terms in a search can be placed into a context in a field of knowledge. Instead of pattern matching a string, the query term can be identified in the semantically structured ontology and results can be returned based on the relationship of the query term with other elements of the ontology. All terms in an ontological neighborhood of the query may provide useful results. Because the concept of an ontology is so general, it is likely that many of the tools and techniques which have been developed for analysis in the GO community have broader application. As well, many of the concepts and methodologies developed in the general ontology community have application within the biological sphere (Gruber, 1995).

11.3 The Gene Ontology

The Gene Ontology (GO) is a highly developed, active, community supported, species independent annotation system for describing genes. Information, tutorials and research tools for the GO can be found at the well-maintained and informative GO website (http://www.geneontology.org) and in articles describing the development of the GO (Ashburner et al., 2000, 2001). The formal structure of the GO is quite simple at first glance, but the GO also has some fairly sophisticated features and it continues to evolve. In this section, we examine the history and motivation for the GO. We also consider the implementation of GO and its various features. Finally, we discuss software tools used to manipulate GO information and to perform calculations with GO data.

11.3.1 History and motivation for GO

To appreciate the GO project one should have an idea of the historical context in which the GO developed. The GO consortium emerged in the late 1990s in a period of enormous productivity but also fragmentation within molecular biology. Several parallel projects to completely sequence the genomes of a number of organisms were ongoing. As well, genome-scale experiments such as gene expression microarray studies had just become technically feasible. In this context, several model organism communities began a large-scale initiative to share information on the genes whose sequence was revealed by the ongoing genome projects and whose expression profiles were being measured with microarrays.

Among the main insights derived from the sequencing work was the clear shared evolutionary history of the genes observed across the various organisms. A large number of genes – the exact numbers and proportions are still debated – reappear with largely the same sequence among highly divergent organisms. The shared gene sequences arise because of the common evolutionary origin of the genomes of these species. This phenomena of common evolutionary origin is termed *homology* in the biological sciences. Homology is a central tool to make sense of the commonalities among the life on earth.

Common evolutionary origins appear at virtually every scale at which a biological system can be described. Common origins underlie developmental patterns of animals and morphological features such as the limb bones of land animals. As well, all vertebrate neurological systems share homologous patterns of development. The maternal care behaviors of mammals are also an example of evolutionary homology. At the molecular scale, homology is manifest in the highly similar nucleotide and amino acid sequences found in the genomes and proteins of various taxa.

For the GO, the fortunate consequence of evolutionary homology is the shared framework that evolution provides for describing what genes do. The various organisms may differ greatly in size, shape and behaviors, yet the common origin of life means that these organisms can be thought of as variations on a theme. The consequence of evolution is that a single vocabulary might be able to characterize the activities of genes and their roles in the underlying life processes. It is thus possible to contemplate a single vocabulary applying *universally* to all organisms.

A central difficulty in creating such a shared terminology is not the nature of biology, but rather the nature of human beings. Despite the unifying principles of evolutionary science, the human communities of molecular biologists and geneticists – the *scientific communities* – have a sociologic tendency to become more specialized and divergent. For this reason the terminology – the names given to genes and their functions – have tended to differ, sometimes considerably. Although a gene might consistently appear in humans, mice and even yeast, that gene would likely have a different name in each taxon, and the jargon used to describe its biological properties would also be somewhat distinct.

The difficulty of divergent vocabularies only became clear in the late 1990s. At that time three separate model organism communities attempted to share the data archived in their species specific databases. The three separate databases were the *Saccharomyces* Genome Database for yeast, FlyBase for fruit flies, and the MGI Database for mouse (Ashburner et al., 2000, 2001). The GO was initially developed in 1998 as a unification strategy to share information contained in just these three databases. The GO rapidly adopted a more lofty ideal to create a unified annotation system for all genes in all organisms. The feasibility of such a unified annotation system rested and continues to depend on the essential commonality and shared evolutionary origins of life on earth.

11.3.2 GO specification and implementation

Creating a shared ontology for describing all genes in all organisms is clearly a daunting challenge. To accomplish this goal, the GO consortium settled on a three-fold collection of structured vocabularies. The three separate vocabularies were adopted to report on three distinct aspects of each gene's properties. These separate vocabularies are Biological Process, Molecular Function and Cellular Component. Genes are annotated into each of these three vocabularies.

Biological Process. The Biological Process ontology aims to capture the biological objective to which a gene's product contributes. The process ontology contains many terms which capture a notion of biological state change or transition. For instance, there are large sections of the ontology dedicated to the cell cycle. The ontology also represents the processes of catabolism and metabolic re-arrangement. This ontology also describes cellular communication processes such as signal transduction and the process of DNA transcription. Importantly, the process ontology does not seek to represent the biochemical events necessary to accomplish a process; rather, the process ontology merely tries to describe the variety of processes available to a cell. This ontology has seen the most development and is the most widely used in computational science.

Molecular Function. Molecular Function represents the biochemical aspect of what a gene product does. The function ontology describes only what is done without attempting to express the context in which reactions occur or what purpose they might serve. Examples of the function ontology include terms like "kinase" or "adenylate cyclase." The terms in molecular function can be more difficult to comprehend without extensive training in the biological sciences.

Cellular Component. The Cellular Component ontology describes the physical localization of a gene product within cells. The terms in this ontology include cellular regions like "cell membrane" or "nucleus." This ontology also includes terms that represent multi-protein complexes such as entities like "ribosomes" or "proteosomes" (Ashburner et al., 2000).

DAG

The terms in each ontology are organized as a directed acyclic graph or DAG. Again, a DAG is a graph where edges have directionality or flow, and a DAG differs from a tree in that a node can have more than one parent. The terms in each ontology DAG are arranged in a pattern of high generality to increasing specificity – from sweeping concepts to refined detail. Figure 11.1 presents a subset of the GO Biological Process ontology.

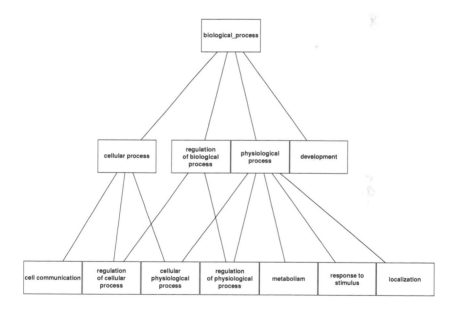

Figure 11.1 *An abbreviated view of the Biological Process arm of the GO, showing how terms are organized from generality to increasing specificity. As well, the figure shows descendant terms with numerous parents. This feature of the GO is essential to represent the semantics of molecular biology. Not all relationships or terms present in the ontology are depicted.*

Table 11.1 *Counts of current (January 2009) annotations to the GO Biological Process ontology in a variety of species. Annotation distinctions are made between those Inferred from Electronic Annotation (IEA) and those which result from all other sources.*

Database	Species	Biological Process	
		All codes	non-IEA codes
SGD	*Saccharomyces cerevisiae*	85,090	44,251
FlyBase	*Drosophila melanogaster*	70,652	54,522
MGI	*Mus musculus*	158,494	58,526
GO Annotation (EBI)	*Homo sapiens*	215,698	64,525

Relationships

There are two main types of relationships between parental nodes and their descendants in the GO. These two relationships are the *is-a* relationship and the *part-of* relationship. The *is-a* relationship indicates that the descendant term is a subclass or subtype of the parental term. The *part-of* relationship is more complex. This relationship indicates that the descendant term is actually a component of the parental term. This relationship is especially important for the Cellular Component branch of the GO. In computational work, the distinction between these relationships is sometimes ignored. All parent-child relationships in the GO are expected to abide the *true-path rule*, which means that the hierarchical structure of the GO must hold from any term to its top level parent. The true-path rule must apply for relationships of either the *is-a* or *part-of* type (see `http://www.geneontology.org/GO.usage.shtml` for more information).

Annotations

Genes are not nodes in the ontology. Rather, genes are annotated to terms within the ontology. When genes are assigned annotation within the GO, those annotations are expected to adhere to the true-path rule. If this property does not hold for some gene, then it is the responsibility of a curator to add the necessary GO terms to make a new path so that a gene may be annotated to the GO in such a way that the true-path rule will hold. Genes can be annotated to any number of different terms within the GO, and thus belong to multiple paths. There is no limit to the number of terms to which a gene may be assigned, nor is there a constraint on the relationships between the terms. Table 11.3.2 gives the number of annotations currently available in several species.

Table 11.2 *Some Gene Ontology evidence codes.*

Code	Description
IC	Inferred by Curator
IDA	Inferred from Direct Assay
IEA	Inferred from Electronic Annotation
IEP	Inferred from Expression Pattern
IGI	Inferred from Genetic Interaction
IMP	Inferred from Mutant Phenotype
IPI	Inferred from Physical Interaction
ISS	Inferred from Sequence or Structural Similarity
NAS	Non-traceable Author Statement
ND	No biological Data available
RCA	Inferred from Reviewed Computational Analysis
TAS	Traceable Author Statement
NR	Not Recorded

Evidence codes

All annotations within GO have evidence codes. There are 17 evidence codes now in use. Some evidence codes refer to "electronically-derived" information (usually meaning sequence similarity). Importantly, a large number of GO annotations are based on real experiments. The number of such experimentally derived annotations continues to grow quite rapidly, and these "real-world"-based annotations are highly valuable. Annotations based on sequence data are often somewhat less trustworthy. Table 11.3.2 lists some of the evidence codes.

Species specificity

The GO is designed to be species independent. This means that the vocabulary should be a common platform for analysis of genes from any taxa. Species independence makes the development of annotation resources faster, so that the effort can leverage work across communities. It also means that the GO can facilitate multi-species comparisons. Species-specific data exist, though, and there must be additional ontologies beyond the GO to analyze such data.

11.3.3 Software tools

There are many software tools in use to analyze GO data, serving a variety of purposes. Some tools, such as AmiGO, are purely for browsing the GO data structure and for comprehensively viewing all gene annotations. Other tools provide methods for enrichment analysis. Three such enrichment analysis tools are FatiGO (Al-

Table 11.3 *Some Gene Ontology software tools.*

Software	Purpose	Language
AmiGO	browse, query, visualize	C, Java, Perl
DAVID	enrichment analysis	ASP, Java
FatiGO	enrichment analysis	undocumented
Ontology Traverser	enrichment analysis	Java and R

Shahrour et al., 2004), `Ontology Traverser` (Young et al., 2005), and `DAVID` (Dennis et al., 2003). Any software that is used for GO content analysis must solve two fundamental problems: (a) the software must represent the GO data structure as a DAG, and (b) the system must be capable of mapping experimentally generated lists to the DAG and computing the necessary count data.

The ontology itself is stored in a relational database. Many people curate the ontology and there are extensive resources to aid in the annotation process. The ontology data structure can be downloaded in at least three formats: MySQL tables, text, and XML. The XML format bears mentioning. XML is the most widely used language for sharing information of the web, and web-based representations of the GO and derivative analyses can facilitate web-based data sharing. Table 11.3.3 lists a few GO software tools.

11.4 Statistical methods

The Gene Ontology is well suited to computational analysis. The fundamental approach we consider is to tabulate experimentally derived sets of genes – gene lists – against the GO data structure. Once the tabulation is accomplished, we analyze the counts of genes annotated at or below each GO node. This procedure can be performed for single lists as well as collections of lists. When analyzing a single gene list, GO tabulation can rapidly summarize the list and reduce its content from a large number of often obscure individual genes to a smaller number of biologically relevant and interpretable categories. This form of GO-directed data reduction is a central benefit of GO analysis.

For analyzing a collection of lists, the GO can generate exploratory comparisons of the lists for data visualization. Exploratory analysis provides a method to compare lists which differ markedly in size, experimental platform, or taxonomic source. In addition to exploratory methods, statistical testing can be used to identify those GO categories and subgraphs with unusual gene counts across the family of lists.

11.4.1 Analyzing a single gene list

The analysis of a single list concerns the count statistics determined when the list is tabulated against the GO structure. The random quantities to consider are the counts of genes annotated at or below each GO node.

Content analysis in a single category

The simplest problem to consider is the analysis of counts at a single GO node. Table 11.4.1 shows the representation of the problem in terms of a two-way table.

Table 11.4 *Two-way table representation of counts at a single GO node. The quantity of interest is X, the number of genes that are both in the list and annotated to the GO term.*

	Annotated to Term	Not Annotated to Term
In Gene List	X	$L - X$
Not In Gene List	$N_C - X$	$N - N_C - (L - X)$

The statistical question revolves around the magnitude of X. Under the null hypothesis that the list is randomly generated, the null distribution for X is usually considered to be the hypergeometric distribution. The hypergeometric distribution is used to model counts of objects drawn from a finite collection without replacement. The hypergeometric is often presented as a description of drawing balls from an urn. The total number of balls in the urn is N, and the balls come in two colors, red and black. Our interest is in the number of red balls we obtain in L sequential draws without replacement. The total number of red balls in the urn is N_C, and the number of black balls is $N - N_C$. The number of red balls in our sample of size L is the random outcome X. The distribution of X is given as:

$$P(X = k | N, N_C, L) \quad = \quad \frac{\dbinom{N_C}{k} \dbinom{N - N_C}{L - k}}{\dbinom{N}{L}}. \qquad (11.1)$$

Quite naturally, this probability mass function is parameterized by the values of N, N_C, and L. In our problem of assessing gene counts at a single GO node, we must make an analogy between the genes derived from the experiment and the imaginary balls drawn from the urn.

In the model, the objects being "drawn" are genes annotated to the GO. The universe of objects is the collection of all genes possibly observable in the assay which have some annotation to the GO. The marked objects are those annotated to the node under consideration. The assumptions we make in applying the hypergeometric model are detailed below:

- N is the total number of genes on the microarray or within the universe of possibly observable genes which have annotations to the GO major branch where the node under consideration resides (e.g., we are considering a node within the Biological Process arm of GO).

- N_C is the number of genes in the universe of possibly observable genes which are annotated to the GO category (GO term) under consideration. Often this is a number far smaller than N.

- L is the number of genes in the list with annotations to the GO branch of the node.

- X represents the number of genes in the list which are also annotated to this GO node.

- We proceed under the null hypothesis that the list was randomly assembled, so that all N genes with GO annotations are *a priori* equally likely to appear in our assembly of L genes.

Under these assumptions, we can suppose that the null distribution for the number of genes annotated at or below a single node follows the probability law given in Equation 11.1. When the number of annotations N grows large and the number of genes L is relatively small, the hypergeometric can be approximated by the binomial distribution:

$$ P(X = k) \approx \binom{L}{k} \left(\frac{N_C}{N} \right)^k \left(1 - \frac{N_C}{N} \right)^{L-k}. \qquad (11.2) $$

Also of great interest is the standardized gene count at a GO node. This standardized count is merely the observed count minus the expected count under randomness, divided by the standard deviation of the count under randomness. For the gene count X this quantity is:

$$ Z = \frac{X - L\left(\frac{N_C}{N} \right)}{\left[L\left(\frac{N_C}{N} \right) \left(1 - \frac{N_C}{N} \right) \frac{(N-L)}{(N-1)} \right]^{\frac{1}{2}}} \qquad (11.3) $$

When L is relatively small with respect to N, the quantity in the denominator is well approximated by the binomial variance estimate. In all situations, the scale change from the binomial variance to the hypergeometric variance is constant across nodes, for it does not depend on N_C but only depends on the total number of annotations for the universe defined by the assay and the size of the list under consideration.

Multivariate distribution of GO counts

There is clearly more information in the tabulation of list counts than the observation at any single GO node. The specification of the joint distribution of counts is therefore of great interest. Under the null hypothesis that the genes in a list are randomly drawn from the collection of all annotated genes – meaning that all genes are equally likely to appear – analysis of the joint distribution follows the combinatorial arguments for the multivariate hypergeometric distribution. However, this joint distribution is complicated by the overlap in the genes annotated to GO nodes in distantly related branches of the GO structure.

The strict inheritance of annotations for nodes arranged in an unbranched path of GO nodes makes analysis of counts in unbranched paths possible. Recall that genes annotated to any GO node are by implication also annotated to their parental nodes according to the true-path rule. The annotations at ancestral nodes propagate up through the ontology by recursive application of the rule. Although distinctions can be made based on the *is-a* and *part-of* parent-child relationships or by evidence codes, in principle all child-node gene annotations are a proper subset of the annotations to each parent node. As will be shown, the analysis of the joint distribution of counts in unbranched paths is an important special case of a tractable joint distribution of GO counts.

Unfortunately, the joint distribution of counts for GO nodes which are linked through branched relationships can be quite complicated. The complication arises because a single gene can be annotated to any number of GO terms throughout the GO topology. When the joint distribution of GO counts in a subgraph containing arbitrarily related nodes is considered, the analysis should take into account the annotation overlap and the consequence for the distribution of counts. The analysis must consider the number of genes which share annotations at or below the nodes under consideration. Although it proves possible to give the joint distribution for pairs of nodes, the complete joint distribution for the entire graph is difficult to specify.

Unbranched paths

We consider first the joint distribution of counts at nodes in an unbranched path stretching from the root GO term to a terminal descendant node. The derivation of the joint distribution for counts in an unbranched path appears below.

1. Denote the list counts for genes annotated at each node along a path of $l+1$ nodes stretching from a root node to a descendant node $l+1$ steps away with the count vector $\mathbf{X} = (X_0, X_1, \dots X_l)^t$. X_0 is the count of the number of genes in the list itself with annotations to the GO branch. For each $i \in 1 \dots l$, the $i-1$ node in the path is a parent of the ith node, so that the inequality $X_i \leq X_{i-1}$ holds. More strongly, because of the true-path rule, the genes being counted at level i are shared by the parent at the level $i-1$.

2. The maximum number of possible annotations at each node – the maximum possible gene counts – along the path follows a similar nested pattern. Denote the vector of the maximum possible annotation counts along the path: $\mathbf{N} = (N_0, N_1, \dots N_l)^t$. Again, we have $N_{i+1} \leq N_i$ because of the nested structure of annotations along the path.

3. The annotation overlap, or genes in common between ancestral nodes and descendant nodes, for genes in our list along the path is denoted $X_{i,i+1}$.

Again, because all annotations at the child are shared by the parents, we have $X_{i,i+1} = X_i - X_{i+1}$.

4. The universe of possible gene annotations along a path also has an overlap structure. For a parent node with count N_i, denote the number of gene annotations not shared by the child node as $N_{i,i+1}^c$.

Using this notation we can write the joint distribution of counts along an unbranched path under the null model that the GO annotated genes in our list are randomly sampled from the universe of genes with annotations:

$$P(\mathbf{X} = \mathbf{x}) = \frac{\binom{N_l}{x_l}\binom{N_{p-1} - N_l}{x_{p-1} - x_l}\cdots\binom{N_0 - N_1}{X_0 - X_1}}{\binom{N}{X_0}}. \quad (11.4)$$

As with the counts at a single node, the without-replacement exact null distribution for counts along an unbranched path can be approximated by a sampling with replacement model. In this case the appropriate distribution to consider is the multinomial distribution. It is still important to consider the overlap in counts between parent and child nodes.

Beyond the joint distribution given above, the exact covariance structure for counts along an unbranched path can also be analyzed. For an ancestral node X_a and a descendant node X_d linked by an unbranched path, the covariance between X_a and X_d is:

$$\mathrm{Cov}(X_a, X_d) = \frac{N_d}{N_a}\sigma_a^2. \quad (11.5)$$

This formula can be arrived at using the definition of covariance and the joint distribution for the node pair. An identity for summing binomial coefficients is important to simplify the joint expectation of counts. In the expression above, σ_a^2 is the variance of X_a, N_a is the count of total annotations to the ancestral node, and N_d is the count of total annotations to the descendant node.

Since both the mean and covariance matrix for counts along an unbranched path are directly calculable, it is possible to arrive at a simple statistic to calculate the total enrichment along an unbranched path. If we consider a single unbranched path of nodes, denoted p, and denote the vector of counts along the path as $\mathbf{X_p}$ with expected counts $\mu_{\mathbf{p}}$, then the enrichment statistic for the full path is the quadratic form:

$$Q_p = (\mathbf{X_p} - \mu_{\mathbf{p}})^t \Sigma_{\mathbf{p}}^{-1}(\mathbf{X_p} - \mu_{\mathbf{p}}). \quad (11.6)$$

The distribution of Q_p can be approximated with a chi-square distribution on l degrees of freedom, where l is the length of the path. It should be noted that in circumstances where GO counts are completely inherited between a parental node and a descendant node so that $\frac{N_d}{N_a} = 1$, a linear dependence will be introduced into the path counts. In this circumstance, one of the completely dependent counts should be removed from the vector $\mathbf{X_p}$ so that the covariance matrix is non-singular.

Bivariate distribution of arbitrary GO terms

Before we proceed to describe methods for analyzing multiple lists, it is useful to discuss the bivariate distribution for counts at a pair of GO nodes related by a branched relationship. Without loss of generality, label the two nodes A and B, and denote the two counts X_A and X_B and denote by X_{AB} the number of genes in the list which are shared by nodes A and B. The distribution can be analyzed by taking expectations over the number of common genes:

$$ f(x_A, x_B) \quad = \quad E_{X_{AB}}(f(x_A, x_B | X_{AB})). \tag{11.7} $$

In this expression $f(x_A, x_B)$ is the bivariate mass function. The difficulty in calculating the expression is the overlap X_{AB}. The overlap is potentially different for all pairs of nodes. Although the bivariate mass function is simple to write, summaries such as the covariance of X_A and X_B must be calculated separately for each pair and cannot easily be expressed in a simple formula such as Equation 11.5.

The joint distribution for an arbitrary pair of nodes is obtained by applying Equation 11.7 (min = $\min(x_A, x_B, N_{AB})$):

$$ \sum_{x_{AB}=0}^{\min} \frac{\binom{N_{AB}}{x_{AB}} \binom{N_A - N_{AB}}{x_A - x_{AB}} \binom{N_B - N_{AB}}{x_B - x_{AB}} \binom{N - (N_A + N_B) + N_{AB}}{X_0 - (x_A - x_B) + x_{AB}}}{\binom{N}{X_0}}. $$

11.4.2 Multiple lists

Now that the tools for analyzing single lists are in hand, the focus can be turned to multiple lists. The GO provides a content-based mechanism to compare lists that differ in size, platform, or species origin. In the multi-list context, a useful starting point is to create visual summaries which describe the relationships between lists based on their GO tabulation. Visual summaries reveal broad scale structure between results and can help to orient further work. An additional goal is to create a testing framework for identifying the GO categories with unusual enrichment across a cohort of lists.

Exploratory data analysis and list distances

The most direct way to compare a collection of lists is to count the overlap in the members of the lists. Unfortunately, counting list overlaps is problematic for several reasons. First, the lists may differ markedly in size, so that list intersection is an asymmetric operation. Second, all lists generated from genome-scale experiments suffer from imperfect statistical power and false negative results. The imperfect power of the list generation mechanism will inevitably result in missed genes which should be included in the results, and these genes will fail to appear in the intersection. Generally speaking, list intersection results in a geometric loss of power because the power to detect genes in both lists is the product of the power of the separate list generation schemes. Finally, all lists suffer from false positive contamination. No matter what rule was used to generate the lists, it is probable that members of the lists have been erroneously included.

Fortunately, the distributional results described in the previous section provide a more coherent approach to compare lists. The bivariate distribution described by Equation 11.7 together with the explicit covariance formula given by Equation 11.5 suggest a distance metric. Denote by $\mathbf{X_{g_i}}$ the vector of counts at all GO nodes for list g_i, and the analogous quantity for list g_j. We can derive both the expectations for the counts, $\mu_{g_i} = \mathbf{E}(\mathbf{X_{g_i}})$, $\mu_{g_j} = \mathbf{E}(\mathbf{X_{g_j}})$, and the covariance matrices for counts, $\mathbf{\Sigma_{g_i}}$ and $\mathbf{\Sigma_{g_j}}$. A natural choice of distance between the lists is then the Mahalanobis distance between $\mathbf{X_{g_i}}$ and $\mathbf{X_{g_j}}$. Setting $\mathbf{Y_{g_i}} = \mathbf{X_{g_i}} - \mu_{g_i}$ and $\mathbf{Y_{g_j}} = \mathbf{X_{g_j}} - \mu_{g_j}$ and $\mathbf{\Sigma_*} = \mathbf{\Sigma_{g_i}} + \mathbf{\Sigma_{g_j}}$, we have

$$D(g_i, g_j) = \left(\mathbf{Y_{g_i}} - \mathbf{Y_{g_j}}\right)^t \mathbf{\Sigma_*}^{-1} \left(\mathbf{Y_{g_i}} - \mathbf{Y_{g_j}}\right). \qquad (11.8)$$

This distance metric is still somewhat difficult to compute. A simpler alternative is to disregard the off-diagonal elements of $\mathbf{\Sigma_{g_i}}$ and $\mathbf{\Sigma_{g_j}}$. In this case we have a simple distance metric:

$$D(g_i, g_j) = \sum_{\text{all nodes}_k} \left(Z_{g_i,k} - Z_{g_j,k}\right)^2. \qquad (11.9)$$

The $Z_{l,k}$ in Equation 11.9 are given by Equation 11.3. Both the distance metrics Equations 11.8 and 11.9 represent an improvement over simple list intersections.

Tests for GO enrichment across a family of lists

Analysis also suggests methods for testing list counts across a family of lists. If we suppose there are k lists under consideration, then analysis of counts across lists should determine whether certain categories are collectively enriched or suppressed across a family of lists.

The first method we consider is a multi-way exact test for counts at a single GO term. A large body of literature exists concerning exact analysis for counts in many two-way tables. With the representation for counts at a single GO term as given in Table 11.4.1, the method described by Mehta et al. (1985) can be used to derive an overall measure of significance of counts across the k tables. Briefly, the method enumerates the total number of two-way tables with marginal totals equal to those observed across the k lists. A network algorithm is employed to determine the total number of such tables. The central quantity of interest is the total number of successes at GO term j:

$$S_j = \sum_{i=1}^{k} X_{i,j}. \qquad (11.10)$$

The work in Mehta et al. (1985) gives a method to compute the mass function of S_j.

For any GO term with an observed total count of s hits across k lists, the p-value for counts is $P(S_j \geq s)$. This p-value should be corrected for the multiplicity of p-values generated when considering many GO terms.

If we recall the formula for measuring total path enrichment given in Equation 11.6, then a simple method presents itself for combining information across k lists. The approach is simply to sum the scores determined by Equation 11.6:

$$T_p = \sum_{i=1}^{k} Q_{i,p}. \qquad (11.11)$$

Under the assumption described for a single path score Equation 11.6, each $Q_{i,p}$ will have an approximately χ_l^2 distribution. If the k lists are derived independently, the quantity T_p will have an approximate χ_{lk}^2 distribution. When multiple lists are derived from the same study, those lists will be dependent, violating the assumptions for the chi-square approximation. As long as the lists are small relative to the number of possible genes, the dependence will be weak and the approximation is useful.

11.5 Application to stem cell data

In this section, we consider meta-analysis of microarray stem cell data from mouse stem cell experiments. The stem-cells being investigated include hematopoetic, skin, hair, neural and cells of embryonic origin. The goal in the analysis is to identify commonalities and differences among the stem cell types.

11.5.1 Background on stem cells

Stem cells are special cells which appear in all multi-cellular organisms and which have the potential to give rise to descendant lineages of more specialized cells. The specialized cells are necessary in multi-cellular organisms to perform particular functions or to participate in distinct organ systems. The ability of cells to specialize is remarkable because, by and large, all cells in any organism share the exact same DNA and therefore have the same genetic information. The process by which cells attain their specific activities is termed *differentiation*. Stem cells are the ancestral cells in the differentiation process. These special cells have the potential to give rise to a variety of different descendant cell lineages. Stem cells are highly abundant in the early growth and development of organisms, but they disappear as organisms mature and then age. Interestingly, reservoirs of stem cells are present in many if not all tissues of adult organisms; these cells are called adult stem cells. Unfortunately, adult stem cells generally have far less potential to generate different types of descendant cells than embryonically-derived stem cells which are, in principle, capable of generating all differentiated cell types. Because of the great potential of stem cells for generating a replacement for damaged or lost tissue, much experimental effort now is directed at characterizing stem cell properties. Interest focuses on finding properties shared by all stem cells, as well as on identifying particular features which make one stem cell population different from another.

11.5.2 Example studies

We consider data from five published studies. A total of 13 gene lists are present in this collection, all derived from Affymetrix microarray experiments on the MG-U74Av2 GeneChip (see Section 1.4.2). The lists range in size from approximately 2,000 genes to as few as about 10. Before we present analysis of these data, we take time to consider the scope of the various individual studies.

Ivanova.   One of the first stem cell papers presented is by Ivanova et al. (2002). This paper describes analysis of transcriptional profiles of stem cells from the hematopoetic system. The paper considers 4 types of cells: long term hematopoetic stem cells (LT-HSC), short term hematopoetic stem cells (ST-HSC), and their descendant cells' early progenitors and late progenitors. Although the results of this study are complex, we consider five lists derived from the experiment which are pertinent to the HSC system: Long term HSC (203 genes), Short term HSC (10 genes), Early progenitors (134 genes), Intermediate progenitors (44 genes), and Late progenitors (182 genes).

Ramalho-Santos.   The paper by Ramalho-Santos et al. (2002) from the Melton laboratory in Boston provides one of the first attempts to identify a common transciptional profile across different types of stem cells. The paper considered three distinct types of stem cells in mice: embryonic stem cells (ESC), neural stem cells (NSC), and hematopoetic stem cells (HSC). The general experimental approach for the NSC and HSC cell types compares the transcriptional profile of the stem cell population to the profile of differentiated cells from the same system. For NSC, the reference population was taken to be lateral ventricle tissue from the brain. For HSC, the reference population was taken to be whole bone marrow. For the ESC, which are ancestral to all mouse cell types, the reference was taken to be the average of the brain and bone marrow differentiated references. Gene lists were derived for each stem cell type based on Affymetrix present/absent calls (Affymetrix, 2002) and on estimated fold changes between the stem cell population and the derived cells. No multiple testing corrections were made, and the lists are relatively large. The ESC list consisted of 1,787 probe sets which are reduced to approximately 1,335 annotated genes in our analysis. The HSC list is also large, consisting of 1,977 probe sets, which are reduced to 1,479 annotated genes for our analysis. The NSC list is largest of all, consisting of 2,458 probe sets, which are reduced to 1,807 genes for our analysis.

Venezia.   The paper by Venezia et al. (2004) concerns HSC only. They perform a time course experiment in which quiescent HSC are stimulated to proliferate by treatment at time 0 with the powerful cytotoxic agent 5FU. The drug 5FU is a commonly used chemotherapy agent used to treat cancer. In this experiment, 5FU is being used to stimulate the normally quiescent HSC to divide; the hope is that the process of stimulation and return to quiescence will reveal important transcriptional properties of the normally quiescent HSC. In this study, mRNA samples are collected from cells at days 0, 1, 2, 3, 6, 10, and 30 after treatment with 5FU. Peak cell division occurs between days 6 and 10. Expression curves were fit to each gene across the time course, and analysis revealed two major types of expression profiles. Some genes demonstrated repressed expression across the proliferative phase of the time course; these genes are termed the *quiescence signature* or Qsig. There are 225 genes assigned to this class. Other genes have elevated expression across the time course; these genes are termed the *proliferation signature* or Psig. There are approximately 265 genes in this cohort.

Morris.   Hair stem cells are of interest in the treatment of hair loss and other disorders of the hair and skin. Morris et al. (2004) describe the isolation of putative stem cells from a distinctive region of the hair follicle in the epithelium. Analysis revealed a gene list of 93 distinct murine genes with GO annotations. These genes are the raw material for GO analysis.

Tumbar.   The paper by Tumbar et al. (2004) is one of the first to describe skin stem cells in adult mice. The paper discusses the ambiguity surrounding skin stem cells, and describes an experimental technique to derive the multi-potent cells from a slowly dividing compartment of the basal epithelium. The stem cells are thought to be found within a specialized area of the basal epidermis called the stem cell niche. The cells isolated in the paper are characterized by their ability to retain a fluorescent label. These cells are named "label-retaining cells" or LRCs. The ability to retain the label shows the cells to be relatively quiescent. Array data were compared against results from mRNAs derived from adjacent basal epidermal tissue. Analysis revealed a gene list of approximately 154 probe sets up-regulated in the LRCs when compared to more specialized, descendant cells from the basal epidermis (BL cells). Analysis shows that approximately 126 characterized mouse genes are represented by these 154 probe sets. We consider the content of these 126 genes when analyzed against the GO data structure.

11.5.3 Meta-analysis of stem cell gene lists

The first step in the meta-analysis is to generate GO tabulations for each of the 13 lists available in the five studies. This step was accomplished using the R package Ontology Traverser developed by Young et al. (2005). Once these results are computed, it is possible to proceed with exploratory analyses and then test for enrichment to identify commonalities and differences between the lists.

Exploratory analysis

To perform exploratory analysis, a 13×13 distance matrix is constructed on the gene lists using the simple distance metric in Equation 11.9. A multi-dimensional scaling method is applied to this distance matrix to generate a bivariate scatter plot of the lists. The scatter plot is presented in Figure 11.2.

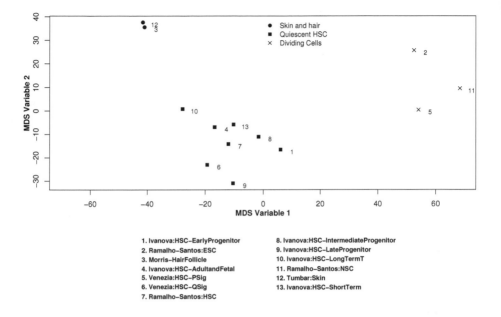

Figure 11.2 *Multi-dimensional scaling analysis of the 13 list distance matrix computed using Equation 11.9. The results suggest the lists might be grouped into 3 families: skin and hair, quiescent HSC, and dividing stem cells.*

Many useful features can be ascertained from Figure 11.2. First, there are three groups of lists. At least two of these groupings correspond to simple biological interpretations. The tightest group consists of the hair and skin. Both of these lists are derived from analysis of epithelial tissue. The other large group corresponding to a simple interpretation consists of the quiescent hematopoetic stem cells. Even lists of widely divergent size are grouped together in this cohort. The final group consists of the ESC list, the NSC list, and the Psig list from the Venezia et al. (2004) study. The common biological thread among these three lists is that the genes in the lists represent actively dividing cell populations. The process of cell division strongly distinguishes this group from the other cohorts.

Commonalities and differences

The statistics developed in Equations 11.10 and 11.11 can be used to analyze the commonalities and differences between lists. To analyze the lists, we first group them into the cohorts made clear by the scatter plot in Figure 11.2. In each of the three groups we then perform the following analyses:

- Multi-way exact tests of each GO term in each of the three groups. This analysis results in a marginal p-value for each GO term in each group. In generating this analysis we also tabulate the total number of gene hits to each term within each group.
- Path enrichment analysis for each unbranched path in each group. Paths are enumerated by identifying all branches of the GO from the root term to terminal nodes.

Once the statistics have been calculated, the remaining task is to identify commonalities and differences between the lists. To identify commonalities, we identify all those GO paths which are significantly enriched in all three groups. To make the analysis even more stringent, we make a further restriction to identify those GO classes which are also significantly enriched by marginal analysis of the term with the multi-way exact test. Figure 11.3 shows the content overlap among the three groups.

The result of the commonality analysis suggests that there are some properties shared by these very different gene lists. All lists share enrichment for the cellular differentiation and cellular communication GO categories, indicating that both the cell differentiation and the cell signaling system is a common feature of the stem cell lists. The regulation of transcription is also a strongly shared property of the lists.

In addition to examining commonalities, the results can also be used to distinguish differences between the groups. To distinguish lists, the following two-step approach is useful:

1. Identify GO paths which are significant in one list cohort, but which are less strongly enriched in the other two cohorts.
2. Among these significant GO paths, identify those nodes which have high gene counts and whose marginal p-values under the exact calculation are very strongly significant.

Analysis of the skin and hair lists (Figure 11.4) shows these to be meaningfully distinct from the other lists. Among the distinctions is the clear enrichment of these lists for the cell-migration and cell-motility system. This characteristic is interesting in light of the clear necessity for cell movement in skin and hair stem cells, as these cells must migrate out of their niche to produce descendants. The skin and hair lists also show enrichment for genes involved in cartilage formation and in early neurogenesis. Both of these characteristics make sense in light of the fact that the skin cells

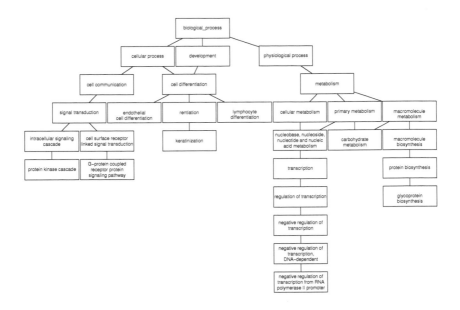

Figure 11.3 *Content overlap among the three groups of gene lists. The analysis presented shows only results from the Biological Process ontology.*

arise from the same developmental lineage that gives rise to connective tissue and brain.

Analysis of the quiescent HSC lists (Figure 11.5) also shows meaningful distinctions from the other lists. Among the distinctions is the clear enrichment of these lists for the cellular quiescence. Interestingly, the motility system is shut off in these cells. This characteristic is interesting in light of the fact that HSC by and large are not believed to migrate in order to generate their progeny. The HSC lists also show clear enrichment for genes known to be involved in specification of the various lineages of cells descendant from the HSC. These GO categories suggest that the HSC cells have the potential to express all these lineage specifying genes.

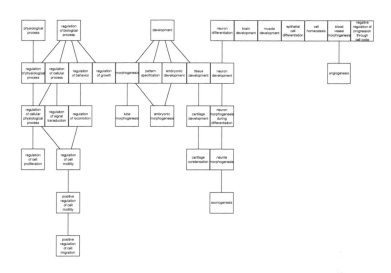

Figure 11.4 *GO subsets which strongly distinguish the skin and hair lists from the other cohorts. Only the Biological Process ontology is depicted.*

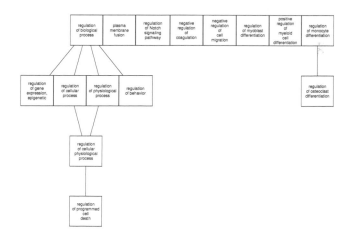

Figure 11.5 *GO subsets which strongly distinguish the quiescent HSC lists from the other cohorts. Only the Biological Process ontology is depicted.*

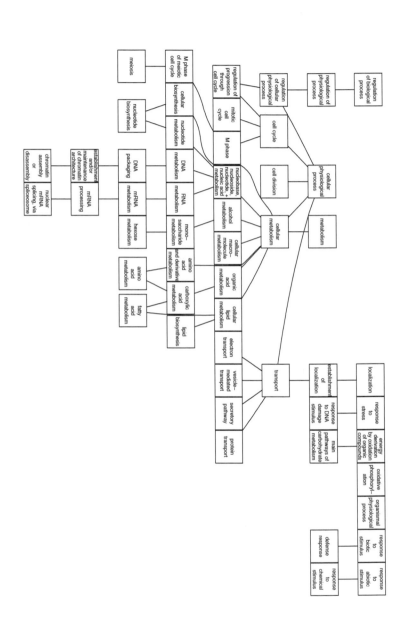

Figure 11.6 *GO subsets which strongly distinguish the proliferating lists from the other cohorts. Only the Biological Process ontology is depicted.*

Analysis of the proliferating cells lists (Figure 11.6) shows a complex set of distinctions from the other groups. First and foremost, the genes in these lists reflect dividing cells, and there is a clear signature of cell division. The other distinctions are more difficult to decipher. Largely speaking, the genes in these lists reflect a heightened metabolic activity compared to those from the other two list groups. The metabolic activity also requires the cellular detoxification system, and these cellular processes are also shown to be enriched. It is not clear why immune response would be elevated in these lists.

11.6 Conclusions

This chapter considers GO-based meta-analysis of genome-scale experiments. The GO is useful in this setting because it permits the joint analysis of lists which differ markedly in size, in experimental platform, or in species origin. The GO is a large controlled vocabulary for describing genes. We provide a variety of statistics for analyzing cross-tabulation of counts against the GO, both for single lists as well as for collections of lists. An example analysis using data from five stem cell studies shows the power of GO-based analysis. Three distinct groups of gene lists are revealed, each of which is biologically meaningful. Deeper analysis using the GO reveals subtle but interesting differences between the lists. GO-based data analysis is an interesting computational arena, with many remaining problems to be addressed.

Part III. Combining Different Data Types

CHAPTER 12

Combining genomic data in human studies

Debashis Ghosh, Daniel Rhodes, and Arul Chinnaiyan

12.1 Introduction

With the development of technology that has allowed for the high throughput minia-turization of standard biochemical assays, it has become possible to globally moni-tor the biochemical activity of populations of cells. This has led to the emergence of cDNA microarrays in medical and scientific research and has allowed for large-scale transcriptional characterization. It should also be noted that microarray technolo-gies would have limited ability without the existence of large-scale genome sequenc-ing projects, such as the Human Genome Project (International Human Genome Se-quencing Consortium, 2001; Venter et al., 2001). Having such sequence data avail-able allows for the characterization of the probes on the microarray. In this chapter, we use the term "genomic data" generically to refer to any genetic data generated using large-scale technologies.

While transcript mRNA microarrays have received much attention in the literature, there has been work on other types of microarrays. Examples include chromatin immunoprecipitation (ChIP) microarrays, which measure transcription factor-DNA binding expression (Lee et al., 2002) and methylation microarrays (Yan et al., 2001), which assess DNA methylation on a global scale. In addition, there has also been much attention on high throughput assays that measure protein-protein interactionsm such as yeast two-hybrid systems (Uetz et al., 2000). There is much interest in at-tempting to integrate these data sets to provide a more complete understanding of the biological mechanisms that are at play. This type of analysis has been given the name "systems biology" in the bioinformatics literature (Ideker et al., 2001).

For the statistician, this area brings many interesting and challenging problems. Al-though the term "meta-analysis" is familiar among most statisticians (Normand, 1999), the term here takes a very different meaning. The most familiar statistical situation involves attempting to combine information from relatively homogeneous data structures from multiple similar experiments. However, in much of the genomic area, the aim is to combine relatively inhomogeneous data structures from multiple experiments that may or may not be similar.

Another complication is that data availability depends on the type of organism studied. In this chapter, we focus on data from human studies. We discuss approaches for combining genomic data in human studies, focusing primarily on methods developed in the cancer setting. Some familiarity with microarray technologies is assumed (see Section 1.4 and also The Chipping Forecast 1999, 2002). Our goal here is to outline the major issues involved in such analyses and describe some solutions that have been proposed. It is not our intent to provide a comprehensive listing of all methodologies that have been used, as the literature is constantly changing. Given the dynamic nature of the field, an important component will be benchmarking of methods to see which should be used in practice.

12.2 Genomic data integration in cancer

12.2.1 Goals

Our group has focused primarily on the analysis of genomic data in cancer studies. There are two broad goals of this research. One is the discovery of new biomarkers that might be used potentially as screening tests or to better predict patient prognosis. Examples of potential promising biomarkers found using gene expression technology include enhancer of zeste homolog 2 (EZH2) in prostate cancer (Varambally et al., 2002). In this study, the transcript mRNA expression EZH2 gene transcript was found to be highly expressed in metastatic prostate cancer. A key point to make at this stage, which we address later, is that mRNA expression does not necessarily perfectly correlate with protein expression. In terms of diseases, the action is happening at the protein level. In protein validation studies done by Varambally et al. (2002), the EZH2 protein was also found to be highly expressed in metastatic prostate cancer. Another example of a potential biomarker found using genomic data technologies is prostasin in ovarian cancer (Mok et al., 2001). In that study, the authors reported a sensitivity of 92% and a specificity of 94% for discriminating ovarian cancer cases from controls using validation by ELISA of serum. Thus, prostasin might serve as a potential biomarker for early detection of ovarian cancer.

The second goal of our research is to better understand the biology of the disease. In the past, cancer was thought of as a heterogeneous collection of diseases. However, a more integrative view of the disease is currently being put forward by many researchers; this view is summarized eloquently in a review article by Hanahan and Weinberg (2000). According to their paradigm, there are six principles that underlie tumorigenesis (the initiation and development of a tumor); equivalently, for a cancer to develop, it must acquire six "hallmark capabilities": self-sufficiency in growth signals, insensitivity to anti-growth signals, evading apoptosis (cell death), limitless replicative potential, sustained angiogenesis, tissue invasion, and metastasis.

With the current availability of large-scale genomic data, we can address the Hanahan and Weinberg (2000) model in two ways. First, we can analyze the data to assess the relative contributions of the six "hallmark capabilities." Second, we can use genomic data to further refine and identify the pathways that comprise each of the individual hallmark capabilities described above.

12.3 Combining data from related technologies: cDNA arrays

The statistical problem closest in spirit to classical meta-analysis involves combining multiple data sets in which the same type of cellular activity was assessed. As an example here, we consider multiple microarray studies in which the same comparison is considered, namely cancer versus normal.

There are several issues that must be considered when attempting such an analysis. First is the problem of study-specific artifacts, such as sampling bias, variations in experimental protocols and differences in laser scanners. However, there are two bigger issues in the analysis of such data. The first is that of matching genes from two studies. This is where the availability of large-scale genomic data figures in hugely. Each spot on a microarray corresponds to a DNA sequence. Spots can be matched to a putative gene in the UniGene Database, a collection of clusters of orthologous genes (http://www.ncbi.nlm.nih.gov/unigene). The UniGene link can then be used to identify common genes across multiple data sets.

This task can be carried out for Affymetrix chips from information at their website (http://www.netaffx.com) or for two-color cDNA microarrays using the SOURCE tool at Stanford (Diehn et al., 2003; http://source.stanford.edu).

A more challenging issue than the mapping involves the fact that the numbers from different microarray platforms represent different things. That is, an expression value of 14 from a cDNA two-color microarray represents relative expression, and therefore is much different from an expression value of 14 measured as an absolute expression on an Affymetrix array. A technique that has proven to be useful as a filtering device to enhance comparability across arrays of different platforms is known as the integrative correlation coefficient or correlation of correlation coefficients (Lee et al., 2002; Parmigiani et al., 2004). The idea underlying this method is that although raw expression values vary from study to study, the inter-gene correlations do not vary as much. Thus, one would consider combining genes that have similar inter-gene correlations across the studies.

In terms of meta-analysis methods put forward, many have been based on the fact that the standardized effect size is combinable across studies. This is the approach advocated by Parmigiani et al. (2004) after filtering based on the integrative correlation coefficient. In Rhodes et al. (2002), the t-statistic is transformed into a p-value, then combined across multiple studies using Fisher's method (Fisher, 1932). By contrast, in Ghosh et al. (2003), the t-statistics themselves are combined directly. An approach more Bayesian in nature is taken by Wang et al. (2004), in which expression values from one study are used to develop a prior distribution for the standardized effect size; data from the remaining studies are used to generate posterior distributions. A fully hierarchical approach was taken by Choi et al. (2003), who then use Markov Chain Monte Carlo methods (Gilks et al., 1995) to sample from the posterior distributions. It should be noted that all of these methods make the assumption that a standardized effect size can be estimated directly for each individual study.

Another approach more in line with classification or supervised learning analyses is to build a classifier or find a gene expression signature on one data set and to see how well it predicts in an independent microarray data set. Such approaches are taken by Beer et al. (2002), Wright et al. (2003), and Jiang et al. (2004). An alternative method using hierarchical clustering, which is an unsupervised learning procedure, was taken by Sørlie et al. (2003). They found a gene expression signature that defined molecular subtypes in breast cancer; they found through interrogation of other data sets that the subtypes were present there as well. Given the growth of publicly available large-scale gene expression data sets, it is increasingly important that results found by one investigator on a particular data set be validated using other data sets as well.

A large-scale comprehensive meta-analysis was performed by Rhodes et al. (2004a), who carried out a meta-analysis of 40 independent data sets (>3,700 array experiments) across 12 tissue sites. They found a universal profile of 67 genes that could differentiate cancer versus noncancer tissue for a variety of cancers. In addition, they identified 36 cancer-specific signatures for determining a tissue-specific cancer. The signatures also demonstrated good discrimination performance on three independent data sets.

A more sophisticated method for meta-analysis is proposed by Shen et al. (2004), based on an idea of Parmigiani et al. (2002). The idea is that for a given gene from a given sample in a given study, it is either over-, under- or non-differentially expressed with respect to a baseline cohort of genes. Each of the three states defines a latent category, which induces a mixture model for gene expression values. The latent states of over-, under- or non-differentially expressed are inferred using a Markov Chain Monte Carlo sampling algorithm. The estimated probabilities of the latent states are then transformed to define a "probability of expression," which is then used as input for a meta-analysis.

There have been many meta-analyses of differential expression across multiple studies. A notable exception is the study by Lee et al. (2004), in which inter-gene correlations across multiple studies are considered. The authors sought to identify pairs of genes that were consistently coexpressed across several data sets. As we describe in the next section, such coexpression is the first step needed in building gene regulatory networks.

12.3.1 Functional and pathway analyses

Gene lists resulting from meta-analyses can be entered into databases representing functional processes. A simple visualization exercise, carried out by Rhodes et al.

(2002), is to find metabolic pathways in which multiple genes exist. One example of such a database is the Kyoto Encyclopedia of Genes and Genomes (KEGG) (`http://www.genome.jp/kegg`). From a list of genes consistently dysregulated across multiple studies comparing prostate cancer to non-prostate cancer, pathways such as the purine biosynthesis were found to have multiple genes. This finding leads to the hypothesis that the purine biosynthesis pathway is dysregulated in prostate cancer. Although the study is only generating a hypothesis and not confirming it, such a computational prediction can help to inform investigators as to the next series of experiments to perform. However, a visual display such as that given by KEGG does not allow for any formal statistical assessment of significance.

More formal statistical analyses for enrichment of functional terms can be done using the hypergeometric distribution. This requires a database of functional annotation terms such as Gene Ontology (GO) (Ashburner et al., 2000, `http://www.geneontology.org`; see also Shaw (Chapter 11 in this volume) for a detailed introduction). The idea behind this procedure is to see if the frequency of certain Gene Ontology terms in a list of genes is similar to or significantly larger than that in an external database. If it is determined that there is statistically significant enrichment of functional annotation terms in a list, then again this generates the hypotheses that certain pathways are dysregulated in the disease process. The null hypothesis is that there is no association between list membership and annotation to a particular GO category; i.e., no functional enrichment of the GO term in the list of genes. This is usually tested using Fisher's exact test.

There are now many publicly available tools for performing such a test (Draghici et al., 2003; Al-Shahrour et al., 2004; Beissbarth and Speed, 2004). Note that these methods are *post hoc* procedures in that the pathway analysis is done conditional on selecting a list of genes. An alternative is to directly model the information contained in the Gene Ontology databases with gene expression data. However, this raises the problem of what constitutes a proper metric by which the heterogeneous information from the two diverse databases can be related; this currently remains an open question. Christian and Guerra (Chapter 14 in this volume) address this issue in the context of expression trait loci (ETL) mapping.

A resource initiated by our group is the ONCOMINE database (Rhodes et al., 2004b; `http://www.oncomine.org`). ONCOMINE represents an effort to systematically curate, analyze and make available all public cancer microarray data via a web-based database and data-mining platform. Within the database, one can perform over 100 types of differential expression analyses based on disease/non-diseased, stage of disease, subtype, etc., reported with study-specific q-values. These analyses are based on standard differential expression analysis with correction for multiple testing using the q-value (Storey, 2002; Storey and Tibshirani, 2003b; for a brief summary see also Sections 7.2.3 and 9.3.2). In addition, one can query individual genes for known available genetic and proteomic information that is stored at other databases (e.g., GenBank, Swiss-Prot, etc.). There are links with pathway databases for visualization and assessing functional enrichment of the gene lists that are found. One can also search for individual genes to see their expression patterns across multiple cancer studies.

12.4 Combining data from different technologies

In the traditional statistical view of meta-analysis, one thinks of attempting to combine information from multiple similar experiments. However, the challenge of bioinformatics is that high throughput functional genomics data are being generated using a variety of platforms and stored in different databases. The challenge then becomes how to integrate these diverse data types.

12.4.1 Bayesian networks

One tool that has been utilized quite heavily for this type of problem has been graphical models (Lauritzen, 1996; Jensen, 2001), also referred to as Bayesian networks or belief networks. The idea of graphical models is to estimate dependencies between random variables through calculation of measures of covariation between them. As a simple example, let us consider three random variables, A, B and C. If we assume that the joint distribution of (A, B, C) is multivariate normal, then assuming the random variables have mean zero, the distribution is summarized by the pairwise correlation coefficients between them. Thus, if we can estimate the correlations, then we have "learned" about the system characterized by A, B and C. There has been a lot of interest in attempting to construct regulatory networks by fitting graphical models to gene expression data only. However, given the amount of experimental variability in such data, this has not turned out to be fruitful, so the focus has turned to building networks with multiple sources of data. As an example, Sun and Zhao (Chapter 15 in this volume) describe a Bayesian framework for regulatory network inference based on combined analysis of gene expression data and protein-DNA interaction data.

One major goal of Bayesian networks has been to predict protein-protein interactions. While much of the genomic data is measured at non-protein levels, actual cellular activity and disease occurs at a protein level. Thus, it is of interest to figure out how well functional genomic correlations predict protein-protein interactions. Other important aspects of protein-protein interaction modeling are discussed by Sun et al. (Chapter 16). This area was first studied in yeast by Jansen et al. (2003). However, they had the advantage of having high throughput protein-protein interaction data available from yeast two-hybrid experiments.

In a recent application, Rhodes et al. (2005b) use Bayesian networks to predict human protein-protein interactions using functional genomic data. Several different types of information are combined in order to develop the graphical model: interactions between orthologs of human proteins, intergene correlations from gene expression profiles, shared functional annotations from Gene Ontology, and shared enrichment domains.

The idea is to create a graphical model using known positive and negative protein-protein interactions in order to develop a scale of evidence for predicting a protein-protein interaction. The positives are defined using the Human Protein Reference Database (HPRD) (Peri et al., 2003; http://www.hprd.org), a bioinformatics resource that contains known protein-protein interactions manually curated from the literature by expert biologists. We queried 11,678 distinct literature-referenced protein-protein interactions among 5,505 proteins. The negatives include all protein pairs in which one protein was assigned to the plasma membrane cellular component and the other to the nuclear cellular component based on Gene Ontology. Based on the fitted model, there are predicted to be approximately 10,000 interactions with a false positive rate of 20% and about 40,000 interactions with a false positive rate of 50%. Several of the predicted protein-protein interactions were verified by subsequent experimentation, while other predictions mimicked what was found in the reported experimental literature. This model is now integrated into ONCOMINE.

Even with the recent successes of the graphical models approach, this area is still in its infancy. One limitation of the graphical model is that it only uses pairwise covariation information. Furthermore, the graphical models used by Jansen et al. (2003) and Rhodes et al. (2005b) involve a binning procedure that seems somewhat *ad hoc*. One interesting alternative has been proposed by Balasubramanian et al. (2004), who propose using a graph-theoretic approach to combining functional genomics data from diverse platforms and test for significance of the nodal connections using permutation testing. Interestingly, there appear to be similarities with the use of graph-theoretic ideas in this area with those in the social network literature (Wasserman and Faust, 1994). This suggests that there may exist techniques from that field that may be of use here.

Another point of the Bayesian networks is that they are bidirectional and do not attempt to impose any directionality. However, we know that activity in biological systems consists of a series of ordered steps. Thus, there might be some advantage to incorporating directionality into the system. Let us take the transcription process as an example. First, there must be binding of DNA to the upstream promoter regions in the genome so that transcription is "turned on." Thus, one could imagine a model for expression as a function of upstream promoter sequence for this scenario. Models like this have been proposed for lower-level eukaryotes (Bussemaker et al., 2001; Conlon et al., 2003) and are referred to as "dictionary models." They take a view that the expression value is a function of a score computed using the sequence data, which is a conditional model. It remains to be seen whether such models could work for human genomic data.

12.4.2 Toward an understanding of regulatory mechanisms

In the previous sections, we have described methods for combining information in order to derive improved gene signatures and to make protein-protein interactions. Another goal of interest is to derive "regulatory" modules. It is likely that some gene expression patterns observed from microarray data represent a downstream readout of a small number of genetic aberrations (e.g., mutations, amplifications, deletions, translocations) that lead to the activation or inactivation of a small number of transcription factors. In some cases, cancer-causing genetic aberrations may not be directly apparent from these downstream gene expression readouts. Recent methods to developing gene expression regulatory modules in human studies are described by Elkon et al. (2003), Segal et al. (2004) and Rhodes et al. (2005a).

The general approach requires a predefined list of genes. The list of genes can come from an external database, such as Gene Ontology (e.g., set of genes involved in a known process), or it may come from a differential expression analysis. Based on the gene list, the Segal et al. (2004) approach is to determine which arrays are commonly induced by multiple gene lists; the gene lists are then combined to form a "core" gene cluster. One then determines which arrays show significant differential expression based on the core gene cluster and assesses whether there is enrichment of clinical annotation in the set of arrays found at the previous step. Through this procedure, Segal et al. (2004) are able to find 456 regulatory modules from gene expression data consisting of measurements of 14,145 genes in 1,917 samples across 22 tissue sites.

The approach taken by Elkon et al. (2003), while similar in spirit, involves a major difference. The difference is that sequence data are integrated with the gene expression profiling data. For the study by Elkon et al. (2003), approximately 13,000 putative promoter start sites were identified based on the NCBI Reference Sequence Database (`ftp://ftp.ncbi.nih.gov/genomes/H_sapiens`). Next, a set of genes determined to be cell cycle-regulated from a human cell cycle gene expression profiling study (Whitfield et al., 2002) were used; of the 874 putative cell cycle genes in that paper, promoter start sites were available for 568 of them. The authors searched for significantly enriched position weight matrices in the entire set of the 568 cell cycle-regulated promoters using the original 13K set as the background set and found enrichment of six binding sets. Thus, this provides a set of candidate transcription factors which may play a role in cell cycle progression.

The study of Rhodes et al. (2005a) is similar to that of Elkon et al. (2003). They derive 265 gene lists from various differential expression analyses using a q-value cutoff of 0.10. Next, they identify putative transcription factor binding sites in the promoter sequences of human genes and come up with a database of 361 transcription factors. Then, enrichment of each transcription factor in each of the gene lists is done; again an adjustment for multiple testing based on false discovery rate calibration is performed. From this analysis, they define 311 regulatory programs that displayed highly significant overlap ($p < 0.00033$) between a gene expression signature and a regulatory signature; these candidate regulatory modules can be tested experimentally.

The crux of the analyses described in this section is that based on defined lists of genes, one calculates overlap measures of enrichment of a certain biological property (here binding sites) with the lists. It is fairly easy to see how other types of biological sequence information (e.g., protein structure information, etc.) might be used here as well. In addition, there are many ways of defining "interesting." It could be differential expression from a two-group comparison, or cell cycle regulated (i.e., periodic expression) in a microarray time-course study. The overlap statistic is a very simple one; many other approaches are possible and may well be more powerful. This area will be a popular one for further study.

12.5 *In vivo/in vitro* genomic data integration

An area that is being considered more frequently in functional genomic studies in cancer is the integration of *in vitro*, i.e., experimental studies, with human gene expression studies, termed *in vivo* data. Integrating results from such experiments with *in vivo* cancer signatures holds the potential both to infer activity of specific oncogenic pathways *in vivo* and to identify relevant effectors of oncogenic pathways. For example, Huang et al. (2003) developed distinct *in vitro* oncogenic signatures for three transcription factors, Myc, Ras and E2F1-3. These signatures were able to predict Myc and Ras state in mammary tumors that developed in transgenic mice expressing either Myc or Ras, suggesting that specific oncogenic events are encoded in global gene-expression profiles.

To begin to understand the mechanisms by which oncogenes cause cancer, investigators have used gene expression profiling to identify downstream targets of oncogenic pathways in cell culture systems. Conceptually, this involves manipulating a gene in an *in vitro* system and measuring a global profile using a gene expression measurement technology and then trying to relate the *in vitro* gene expression profile to an *in vivo* gene expression profile. Such an approach was taken by Lamb et al. (2003) to determine the direct transcriptional effects of oncogene Cyclin D1. *In vitro* experiments were performed in which the Cyclin D1 was both over and underexpressed, and global gene expression profiles were determined. Lists of differentially expressed genes were then generated. To correlate the lists with in vivo gene expression data, a two-step process was utilized in which genes were first ordered based on correlation with Cyclin D1. Then, a Kolmogorov-Smirnov statistic was used to determine if the lists clustered within the ordered list based on correlation. Since there was significant evidence of clustering, Lamb et al. (2003) found that the *in vitro*-defined targets of Cyclin D1 were correlated with Cyclin D1 levels *in vivo*. This suggests that the direct regulatory effects of Cyclin D1 may play an important role in tumorigenesis. The statistical problem brought by up this type of analysis is determining clustering of a list of genes within an ordered list of genes. Although a Kolmogorov-Smirnov statistic has the advantage of being a nonparametric statistic, the potential disadvantage to the use of such a method will be a loss of efficiency. Determining alternative methodologies for this type of problem will be important.

Another setting that leads to joint consideration of *in vitro* and *in vivo* genomic data is when the *in vitro* experiment is performed in a model organism system. For example, Sweet-Cordero et al. (2005) defined a signature by comparing lung tumors generated from a spontaneous KRAS mutation mouse model to normal mouse lung and correlating it with gene expression profiles in human lung cancer studies. The major issue in such an analysis is mapping mouse genes to orthologous human genes. Sweet-Cordero et al. (2005) found that the mouse signature shared significant similarity with human lung adenocarcinoma but not with other lung cancer types. Next, they looked for evidence of the KRAS signature in human tumors carrying activating KRAS mutations relative to wild type tumors. Although no individual genes were significantly associated with KRAS mutation status in human tumors, the mouse KRAS signature was significantly enriched among genes rank-ordered by differential expression in human tumors with a KRAS mutation.

It is expected that experiments such as those described above will become much more commonplace in the future. Thus, it will be critical to address issues and to develop methods for integrating *in vivo* and *in vitro* genomic data so that inferences regarding transcriptional regulatory pathways in cancer can be generated.

12.6 Software availability

Public use software programs implementing these recent methods have been developed and are still being improved. As mentioned earlier, our group has developed the publicly available ONCOMINE database (http://www.oncomine.org). This database is geared toward biologists and therefore carries out several types of automated data analyses. Examples include differential expression analyses , analyses for functional enrichment of GO terms and Kolmogorov-Smirnov analyses in the spirit of Lamb et al. (2003). In addition, links to the protein-protein prediction project of Rhodes et al. (2005b) are available.

Many primary analysts use software languages such as MATLAB and R (R Development Core Team, 2005) for the analysis of genomic data. A major benefit of R is that it is a high-level interpretable language that allows for relatively fast development of methods. In addition, it has a nice ability for packaging related components.

The world-wide, open source and open development BioConductor project encourages development of bioinformatics software packages in R (Gentleman et al., 2004). The goals of the BioConductor project include:

1. fostering collaborative development and widespread use of innovative software,
2. reducing barriers to entry into interdisciplinary scientific research, and
3. promoting the achievement of remote reproducibility of research results.

More information about BioConductor is available at the web site http://www.bioconductor.org.

Another language that is important in this type of bioinformatics research is Perl (http://www.perl.org). Given that many of the databases are text databases, it is very important to be able to manipulate such databases relatively easily. Perl is a very useful language for such text manipulations.

12.7 Discussion

In this chapter, we have provided an overview of the area of functional genomic analyses. Because there are many different types of functional genomic data sets being generated, we can define a generalization of the statistical concept of meta-analysis. Now, analysts are faced with the prospect of combining different sources of information from a wide variety of platforms based on different technologies.

One of the techniques described earlier, graphical models, is a tool from the area of machine learning. Machine learning algorithms tend to be black-box algorithms that are useful for predictive inference. While the application of machine learning algorithms to high-dimensional genomic data sets will lead to some predictions that will be borne out, it is also important to attempt to build in biological information as much as possible into the analyses. As an example, a central tenet of biology is that binding of DNA to the binding site's transcription factors leads to activation of gene expression. It would seem sensible that a model in which transcription factor information is the independent factor and gene expression is the dependent variable should be a better model for the system than a graphical model that assumes no directionality.

Finally, an important non-statistical issue that needs to be addressed is how to store information from these types of analyses such that they themselves can be combined. One can imagine that lists of genes from different analyses can be used to make inferences about various biological aspects in cancer studies. It then may be of interest to compare the lists themselves in another type of meta-analysis so that higher-order inferences about the biological network can be made. However, to do this will require work to develop database requirements and standardization, much as was done in the case of microarrays (Brazma et al., 2001).

Acknowledgments

Debashis Ghosh would like to acknowledge the support of grant GM72007 from the Joint NSF/NIGMS Biological Mathematics Program.

CHAPTER 13

An overview of statistical approaches for expression trait loci mapping

Christina Kendziorski and Meng Chen

13.1 Introduction

Karl Sax was a pioneer in the field of quantitative trait loci (QTL) mapping. In his ground breaking 1923 paper (Sax, 1923), Sax identified a quantitative trait locus (QTL) for seed weight by associating the trait with seed color (a "marker" for which genotype information could be inferred). The next 60 years saw only a handful of similar studies, due mainly to limitations imposed by the difficulty in arranging crosses with a large number of genetic markers. This changed in the 1980s following the discovery that abundant, highly polymorphic variation could be used to derive molecular markers densely spaced throughout the genome (Botstein et al., 1980). This advance, combined with statistical methods for QTL mapping (Lander and Botstein, 1989), led to hundreds of QTL mapping studies.

A recent advance of comparable significance has been made in the area of phenotyping. With high throughput technologies now widely available, investigators today can easily measure thousands of traits for QTL mapping. Gene expression abundances measured via microarrays (see Section 1.4) are particularly amenable to QTL mapping, and most scientists agree that the mapping of gene expression has the potential to impact a broad range of biological endeavors (Cox, 2004; Broman, 2005).

The optimism is based largely on the first expression trait loci (ETL) studies, which have demonstrated utility: in identifying candidate genes (Schadt et al., 2003; Hubner et al., 2005; Bystrykh et al., 2005), in inferring not only correlative but also causal relationships between modulator and modulated genes (Brem et al., 2002; Schadt et al., 2003; Yvert et al., 2003), in elucidating subclasses of clinical phenotypes (Schadt et al., 2003; Bystrykh et al., 2005; Chesler et al., 2005; Hubner et al., 2005), and perhaps most importantly, in identifying "hot spot" regions, genomic regions where multiple transcripts map (Schadt et al., 2003; Brem et al., 2002; Morley et al., 2004; Bystrykh et al., 2005; Chesler et al., 2005; Hubner et al., 2005). Hot spot regions are attractive for follow up studies as they putatively contain master regulators that affect transcripts of common function. The identification of master regulators could give critical information on mechanisms of regulation that remain poorly characterized, ultimately leading to targets of gene therapies (Cox, 2004; Schadt et al., 2003). As a result of these successes, a number of efforts are now underway to localize the genetic basis of gene expression.

It is clear that the experimental setup in an ETL mapping study is structurally similar to a traditional QTL mapping study, but with thousands of phenotypes; and, as a result, most published studies to date have used QTL mapping methods in the ETL setting. Lan et al. (2003) reduced the expression measurements to a few summary scores using a principal components analysis and then used single-trait QTL mapping methods to map the summary phenotypes. Doing so proved useful; however, transcript-specific information could not be recovered. Others have used a "transcript-based" approach. In a transcript-based approach, each transcript is treated separately as a one-dimensional phenotype for QTL mapping. Single-trait QTL analysis is then carried out thousands of times (once for each transcript). Notably, although adjustments are made for multiple tests across the genome, no adjustments are made for multiple tests across transcripts. This leads to a potentially serious multiple testing problem and an inflated false discovery rate (FDR; Benjamini and Hochberg, 1995; see also Section 7.2.3).

An alternative approach recognizes the similarities between ETL mapping and the problem of identifying differentially expressed (DE) transcripts in a standard microarray experiment. By grouping animals with similar marker genotypes, the ETL mapping problem at a particular marker reduces to identifying DE transcripts across the genotype groups. Any method developed for identifying DE transcripts could be applied. Similar to the transcript-based approach, this "marker-based" approach is also subject to inflated FDR as here multiplicities across markers are not accounted for. For some labs, an inflated FDR is tolerable as many genes can be tested quickly for certain properties and discarded if found to be false positives. However, for many labs, validation tests are prohibitively expensive and statistical methods that control error rates across both markers and transcripts are needed. Kendziorski et al. (2006) proposed such an approach, the mixture over markers (MOM) model.

In this chapter, we review transcript-based approaches, marker-based approaches, and the MOM model approach to ETL mapping. The advantages and disadvantages of these approaches are discussed in Sections 13.2 and 13.4. Utility is evaluated using simulated data and data from two case studies (Section 13.3).

13.2 ETL mapping data and methods

13.2.1 Data

The general data collected in an ETL mapping experiment consists minimally of a genetic map, marker genotypes, and microarray data (phenotypes) collected on a set of individuals. A genetic marker is a region of the genome of known, or estimated, location. These locations make up the genetic map. At each marker, genotypes are obtained. ETL mapping studies take place in both human and experimental populations; we focus here on the latter. For these populations, the number of possible marker genotypes is relatively small.

Studies with experimental populations most often involve arranging a cross between two inbred strains differing substantially in some trait of interest to produce F_1 offspring. Segregating progeny are then typically derived from a B_1 backcross ($F_1 \times$ Parent) or an F_2 intercross ($F_1 \times F_1$). Repeated intercrossing ($F_n \times F_n$) can also be done to generate recombinant inbred (RI) lines. For simplicity of notation, we focus on a backcross population. This is not required and is relaxed in the simulation and case studies sections. Consider two inbred parental populations P_1 and P_2, genotyped as AA and aa, respectively, at M markers. The offspring of the first filial generation (F_1) have genotype Aa at each marker (allele A from parent P_1 and a from parent P_2). In a backcross, the F_1 offspring are crossed back to a parental line, say P_1, resulting in a population with genotypes AA or Aa at a given marker. We denote AA by 0 and Aa by 1.

For each member of the backcross population, phenotypes are collected via microarrays. For the kth animal, let $y_{t,k}$, $t = 1, 2, \ldots, T$, denote the expression level for transcript t and $g_{m,k}$, $k = 1, 2, \ldots, n$, denote the genotype at marker m. To avoid confusion when referring to genes on a genetic map and gene expression levels measured on a microarray (where the physical location of the gene is often not known) we use the term "gene" when referring to the former, and "transcript" or "trait" when referring to the latter.

Most questions addressed in an ETL mapping study rely on the ability to identify a list of significant linkages between transcripts and markers. To be precise, a transcript t is linked to marker m if $\mu_{t,0} \neq \mu_{t,1}$, where $\mu_{t,0(1)}$ denotes the latent mean level of expression of transcript t for the population of animals with genotype $0(1)$ at marker m. Suppose observations $y_{t,k}$ have density $f_{obs}(y_{t,k}|\mu_{t,g_{m,k}}, \theta)$, where θ denotes any remaining unknown parameters. Assuming independence across animals, under the null hypothesis of no linkage, the data are governed by

$$\prod_{k=1}^{n} f_{obs}(y_{t,k}|\mu_{t,0} = \mu_{t,1}, \theta)$$

and under the alternative by

$$\prod_{k=1}^{n} [f_{obs}(y_{t,k}|\mu_{t,0}, \theta)]^{1-g_{m,k}} [f_{obs}(y_{t,k}|\mu_{t,1}, \theta)]^{g_{m,k}} .$$

As discussed below, a main difference between the transcript-based (TB) and marker-based (MB) approaches arises from different assumptions regarding the latent means.

13.2.2 Transcript-based approach

A TB approach refers generally to the repeated application of any single phenotype QTL mapping method to each mRNA transcript, with locations identified as important if the test statistic of interest exceeds some critical value. The lod score

$$\log_{10}\left(\frac{\prod_{k=1}^{n} f_{obs}(y_{t,k}|\hat{\mu}_{t,0}, \hat{\mu}_{t,1}, \hat{\theta})}{\prod_{k=1}^{n} f_{obs}(y_{t,k}|\hat{\mu}, \hat{\theta})}\right)$$

is often used as the statistic measuring evidence in favor of linkage, where $(\hat{\cdot})$ denotes the maximum likelihood estimate of the associated parameter(s) and μ denotes the mean common across genotype groups (Lander and Botstein, 1989). Critical values that adjust for multiplicities across genome locations can be obtained theoretically (Dupuis and Siegmund, 1999) or via permutations (Churchill and Doerge, 1994).

The specific TB approach considered here assumes a Gaussian density for f_{obs} with critical values determined by the formulas given in Dupuis and Siegmund (1999). We consider the output from this approach at markers and refer to this as a TB marker regression (TB-MR) approach. The restriction to consider output only at markers is done to facilitate comparisons with MB methods, discussed below. For TB-MR, the genome-wide Type I error rate per transcript is controlled at 5% (Dupuis and Siegmund, 1999).

13.2.3 Marker-based approaches

To identify transcripts significantly linked to genomic locations, instead of testing each transcript for significant linkage across markers, one could test at each marker for significant linkage across transcripts. This amounts to identifying DE transcripts at each marker, with groups determined by marker genotypes. The MB approach refers generally to the repeated application, at each marker, of any method for identifying DE transcripts. In this setting, a number of approaches could be used; we consider four.

The first is an empirical Bayes approach, *EBarrays*, with the log-Normal Normal model (LNN) described in detail in Kendziorski et al. (2003, 2006). This approach calculates the posterior probability of differential expression for every transcript. Thresholds can be chosen to control the expected posterior FDR across transcripts. For example, by specifying the threshold to be the smallest posterior probability such that the average posterior probability of all transcripts exceeding the threshold is larger than $1 - \alpha$, the posterior expected FDR is controlled at $100\alpha\%$ (Newton et al., 2004). This marker-based empirical Bayes approach will be referred to as MB-EB. As in TB-MR, the LNN model assumes a Gaussian density for f_{obs}.

The second marker-based approach consists of obtaining p-values from a Student t-test followed by p-value adjustment; and the last two approaches consider moderated t-statistics followed by p-value adjustment. The details of the moderated statistic construction are given in Smyth (2004) and Tusher et al. (2003), respectively (see also Section 8.2.2). Adjustment for these last three methods is done using q-values to control the overall false discovery rate (FDR). In particular, to control the FDR at α, transcripts with q-values $<= \alpha$ are considered significant (Storey and Tibshirani, 2003b; see also Section 9.3.2). MB-Q, MB-LIMMA, and MB-SAM will denote the three marker-based approaches, respectively.

13.2.4 Other approaches

Although the TB and MB approaches are in many ways fundamentally different, they share an important flaw. Separate tests are conducted for each transcript-marker pair, and each measures evidence that the transcript maps to that marker relative to evidence that it maps nowhere. Since a transcript can map to any of many marker locations, the evidence that a transcript maps to a particular marker should not be judged relative only to the possibility that it maps nowhere, but rather relative to the possibility that it maps nowhere *or* to some other marker. This idea motivates the mixture over markers (MOM) model (Kendziorski et al., 2006). Briefly, MOM assumes a transcript t maps nowhere with probability p_0 or to marker m with probability p_m where $p_0 + \sum_{m=1}^{M} p_m = 1$ and M denotes the total number of markers. The marginal distribution of the data $\mathbf{y}_t$ is then given by

$$p_0 f_0(\mathbf{y}_t) + \sum_{m=1}^{M} p_m f_m(\mathbf{y}_t), \tag{13.1}$$

where f_m describes the distribution of data if transcript t maps to marker m (f_0 describes the data for non-mapping transcripts). The component densities are predictive distributions that can be derived under different parametric assumptions. For comparison, we take Gaussian observation components for the log measurements with Normal priors on the latent expression levels.

13.3 Evaluation of ETL mapping methods

The methods discussed above were evaluated using simulated data and data from two case studies. The simulations are in no way designed to capture the many complexities of ETL mapping data. Nevertheless, they do provide some insight into operating characteristics of each of the approaches. The first case study concerns an experiment in yeast and the second a study of diabetes in mouse.

13.3.1 Simulation

Recall that for a backcross population, a subject has one of two genotypes (AA or Aa) at each marker locus. For an F_2, three genotypes are possible (AA, Aa, or aa) and, as a result, a given transcript may be equivalently expressed (EE) or may be in any one of 4 DE patterns ($AA|Aa, aa$; $AA, Aa|aa$; $AA, aa|Aa$; $AA|Aa|aa$). Here $|$ denotes inequality among the latent genotype group means. We performed a simulation of an F_2 population in which pattern membership was determined by a multinomial where the expected proportion of transcripts in each DE pattern was specified at 3%, 3%, 1% and 3%, respectively (1% is used for the pattern that is least biologically plausible).

Care was taken to protect against biasing the results in favor of any of the methods considered. The details are given in Kendziorski et al. (2006). In short, a major difference among methods lies in the estimation of transcript variance σ_t^2. To set the variance for a simulated transcript t, we used the posterior mean of σ_t^2, given by

$$\frac{\sum_{k=1}^n \left(y_{t,k} - \bar{y}_{t,\cdot}\right)^2 + \nu_0 \sigma_0^2}{\nu_0 + n - 2},$$

derived assuming the transcript-specific variance is distributed as scaled inverse chi-square: $\sigma_t^2 \sim$ Inverse $\chi^2 \left(\nu_0, \sigma_0^2\right)$. As $\nu_0 \to 0$, the posterior mean approaches $[(n-1)s^2/(n-2)] \approx s^2$, the transcript-specific sample variance, which is the naive estimate of σ_t^2 for an EE transcript under TB-MR assumptions. Data simulated with small ν_0 are therefore consistent with assumptions made in TB-MR. As $\nu_0 \to \infty$, the posterior mean approaches a constant value σ_0^2, which is assumed in MB-EB (note that this assumption implies a constant coefficient of variation on the raw gene expression scale). By varying ν_0, operating characteristics could be evaluated without biasing the results in favor of one method. Data simulated by this empirical method had marginal distributions that were virtually indistinguishable from the observed data.

We consider a single ETL simulation with 100 animals and 2 chromosomes. Marker genotype data were obtained from chromosomes 2 and 3 of the F_2 data described in the next section. Chromosome 2 (3) contained 17 (6) markers with an average intermarker distance of 7.6 (17.7) cM. An ETL at marker 5 on chromosome 2 was simulated; no ETL was simulated on chromosome 3. Seven sets of simulations were obtained for ν_0 between 5^{-5} and 5^5 (ν_0 for the actual F_2 data was estimated near 5). For each value of ν_0, 20 simulated data sets were generated. At each fixed ν_0, the profile marginal MLE was obtained for σ_0^2.

FDR gives the proportion of transcripts identified incorrectly as mapping to chromosome 2; i.e., they were EE or they were DE but mapped outside the region flanking the true ETL. Table 13.1 reports the operating characteristics. FDR is well above the target level of 0.05 for most methods and most values of ν_0. MOM is the only one of the approaches capable of FDR control in this simple simulation setting. Power in this context measures the ability to identify the DE transcripts exactly at marker 5 or either of the flanking markers which are 16.5 and 5.8 cM away, respectively. There is little variation in power across ν_0. MB-Q is the most powerful method, followed by TB-MR, MB-EB, and MOM. The difference in power between MOM and the others is statistically significant, but perhaps not *practically* significant as power is still near 80%.

As shown in Table 13.1, the results from MB-Q, MB-LIMMA and MB-SAM were

Table 13.1 *Average operating characteristics (OCs) for TB-MR, MB-EB, MB-Q, MB-LIMMA, MB-SAM, and MOM. Averages are calculated over 20 data sets; standard errors were less than 0.005. OC definitions and details of the simulation are given in the text (see Section 13.3.1).*

OC	Method	ν_0						
		5^{-5}	5^{-3}	5^{-1}	5^0	5^1	5^3	5^5
FDR	TB-MR	0.286	0.286	0.293	0.285	0.286	0.280	0.301
	MB-EB	0.282	0.281	0.285	0.279	0.269	0.117	0.034
	MB-Q	0.240	0.246	0.246	0.240	0.245	0.230	0.226
	MB-LIMMA	0.238	0.236	0.232	0.237	0.235	0.237	0.229
	MB-SAM	0.233	0.238	0.235	0.232	0.238	0.236	0.221
	MOM	0.038	0.041	0.046	0.037	0.036	0.005	0.002
Power	TB-MR	0.884	0.886	0.887	0.886	0.889	0.919	0.868
	MB-EB	0.820	0.817	0.815	0.823	0.833	0.895	0.837
	MB-Q	0.911	0.912	0.913	0.912	0.917	0.949	0.918
	MB-LIMMA	0.900	0.910	0.909	0.900	0.914	0.935	0.899
	MB-SAM	0.897	0.908	0.906	0.898	0.913	0.933	0.899
	MOM	0.848	0.851	0.853	0.850	0.856	0.860	0.811

very similar, most likely because the relatively large sample size (100 animals) yields statistics in MB-LIMMA and MB-SAM that have been "moderated" only slightly. A similar result was reported in Smyth (2004), where an experiment with 16 animals was considered. For this reason, only results for MB-Q are discussed.

13.3.2 Case studies

To further compare these approaches, we consider ETL mapping data from the yeast experiment described in Brem et al. (2002). It is structured as a backcross between a standard laboratory strain (BY) and a wild isolate from a California vineyard (RM). There are 6,215 transcripts and 3,312 markers. With only 40 segregants in the cross, recombinants are limited. We removed pairs of markers with fewer than 10 recombinants between them, leaving 88 markers.

Brem et al. (2002) identified eight regions enriched for linkage across the genome. Many transcripts in these hot spot regions have been at least partly validated using independent experiments. As noted in the Introduction (Section 13.1), these regions are of much interest as they may contain a master regulator responsible for the control of transcripts sharing common biological function. A statistical test for enrichment of common function can be done via the GOHyperG package in BioConductor (Gentleman et al., 2004). GOHyperG uses data from Gene Ontology (GO), where transcripts are categorized at varying levels of biological detail (the three broadest levels are molecular function, cellular component, and biological process – there are many subcategories within each). A more detailed GO description is given by Shaw (Chapter 11).

For a given set of mapping transcripts and a given function, a hypergeometric calculation is performed to test for enrichment of that function across the transcripts. Interpretation of resulting p-values is not straightforward due to dependencies in the tested hypotheses. Furthermore, the hypergeometric calculation tends to result in small p-values when GO nodes with few transcripts are considered. For these reasons, it has been suggested that one consider only interesting small p-values obtained from a relatively large set of transcripts (> 10) (Gentleman, 2004). Applying this criterion to the results from Brem et al. (2002) gives five regions, shown in Table 13.2.

Table 13.3 shows information similar to that in Table 13.2, for the five regions with the largest number of mapping transcripts identified by MOM, TB-MR, and MB-Q. We see that TB-MR identifies three of the five regions identified by Brem et al. (2002) on chromosomes 3, 12, and 14. The location identified by Brem et al. (2002) on chromosome 2 is missed by TB-MR; and the location identified by TB-MR on chromosome 9 is not found using any other method and shows little evidence for enrichment of common function. This is likely a false positive. Similar results are obtained from MB-Q, with three of the five regions identified, and one potentially spurious identification on chromosome 8. Note that the region identified by all methods on chromosome 15 is one of the eight originally identified by Brem et al. (2002). It was excluded when constructing the list of five due to a relatively large p-value (0.02). It is difficult to judge whether or not this region is a false positive. Considering that all methods point to this region, perhaps it is not.

Table 13.2 *Results reproduced from Brem et al. (2002). Chromosomal locations, number of transcripts mapping to each region, biological function common to these transcripts, and p-values from* GoHyperG *are shown. BP gives the number of kilobases from the 5' end of the chromosome.*

Chromosome (BP)	# Mapping Transcripts	Common Function	p-value
2 (550)	18	Cell Separation	$\sim 10^{-7}$
3 (90)	21	Leucine Biosynthesis	$\sim 10^{-7}$
3 (190)	28	Mating	$\sim 10^{-10}$
12 (670)	28	Fatty Acid Metabolism	$\sim 10^{-7}$
14 (490)	94	Mitochondrial Induction	$\sim 10^{-6}$

The MOM model performs better: four of the five regions identified by Brem et al. (2002) (on chromosomes 2, 3, 12, and 14) are also identified by MOM. The one region identified by Brem et al. (2002) but not MOM is a second location on chromosome 3. There are not enough markers (the selected 88) to distinguish between these two regions using MOM. In addition to improved hot spot localization, MOM is generally more sensitive than the other methods. We suspect that the increased

Table 13.3 *Top five regions identified by TB-MR, TB-Q, and MOM. For each method and region, chromosomal locations, number of transcripts mapping to each region, biological function common to these transcripts, and p-values from* GoHyperG *are shown. BP gives the number of kilobases from the 5′ end of the chromosome.*

Method	Chromosome (BP)	# Mapping Transcripts	Common Function	p-value
TB-MR	3 (75)	29	Leucine Biosynthesis	$\sim 10^{-6}$
TB-MR	12 (607)	21	Fatty Acid Metabolism	$\sim 10^{-7}$
TB-MR	14 (502)	644	Mitochondrial Induction	$\sim 10^{-6}$
TB-MR	15 (1)	27	Glucan Metabolism	> 0.2
TB-MR	9 (99)	19	Iron Transport	0.03
MOM	2 (602)	56	Cell Separation	$\sim 10^{-5}$
MOM	3 (75)	56	Leucine Biosynthesis	$\sim 10^{-6}$
MOM	12 (872)	55	Fatty Acid Metabolism	$\sim 10^{-8}$
MOM	14 (502)	94	Mitochondrial Induction	$\sim 10^{-6}$
MOM	15 (1)	288	Glucan Metabolism	$\sim 10^{-3}$
MB-Q	3 (75)	31	Leucine Biosynthesis	$\sim 10^{-5}$
MB-Q	12 (607)	36	Fatty Acid Metabolism	$\sim 10^{-7}$
MB-Q	14 (502)	78	Mitochondrial Induction	$\sim 10^{-5}$
MB-Q	15 (1)	29	Glucan Metabolism	10^{-1}
MB-Q	8 (80)	81	Response to Pheromone	0.001

number of identifications made by MOM are not false discoveries as the additional transcripts maintain evidence for enrichment of the common function.

It is insightful to check the results from these approaches when control of particular error rates is not used for hot spot identification. For example, instead of defining hot spots in terms of the number of mapping transcripts (which depends on particular thresholds to generate binary calls), one could consider average evidence (across transcripts) of mapping at each location (average lod score, average posterior probability, or the average of $1 - q$-value). Given hot spots identified in this way, one can simply rank transcripts at each hot spot by this evidence measure and then consider the top N transcripts for some N. In terms of regions identified and tests for enrichment of common function, we find results similar to those shown in Table 13.3 for N of 50 and 100.

The ETL mapping approaches were also evaluated using data from a study of diabetes in mouse. For details on the experiment, see Kendziorski et al. (2006). Briefly, it is well known that the *ob* mutation in the C57BL/6J mouse background (B6-*ob/ob*) causes obesity, but only mild and transient diabetes (Coleman and Hummel, 1973), while the same mutation in the BTBR genetic background (BTBR-*ob/ob*) causes severe Type 2 diabetes (Stoehr et al., 2000). To gain insight into the genetic basis of these differences, a (B6 x BTBR) F_2-cross was generated yielding 110 animals. Selective phenotyping (Jin et al., 2004) was employed to identify 60 F_2 *ob/ob* mice. For each of the 60 mice, liver tissue was isolated and 45,265 mRNA abundance traits were collected at 10 weeks of age using Affymetrix GeneChips MOE430A and MOE430B (see Section 1.4.2). The probe level data were processed using Robust Multi-array Average (RMA) to give a single, normalized, background corrected summary score of expression for each transcript (Irizarry et al., 2003b; Section 1.4.2). Low abundance transcripts, defined as transcripts with average expression level below the tenth percentile, were removed, leaving 40,738 traits. Genotypes for 145 markers were also obtained (over 90% of the animals provided genotype data at any given marker).

Each method was applied to identify ETL. Hot spot regions are shown in the top panel of Figure 13.1. The first marker, D2Mit241, is adjacent to D2Mit9, which has recently been identified as an obesity modifier locus (Stoehr et al., 2004). Two additional regions identified by four of the five methods (on chromosomes 4 and 10) are not yet known to be involved in diabetes, although we note that the region identified on chromosome 4 has been implicated in other analyses done in the Attie lab. The two regions identified by MOM alone on chromosomes 5 and 8 have been identified by other groups in earlier studies: D5Mit1 is a location known to affect triglyceride levels (Colinayo et al., 2003) and D8Mit249 is the marker on our map closest to the "fat" gene which is known to affect both diabetes and obesity (Naggert et al., 1995). This provides some evidence for the success of the MOM approach, but much more biological validation is required.

It is interesting to note that the agreement between FDR-controlled and rank-based inferences observed for the yeast study was not observed here. Figure 13.1 (bottom panel) gives results from the diabetes case study using the binary scores. As shown, there is much less agreement across methods when the binary scores are used. We expect there are conditions under which averaging evidence across transcripts is more advantageous than reducing to a binary score (and vice versa). This is currently an area under investigation.

13.4 Discussion

The field of QTL mapping was reignited in the 1980s by advances that allowed for the relatively easy identification of genetic markers and their genotypes. Today, with major developments in high throughput technologies, a similar advance has taken place that allows for measurement of thousands of phenotypes. The number and nature of these phenotypes are what distinguish QTL from ETL mapping. In fact, ETL mapping is exactly traditional QTL mapping, but with thousands of expression traits considered as phenotypes. The simplicity with which this difference can be stated perhaps obscures the resulting challenges posed for the statistical analysis of ETL data.

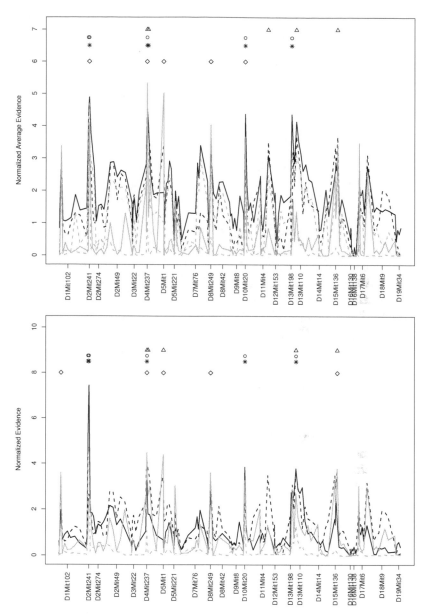

Figure 13.1 *Evidence of linkage across the genome for each approach (lod score for TB-MR, posterior probability for MB-EB and MOM, and* $1 - q$*-value for MB-Q). TB-MR, MB-EB, MOM, and MB-Q are shown in black solid, black dashed, grey solid, and grey dashed lines and represented by the circle, triangle, diamond, and star, respectively. The top panel averages evidence of mapping over transcripts; the bottom panel gives normalized totals of mapping transcripts based on thresholding to control FDR. The five markers with the strongest evidence of mapping transcripts are indicated by the corresponding method symbol.*

When faced with just about any statistical problem, it is often best to first consider methods that are currently available. This was done for ETL mapping. The earliest ETL papers applied traditional QTL mapping methods to each transcript in isolation. Doing so does not account for multiple tests across transcripts, however, and we found this to have a real impact on increased FDR even in very simplified simulation settings. For some labs, an inflated FDR is tolerable as many genes can be tested quickly for certain properties and discarded if found to be false positives. However, for many labs, such tests are prohibitively expensive and more appropriate statistical methods are needed.

More recent ETL studies have made attempts at adjusting for multiplicities across both markers and transcripts using a two-stage approach (Chesler et al., 2005; Hubner et al., 2005). The first stage obtains a single p-value for each transcript that is adjusted for multiple tests across markers; stage two controls the FDR across transcripts by calculating q-values from these p-values. With this approach, mapping transcripts are identified, along with the single most likely location to which these transcripts map. Preliminary simulation results (not shown) show very low power if attempts are made to control the FDR at 5%. This is consistent with the results reported in Chesler et al. (2005), where an FDR cutoff of 25% is used so that 101 transcripts can be identified (out of $12, 422$ total transcripts).

Our general conclusion is that a clever application of statistical methods developed in the context of QTL mapping and/or multiple testing is not sufficient to address the complexities of the ETL mapping problem. As a result, we continue to investigate MOM. The MOM approach was designed explicitly to address the ETL mapping question. Operating characteristics evaluated via simulations as well as results from case studies are encouraging. Another nice feature of the MOM framework is that it can be extended to account for interval and multiple ETL mapping. This work is underway.

In summary, much more work is required before the analysis of ETL data becomes routine. In practice, we suggest an investigator apply a number of tools and focus initially on genomic locations at which most methods agree (such as the four regions shown in the top panel of Figure 13.1), keeping in mind that assumptions across different methods are often very similar and therefore by no means are the results of different methods independent confirmations. Statisticians can contribute to the ETL mapping effort by method development, evaluation, and validation, and by carefully considering those genomic regions that *do not* agree across methods. Such regions can provide valuable insights so that specific conditions under which different methods work best can be identified. Advances in each area and communication between the two are required to maximize the amount of information that can be derived from ETL mapping studies.

CHAPTER 14

Incorporating GO annotation information in expression trait loci mapping

J. Blair Christian and Rudy Guerra

14.1 Introduction

Microarrays allow simultaneous measurements of gene expression of almost every gene in the human genome. Understanding the relationship between a particular genetic location and its expression is fundamental to elucidating the relationships among genes, transcripts of other genes, and proteins translated from those transcripts. For example, in the nuclear hormone receptor superfamily, the Peroxisome Proliferator-Activated Receptor Gamma (PPARγ) protein is one whose variants have different effects on the production of a large group of transcripts (Bush et al., 2007).

Currently, there are few statistical approaches that use all available biological information beyond the expression data in analyzing mRNA transcript experiments. However, it is expected that incorporating additional data or knowledge can help to provide better understanding of gene expression signatures.

To help improve current approaches and provide further understanding in using additional biological information in expression studies, we discuss the use of Gene Ontology (GO) data in expression trait mapping via genotype data. A key feature of our investigation is analyzing genes that are correlated as a function of GO distance. In this sense we examine the incorporation of "biological distance" into expression trait loci mapping. Several authors have proposed incorporating GO data in expression studies, for example, clustering expression profiles (Pan, 2006). However, most do not consider the possible adverse effects of using GO data, especially GO information based on weak evidence. In some cases, the incorporation of GO information is done rather informally. Here we return to more basic principles and issues, focusing instead on a proof of concept investigation addressing how to include the GO information and circumstances under which it may be helpful to do so.

Simulations are used to compare ETL mapping with and without the use of GO data.

We use Receiver Operating Characteristic curves to compare these approaches, and give recommendations for when it is advantageous to include annotation information into gene mapping. The greatest benefit arises in pleiotropic relationships where each transcript has low to moderate heritability and the correlation among the transcripts is not too high, although using excessively noisy annotations can lead to erroneous results.

14.2 Expression trait loci mapping

Quantitative trait loci (QTL) mapping is a statistical approach to identify correlations between genetic loci and a quantitative trait, such as low density lipoprotein cholesterol. The basic approach is to genotype a number of loci and test for associations between genotype and phenotype. A given quantitative trait may be determined by anywhere from one to many genetic loci. In the latter case, the multiple loci may act additively or interact with each other.

Measured gene expression may be viewed as a quantitative trait; loci contributing to the trait are called expression (quantitative) trait loci (ETL). ETL mapping is an active area of research. A study begins with the collection of both marker and transcript data on the same subjects. Thus, the measured data are typically represented as two matrices: n_{RNA} transcripts measured for each of n_{sub} subjects, as well as n_{SNP} biallelic SNPs genotyped for these same n_{sub} subjects. Brem et al. (2002) published the first ETL mapping paper in the bioinformatics literature, describing a collection of two-channel RNA measurements and genotypes from a yeast study. Interest in this problem has been growing ever since (Cheung et al., 2003; Schadt et al., 2003; Broman, 2005; Li and Burmeister, 2005; Kendziorski et al., 2006, 2004). Chapter 13 by Kendziorski and Chen also discusses ETL mapping.

The primary focus of this chapter is to explore the possibility of improving ETL analysis by incorporating biological annotation data. Instead of treating genes as mathematically independent units, we incorporate into the modeling process both known and inferred biological relationships between genes and their products. Thus, we combine the following data types: RNA measurements (gene expressions), single nucleotide polymorphism (SNP) genotypes, gene ontology (GO) information, and the annotation mappings of genes to the ontology.

Historically, countless tedious experiments have been performed to understand the relationships between genes, RNA, and protein. This effort has yielded a relatively small number of very high quality estimates of interactions between the genotype at one gene and expression of another gene. The fruit of this labor is seen in graphs such as the protein-protein interactions in the Kyoto Encyclopedia of Genes and Genomes (KEGG; `http://www.genome.ad.jp/kegg`) database. KEGG does not explicitly show relationships between genotypes and gene expressions, but there are implicit relationships indicated by the protein interactions. In this sense, there is prior knowledge about possible relationships between genotype and gene expression. This prior information could make finding these relationships an easier task than is currently possible. Although new technologies can generate many hypotheses about genotype-expression relationships, they have changed the paradigm from encouraging high quality estimates about a few components to one of lower quality estimates about very many components.

In the ETL mapping context there are many (univariate) traits, each of which can be tested for association with each of many genetic loci. With hundreds of thousands of traits and hundreds of thousands of loci, the multiple testing problem is quite challenging. It may, however, be of benefit to use known or inferred biological information that generates homogeneous clusters of the univariate traits, for which higher statistical power may be achieved by conducting the genotype-phenotype analysis within clusters. Alternatively, the biologically correlated univariate traits may be viewed as a (multivariate) response in a multivariate analysis of variance. In statistical language, this is an instance of data dimension reduction for high-dimensional problems. Here we integrate GO annotation information to reduce the number of phenotypes (expression) tested, from the observed number of univariate gene expressions to a relatively much smaller number of multivariate gene expression phenotypes.

The biological justification for using GO data in the ETL mapping problem rests on the dynamical systems framework (Kacser and Burns, 1973), whereby genes with similar function have similar expressions. This motivates grouping genes into biologically meaningful units that are then analyzed as a multivariate phenotype. In this chapter, we examine whether the ability to identify accurately biologically correlated univariate expressions (and subsequently the multivariate phenotypes) is sufficient to allow the theoretical benefits of multivariate approaches to be realized. Indeed, it is known that the power to detect genetic linkage via multivariate traits can be better than that based on the individual components of the multivariate trait (Amos et al., 2001). We give evidence in this work that under certain conditions, the same is true for multivariate expression traits identified by GO information.

From a biological perspective we are working under a "pleiotropy hypothesis," where one gene (via a SNP genotype) can affect the expression of other genes. In practice, of course, the pleiotropies are largely unknown, so the candidate pleiotropic transcript clusters must be estimated. Very little is known about these pleiotropies in the ETL mapping context. In related work on clustering of expression profiles, Pan (2006), Huang and Pan (2006) and Huang et al. (2006) demonstrate that incorporation of annotation data can improve the biological quality of the clusters. In particular, Pan (2006), using model-based clustering on microarray data, found the best results when combining annotation information with the transcript data. Thus, if the biologically meaningful clusters are estimated accurately and they are indeed part of a pleiotropy, then genotype-phenotype inference may be improved in the context of ETL mapping. Incorporating incorrect external data and/or incorporating external data incorrectly, however, can damage the quality of the inference. This issue may seem obvious, but the extent or nature of the potential problems are not clearly understood. The issue has received almost no attention, but it is important to know the impact of the underlying assumptions on performance. We provide some guidance as part of this investigation of ETL mapping with annotation data.

14.3 Data

The different types of data are related by referring to common genomes or to common homologies. The primary data we use here consists of four types:

1. RNA (gene expressions) observations on each subject;
2. SNP genotype observations on each subject;
3. a gene ontology, which is a structured description of various biological contexts;
4. the annotations of the particular genes (and their products) being studied to the ontology.

Note the distinction between a gene ontology and annotations to the ontology. The annotations may be viewed as the "GO data" for our ETL mapping problem. As described below, the GO is the same for all users and studies. The particular annotations for a given study depend on the genes under investigation and thus an annotation data set will vary from one study to another. We can think of the ontology as a word dictionary. Different people read different books, but they all refer to the same dictionary to look up different words. In this work, all data types depend directly on the current estimated version of the genome. Without loss of generality, we have chosen to use the human genome, which is updated quarterly.

We assume a sample of n_{sub} subjects. Each subject is genotyped at each (SNP) marker, and gene expression (transcript) levels are measured for a particular tissue of interest. The SNP genotypes are the independent variables in ETL mapping. The measured gene expressions are the dependent (response) quantitative variables.

14.3.1 Gene Ontology

Independent of the marker and gene expression data at hand, there are relationships inside the cell that involve the genes of interest. One formal way of describing the relationships is through the use of an ontology, which is a taxonomy for describing relationships in a structured way. The taxonomy is presented using the language of graph theory. Knowing these relationships among the genes may help improve inference about genotype associations with gene expression.

Annotation is information about a gene within a particular context. When referring to "annotation data" it is helpful to think of a pair: (1) a graph or ontology and (2) mappings from another data set to this graph. Each gene transcript can be annotated to zero or more nodes of the graph; the nodes represent the biological information on the genes.

Graphs provide a mathematical representation of the relationships between biological concepts in different domains. One domain is the cellular component or components where a particular protein is found. In this case, places where a protein can be found in a cell are the *nodes* in the graph, such as "cytoplasm", "cell membrane" or "Golgi body." What makes this an ontology is the way in which these nodes are connected. An *edge* is a connection between two nodes. In an ontology, the edges have a direction, which is used to create the structure between nodes. The standard used is that edges join two nodes that are involved in hierarchical relationship. For example, given two nodes, "cell" and "nucleus," the ontology has an edge connecting them, "cell" ← "nucleus", which can be read as a cell *has a* nucleus. Further, we can add an edge for "chromosome" to indicate that a nucleus has a chromosome, denoting this set of relationships "cell" ← "nucleus" ← "chromosome."

Annotation data are any data about the biological context of the marker or gene. One database of annotation information is the Gene Ontology (GO), which annotates genes to three different contexts: the cellular components (CC) where the protein is expressed; the biological processes (BP) in which the expressed protein is involved; and the molecular functions (MF) of the expressed protein (Ashburner et al., 2000). Other similar databases exist, such as KEGG; however, we restrict attention to the GO database.

We can now be precise about the "annotation data" used in the ETL mapping problem. It is the collection of each gene's *annotations*. For a given ontology, each gene may or may not have available information for mapping it to nodes on the GO graph. If information is available, a gene will map to one or more nodes in the GO graph. If a gene were annotated to a particular node (N), it would also be biologically related to the ancestor nodes shown in the ontology, referred to as the ontological *subgraph* induced by node (N).

When a gene or gene product is labeled with a GO ID, the annotation of the gene or gene product to a given node is given an *evidence code* to denote the origin of the annotation, an indication of its quality. Examples of evidence codes to label proteins are given by Shaw (Chapter 11 in this volume, Table 11.3.2).

The annotation information on each gene transcript comes from databases built by the research community (Ashburner et al., 2000; Kanehisa and Goto, 2000; Mewes et al., 2004). We use the Biological Process ontology in this work. In the simulations discussed below, we assume that a gene maps to at most a single node. For an introduction to working with GO data, see Chapter 11 by Shaw or Gentleman (2008). For a computational perspective, see the Bioconductor annotation vignette: `http://bioconductor.org/packages/bioc/html/annotate.html`. The GO tree is updated monthly, with archives available at `ftp://ftp.geneontology.org/pub/go/ontology-archive/`.

14.4 Methodology

The standard data in ETL mapping consists of n_{RNA} transcripts and n_{SNP} SNP marker genotypes. Thus, the genotype matrix X of dimension $n_{sub} \times n_{SNP}$ and the transcript matrix Y is of dimension $n_{sub} \times n_{RNA}$. Most studies do not use biological meta-data (such as Gene Ontology) related to the markers and transcripts. Currently, a major limitation to carrying out more complicated ETL mapping techniques is insufficient sample size. These more complicated techniques may involve non-trivial testing of multiple SNPs and transcripts simultaneously, estimating covariances between transcripts, and identifying epistatic interactions or other interactions. In all cases to date both $n_{sub} \ll n_{RNA}$ and $n_{sub} \ll n_{SNP}$. More specifically, n_{sub} is typically in the 100s, while n_{RNA} and n_{SNP} are in the 10,000s or even 100,000s. This situation has led to the primary use of simple, basic models with fewer variables than those that would be used if $n_{sub} \gg n_{RNA} + n_{SNP}$. This situation motivates the use of techniques such as dimension reduction and other methods for sparse data.

Treating each marker and transcript independently is a basic approach to ETL mapping. This approach performs many one-way ANOVAs, whose results may or may not have been corrected for multiple testing (e.g., using the false discovery rate (FDR) Benjamini and Hochberg, 1995).

14.4.1 Incorporating GO annotation in ETL mapping

In this section, we describe a simple multivariate analysis of variance (MANOVA) approach for incorporating GO annotation information in ETL mapping. A more sophisticated approach could be developed, but a primary objective is simply to determine whether or not incorporating GO data can improve the genotype-phenotype inference. In particular, we are interested in those situations that do not improve the inference (as estimated with ROC curves, see below). In order to maintain focus on the inclusion of GO data, using a simple procedure makes it easier to pinpoint GO problems specifically. There are several ways in which GO data can be integrated into the analysis. Here, we opt for a simple nonparametric approach. We thus have the following two-stage method:

1. Cluster the transcripts based on the distance between their annotations. The GO distance between two genes is defined as a distance between the subgraphs induced by each gene's annotation.
2. Each cluster of transcripts is then viewed as a multivariate phenotypic response. For clusters with more than one transcript, we use MANOVA to test for associations; otherwise we treat the transcripts as singleton clusters and perform a univariate one-way ANOVA on them.

The F-test is used for association in the ANOVA case and Wilks' Λ in the MANOVA case. We adjust for multiple testing controlling FDR. When changing paradigms from many univariate tests to fewer multivariate tests, the hypotheses are different.

Also, the meanings of false negative and false positive rates change when moving to the multivariate setting. The null hypothesis for the (univariate) ANOVA setup at transcript i and SNP j is that expression of mRNA i is not associated with the polymorphism at a particular SNP j. However, in the multivariate case the null hypothesis is that none of the transcripts $\{i_1, \ldots, i_{n_{clust}}\}$ is associated with SNP j.

To accommodate the change in dimension in going between the univariate and multivariate settings, we compare the Receiver Operating Characteristic (ROC) curves for each method. Specifically, we compare the area under each ROC curve (AUC). The ROC is a plot of the true positive rate (y-axis) versus the false positive rate (x-axis), equivalent to a plot of sensitivity versus ($1-$specificity). The ROC curve is estimated, so there is error in the AUC. The AUC is calculated using the Wilcoxon U-statistic; its standard error is estimated using a normal approximation.

14.4.2 Distances between annotated transcripts

The first step in moving from biological annotations to a notion of biological distance is the creation of a distance metric between the subgraphs induced by the annotations of two transcripts. Figure 14.1 shows a simple example where each gene has exactly one annotation. Two distances are represented, adapted from similarity scores of Gentleman (2004). Zhou et al. (2002) was the first to discuss the concept of a distance matrix in relation to transcript data, but in a different context and only implicitly. Zhou et al. (2002) estimated the GO annotations for different organisms by using sequence homologies to annotated genes in other organisms.

The distance between two subgraphs $G_1 = \{N_1, E_1\}$ and $G_2 = \{N_2, E_2\}$ of a graph $G = \{N, E\}$ can be defined in many ways (N denotes nodes, E denotes edges). The two metrics we consider here are motivated by their biological relevance (Gentleman, 2004). The first is based on the "union-intersection" (UI) similarity metric. The UI distance D_{UI} is constructed in the following way ($card$ denotes set cardinality):

$$D_{UI}[G_1, G_2] \equiv 1 - \frac{card\left(N_1 \bigcap N_2\right)}{card\left(N_1 \bigcup N_2\right)},$$

which creates a distance between zero and one. The "longest path" (LP) similarity between two directed acyclic graphs (DAGs) G_1 and G_2 is defined as the longest path shared by both of the graphs. The LP similarity between two graphs, $LP(G_1, G_2)$ is constructed using Algorithm 1 (p. 233).

From this LP similarity, we can create a distance metric in several ways. The most natural way to create a distance in $[0, 1]$ is:

$$D_{LP}[G_1, G_2] \equiv 1 - \frac{LP(G_1, G_2)}{LP(G, G)}.$$

Note that both distances have an unnormalized version that is equivalent (just multiply by the denominator). In order to take into account other factors, it may be worth normalizing distances in other ways, for example by:

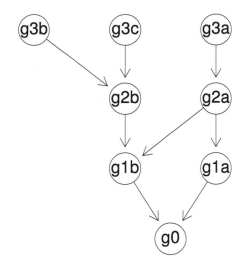

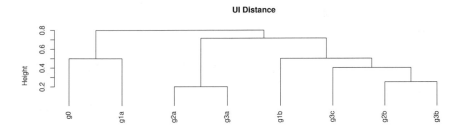

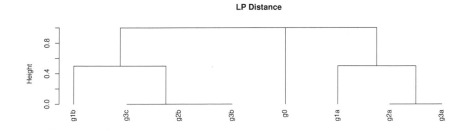

Figure 14.1 *(a) At top, an example graph under the assumption that one and only one gene is annotated to each node on the graph. (b) Hierarchical clustering dendrograms showing biological distances between nodes. Both the union intersection (UI) and longest path (LP) graph distance metrics are shown.*

Algorithm 1 Calculating Longest Paths Between Two Graphs.

Begin at the root node; set $LP = 0$
for each of the root node's children $n_i \in children\,(n_{root})$ **do**
    set $LPC = 0$, the number of edges both graphs share from the current node to
    the root node
    if both $n_i \in N_1$ and $n_i \in N_2$ **then**
        $LPC + +$, repeat for $n_j \in children\,(n_i)$
        if $LPC > LP$ **then**
            $LP = LPC$
        end if
    end if
end for

$\min\,(LP\,(G_1, G_1)\,, LP\,(G_2, G_2))$; $\max\,(LP\,(G_1, G_1)\,, LP\,(G_2, G_2))$;
$\min\,(card\,(N_1)\,, card\,(N_2))$; or $\max\,(card\,(N_1)\,, card\,(N_2))$.

Biologically, the UI distance is intuitive for genes with many annotations, and thus very large induced subgraphs. This gives a general idea of their similarities. The LP distance seems intuitive for comparing genes with singleton annotations, as it is directly proportional to the depth of the deepest shared node between two nodes. The deeper the shared nodes, the more specific the biological functional similarity.

Formally, a distance metric $d(x, y)$ must satisfy three properties:

1. Nonnegativity: $d(x, y) \geq 0$, with $d(x, y) = 0$ if and only if $x = y$.
2. Symmetry: $d(x, y) = d(y, x)$.
3. Satisfies the triangle inequality: $d(x, z) \leq d(x, y) + d(y, z)$.

These properties hold for both UI and LP.

14.5 Simulations

In this section, we report on simulations designed to investigate the influence of GO data in ETL mapping. These simulations are structured in order of increasing reality to examine the effects of various factors, including: heritability, correlation within biological clusters, and how well the annotations are determined.

Example 1: We assume that the true annotations are known for each transcript. We also assume that we know cluster memberships; i.e., we know which multivariate phenotypes we should be analyzing. This example will provide benchmark results under ideal assumptions. Questions of interest include:

1. How does the GO correlation within multivariate phenotype affect performance?
2. How does the degree of heritability of the ETL affect performance?

Example 2: We extend Example 1 by fixing the heritability and correlation within multivariate phenotypes, but varying the correlation between transcripts in the multivariate phenotype and those transcripts that are outside the multivariate phenotype.

Example 3: Here we assume that the true annotations are unknown and must be estimated. We are interested in determining which factors in the estimation of the annotation structure most affect performance. This scenario allows us to examine the hypothesis that the GO (biological) distance effect is more important than the clustering effect.

The simulation assumptions apply to both the mRNA measurements and the genetic context. Without loss of generality, we use the following genetic assumptions: an additive model for ETL; a two allele setting in Hardy-Weinberg Equilibrium, $p = q = .5$, where p and q are the population frequencies of the major/minor alleles; independent genetic markers; no interaction between the genotypic and the environmental variances, i.e., $\text{Cov}(V_G, V_E) = 0$; and no dominance or epistatic effects, i.e., $V_D = V_I = 0$. We assume that after suitable transformation each gene expression trait is normally distributed. For the simulation we generate two data matrices: an $n_{sub} \times n_{RNA}$ matrix of transcripts, and an $n_{sub} \times n_{SNP}$ matrix of genotypes.

14.5.1 Simulation Example 1

As a motivation for Example 1, we consider the question: What is the best case scenario for use of GO data? Specifically, when does it work better than, worse than, or about the same as many one way ANOVAs? Therefore, we begin by assuming that perfect information about the biological annotation is known for each gene, and that there is a perfect relationship between biological distance and pleiotropic effects. That is, we assume that when more than one RNA are (in part) affected by variation at the same SNP, then all of these RNAs are "close" in the biological distance sense. These ideal conditions provide a benchmark for further analyses. In the other two examples, we introduce more realistic scenarios, for example, uncertainty into the biological distances between RNA.

The simulation configuration is as follows. For each of 1,000 runs, an experiment is conducted over 100 subjects. Each subject has one SNP measured and eight RNA transcript levels measured. Further, the biological GO distances among the RNAs are assumed known. There is a pleiotropy so that four RNA transcripts are GO-correlated and influenced by the same SNP, and four biologically distant RNA transcripts are generated independently of each other, and of the SNP. This SNP is taken as an ETL for the four pleiotropic RNA. We assume that the genes where the SNP genotypes are measured do not include any genes where RNA transcripts are measured.

Two factors were altered for each set of 1,000 runs: (i) heritability of gene expression due to polymorphism at the SNP locus; (ii) biological GO distance within the pleiotropic RNA. The correlations due to annotation distance within the cluster are in $\{0, 0.25, 0.5\}$, and the levels of heritability used are in $\{0.01, 0.1, 0.2\}$. For each set of experiments, the results were produced using two methods. The base analysis consists of eight one-way ANOVAs (i.e., eight significance tests). For the alternative analysis, which takes account of the known clustering, we use one 4-dimensional MANOVA and four one-way ANOVAs, a total of five tests. We refer to the combination of ANOVA and MANOVA tests as the *hybrid method*. In both cases, results are corrected for multiple testing. After 1,000 runs, there are thus 8,000 p-values for the one way ANOVAs and 5,000 p-values for the hybrid method. For each method, the false negative and false positive rates are calculated and ROC curves are constructed from these. Estimates of the areas under the curve (AUC) were obtained for each triplet of heritability, biological distance, and analysis method. Additional low level details of the data generation may be found in Christian (2009).

14.5.2 Simulation Example 2

The set-up is the same as in Example 1 except that there is only one factor that is varied in the simulation. We vary the GO correlation between the transcripts in the multivariate phenotype and those outside it. The levels of correlation due to annotation distance between the set of expressions in the multivariate phenotype and the set outside the phenotype are in $\{0, 0.1, 0.2\}$, while the correlation due to annotation distance within the multivariate phenotype was fixed at 0.25. The level of heritability is fixed at 0.2. These values of correlation within multivariate phenotypes and heritability are chosen because these settings produce the most variation in Example 1.

14.5.3 Simulation Example 3

Example 3 mimics reality more closely than the previous two examples by making the pleiotropies unknown. It is thus required to estimate the cluster of GO-correlated gene expressions. By using estimated pleiotropies, we gain some idea regarding the amount and quality of annotation data needed to improve the quality of ETL mapping relative to results obtained from many one-way ANOVAs based only on expression data.

As in Examples 1 and 2, each of 100 subjects has one SNP measured and eight RNA transcript levels measured. We continue with four biologically distant RNA transcripts that are generated independently of each other and of the SNP, and four GO-correlated RNA transcripts affected by the same SNP. The biological distance and pleiotropy is created in the following way. The ETL defines the pleiotropy, i.e., the multivariate phenotype. For each of the 1,000 runs of the experiment, annotations are generated which reflect the desired relevant biological distances. This is done by randomly sampling annotations from areas of the ontology that are biologically close or far, as defined in the pleiotropy. Specifically, all of the annotations in the ontology are clustered, then the four closest transcripts are assigned annotations which are close to each other, while the other four transcripts' annotations are assigned to elsewhere on the ontology. From this annotation information, three clustering methods are used to estimate a pleiotropic multivariate phenotype:

1. nonparametric hierarchical clustering using expression data only;
2. nonparametric hierarchical clustering using annotation data only;
3. normal mixture model-based clustering (Pan, 2006) using both expression and annotation data.

We consider different clustering methods in order to clarify poor performance factors.

Estimating the cluster of expressions depends on GO distances between genes; we use the UI distance in the clustering algorithms. In each experiment, there are three estimates of pleiotropies. Each estimate may contain a different number of transcripts. For each experiment, we have five sets of p-values: one set of eight one-way ANOVAs, and three sets of varying dimensional MANOVAs from each method for estimating clustered transcripts.

Simulating GO-correlated data

Since GO-based simulations are relatively new in research, we provide some details on generating simulated annotations. Assume a desired pleiotropy is stated, e.g., $\{SNP_j \rightarrow RNA_1, SNP_j \rightarrow RNA_2, SNP_j \rightarrow RNA_3, SNP_j \rightarrow RNA_4\}$. First, an ontology is needed. Here, a subgraph of the (real) GO Biological Process ontology is used. The subgraph was induced by randomly sampling 50 nodes from the BP ontology. The goal is to create annotations which are biologically close for transcripts in the pleiotropy (cluster) and which are biologically far from transcripts not in the pleiotropy. There is no unique way to do this; the approach taken here is to make the assumption that each transcript will be assigned exactly one annotation on the ontology. Then the distances between all of the ontology nodes are calculated and nodes clustered using hierarchical clustering. The clustering algorithm continues until a group of ten biologically close annotations are found. The annotations for the desired pleiotropy of four transcripts are randomly sampled from the group of 10 close nodes. For each iteration, the annotations for the four RNA transcripts in the pleiotropy are sampled without replacement from a set of ten close (new) nodes. The transcripts not involved in the pleiotropy have their annotations sampled without replacement from outside of the ten biologically close annotations. This ensures that the annotation for the pleiotropy follows the assumption of coexpression within the pleiotropy. Figure 14.2 shows an example of a set of close nodes as identified by hierarchical clustering with complete linkage and the union-intersection distance metric.

Once both the annotations and SNP genotypes have been created, the RNA transcript data can be generated. The expression data (y_{ij}) for transcript j of subject i is

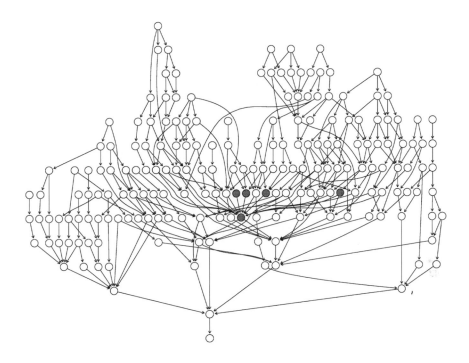

Figure 14.2 *Example 3, shading indicates biologically close annotation nodes as identified by hierarchical clustering with complete linkage and the union-intersection distance metric.*

generated according to the model

$$y_{ij} = \mu_j + \alpha_{ij} + \epsilon_{ij},$$

where μ_j is a mean; α_{ij} the ETL (SNP) effect, whose variance is denoted by σ_g^2; and ϵ_{ij} is an error term. The error terms are generated with respect to all transcripts according to a multivariate normal distribution, $\epsilon_i \sim N_{n_{RNA}}(0, \Sigma)$, with the covariance matrix representing GO-based distances (see below). The heritability (h^2) of expression (Y) attributable to polymorphism at the SNP locus is defined as $h^2 = \sigma_g^2 / (\sigma_g^2 + \sigma_e^2)$.

The main issue is ensuring that both the covariance in the transcripts and the relevant genotype effects are implemented correctly. With the heritability, allele frequencies, and variance component of the ETL fixed, the SNP effect (α_{ij}) can be easily determined for each of the three genotypes. The remaining component needed to calculate the RNA transcript values is the multivariate error term for each subject. In order to do that, it is necessary to calculate the covariance between the error terms, which reflects the biological distance in the annotations. The covariance structure is due to distances between transcript annotations. With h^2 fixed and σ_g^2 set without loss of generality to 1, we can determine σ_e^2 as

$$\sigma_e^2 = \begin{cases} \frac{\sigma_g^2}{h^2} - \sigma_g^2 = \frac{1-h^2}{h^2} & \text{when } h^2 > 0 \\ 1 & \text{when } h^2 = 0. \end{cases}$$

Since the transcripts are not independent, the correlation due to the biological information must be taken into account when generating the transcript measurements for each subject. This is done by first creating a correlation matrix based on the annotations, then transforming it to a covariance matrix using the variance of each transcript, σ_e^2.

This is still an open area of research, and here we present one approach. The covariance of biological entities located d GO units apart is denoted $C(d)$. For a full description of covariance functions, consult Schabenberger and Gotway (2005, Chapter 4). The family of covariance functions we use is limited to the Gaussian model,

$$C(d) = \sigma^2 \exp\left\{-\theta d^2\right\}.$$

The parameters can be estimated from true annotations. Note that θ is related to the biological distance at which there is no longer covariance. To create the correlation matrix among a set of transcripts, we apply a Gaussian covariance function to the biological distances between the annotations for each transcript, yielding an $n_{RNA} \times n_{RNA}$ correlation matrix, R:

$$R[i, j] = \begin{cases} 1 & \text{when } i = j \\ C(d) & \text{when } i \neq j, \end{cases}$$

where $C(d)$ is as given above. Given the correlation matrix and the variance, the covariance matrix is created as $\Sigma = \sigma_e^2 R$. A set of GO-correlated expression data can now be simulated.

14.6 Results

Each simulation experiment yields an ROC curve for each mapping method. By comparing the AUC values from the ROC curves, we can describe the hybrid method as statistically better, worse, or no different than the multiple one-way ANOVA method. The result consists of sets of scenarios where both methods work poorly, both methods work well, and where one method is better than the other.

14.6.1 Results for Example 1

Table 14.1 summarizes the AUC results by method (one-way ANOVAs and hybrid) and GO correlation. For the case with $h^2 = 0.01$, there is no difference between the one-way ANOVAs and the hybrid confidence intervals at each level of GO correlation as indicated by their standard errors. Moreover, neither method is able to find the ETL better than random chance, yielding near linear ROC curves with AUCs whose confidence intervals include 0.5. The results are not surprising as the level of heritability in this configuration is practically negligible at $h^2 = 0.01$. In this sense, this is a negative control case in which we did not expect to see detectable ETL.

Table 14.1 *Example 1. Mean AUC for ANOVA and Hybrid methods, along with standard error for the difference in means (SE Diff), for different levels of heritability and annotation correlation within GO cluster.*

Method; $\rho =$	$h^2 = 0.01$			$h^2 = 0.1$			$h^2 = 0.2$		
	0.0	0.25	0.5	0.0	0.25	0.5	0.0	0.25	0.5
ANOVA	0.50	0.49	0.50	0.61	0.60	0.60	0.87	0.86	0.85
Hybrid	0.50	0.49	0.50	0.70	0.60	0.58	0.98	0.90	0.82
SE Diff	0.01	0.01	0.01	0.01	0.01	0.01	0.005	0.006	0.007

As shown in Table 14.1 and Figure 14.3(a), gains from incorporating GO data begin to appear with a higher level of heritability of $h^2 = 0.1$. Although higher than the previous case of $h^2 = 0.01$, a 10% heritability is still low and finding genetic loci for such quantitative traits is difficult. Nevertheless, there is a noticeable difference between the one-way ANOVAs and the hybrid method. At the lowest level of within-cluster GO correlation the hybrid method has an average AUC of 69.5% compared to 60.7% with the one-way ANOVAs. At the two other levels of within-cluster correlation the two methods are practically equivalent. One other difference between the methods is that the AUC values for the one-way ANOVAs are constant over the correlations while the hybrid method has a decreasing AUC with increasing correlation. The constancy of the ANOVA AUC is expected since the the F-tests for each ANOVA are independent of one another across the three correlations. The decreasing power of the MANOVA test with increasing correlation among the multivariate component traits is also not surprising. It is known that the likelihood ratio Wilks test is sensitive to correlation in the dependent variables (Cole et al., 1994), with power decreasing (increasing) with increasing positive (negative) correlation among the variables. For moderate heritability of 20% (Table 14.1 and Figure 14.3(b)) the AUC for the hybrid method is higher at correlations of 0 (0.980 vs 0.867) and 0.25 (0.896 vs 0.856). The trend of decreasing performance as within-cluster correlation increases holds at $h^2 = 0.2$. Finally, for heritabilities above 0.5, both methods work well, yielding AUC values above 0.99 in most cases (data not shown).

14.6.2 Results for Example 2

In Example 2, the heritability and GO correlation within the cluster are fixed at $h^2 = 0.2$ and $\rho_{within} = 0.25$, respectively. We vary the correlation, $\rho_{between}$ in {0,0.1,0.2}, between the transcripts in and out of the pleiotropic cluster. As in Example 1, each of the three experiments is based on 1,000 runs, resulting in 8,000 p-values

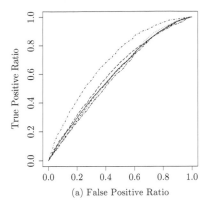
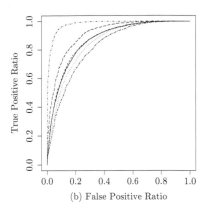

(a) False Positive Ratio   (b) False Positive Ratio

Figure 14.3 *Example 1: ROC curves for different levels of annotation correlation. (a)* $h^2 =$ 0.1. *(b)* $h^2 = 0.2$. *Most curves overlap; in both cases the highest curve corresponds to the hybrid method for* $\rho = 0.0$.

for the one way ANOVAs, and 5,000 p-values for the hybrid method. Across correlation values, mean AUC values for the ANOVA method ranged from 86.2%–86.9%, while those for the hybrid method were 90.4%–90.8%. The SE is 0.004 in all cases.

The ROC curves within each class of mapping method are virtually identical across correlation values $\rho_{between}$ (not shown). At each value of $\rho_{between}$, the hybrid method performs better than the baseline one-way ANOVAs. Unlike Example 1, the difference between the ANOVA and the hybrid methods is independent of the levels of $\rho_{between}$. The hybrid method shows a constant AUC of about 90% and the one-way ANOVAs at about 87%. The AUC is constant since the within-GO correlation is fixed in this simulation configuration.

14.6.3 Results of Example 3

Example 3 introduces uncertainty into the process by making the true GO clusters unknown and thus estimated as part of the analysis. Three methods were used to estimate the GO clusters. This results in eight one-way ANOVAs for each experiment, and four sets of p-values of varying dimensions from the hybrid method.

The various methods of estimating GO clusters are compared for $n_{sub} = 100$ simulated subjects replicated for $n_{runs} = 1,000$. Distance-based clustering methods use the union-intersection metric. Results for the case $h^2 = 0.2$ are summarized in Table 14.2. All methods yield an AUC greater than 99%. The standard errors of the non-ANOVA methods are 0.00075 for expression-only based clustering, 0.00123 for annotation and expression based clustering, and 0.00193 for annotation-only based clustering. The ordering indicates that annotation-only based clustering leads to higher variability in AUC than expression-only clustering. Annotation-only clustering is even more variable than when using both annotation and expression data together.

Transcript Clustering Method	AUC	SE
ANOVA	0.9986	0.00078
Annotation only	0.9978	0.00193
Expression only	0.9987	0.00075
Annotation and expression	0.9983	0.00123

Table 14.2 *Example 3 with $h^2 = 0.2$. Annotation-only and expression-only results based on nonparametric hierarchical clustering under complete linkage. Clustering based on both annotation and expression based on a stratified normal mixture model method (Pan, 2006).*

14.7 Conclusions

Previous studies in the analysis of gene expression data, but not ETL analysis, have demonstrated that there is merit in using GO annotation data. Clustering of gene expression profiles (Pan, 2006) results in more biologically meaningful clusters with GO data than without. However, previous studies have provided little discussion or guidance on how best to combine the GO data with expression data and how much relative improvement can be expected. Not all GO annotations, for example, are equally substantiated. Therefore, it is possible that one can introduce more noise into the analysis than is necessary.

In this chapter, we have considered the possible benefits of incorporating GO data in the statistical analysis of ETL mapping. A simple approach to QTL mapping is based on analysis of variance models. Motivated by the use of multivariate phenotypes in linkage analysis to increase power over univariate traits, we have used a similar approach. We use a MANOVA approach to ETL mapping whereby the multivariate phenotype is composed of GO-correlated gene expressions. Using simple MANOVA models allows us to avoid potentially confounding factors in more complicated approaches. In this sense, although the MANOVA approach can be used in practice, our work is more in line with a proof of concept investigation. Our simulations indicate that there can be merit to integrating GO data in an ETL analysis. Under low to moderate heritability and known or well-estimated clusters of gene expressions (the multivariate pleiotropic phenotype), a MANOVA approach can perform better than a more standard approach of multiple one-way ANOVAs. However, as the pleiotropy estimates become increasingly noisy because of weakly substantiated GO information, the performance of the MANOVA method can decline. In practice, since the truth of the GO cluster and the associated within-GO correlation are not known, we advise both the singleton one-way ANOVAs and MANOVA analyses. With highly (positively) correlated traits, the MANOVA approach may lose power.

We have also highlighted some of the issues related to working with GO data. The issue of correlating genes via biological information is a nontrivial task. Perhaps the most common way to do so is to consider the corresponding protein-protein interactions reported in assorted databases. Using Gene Ontology is another approach. A basic question, however, is how one defines correlation (or association or closeness) in this context. Here, we have discussed ideas of distances between genes in relation to their GO annotations. In turn, the distances can be used to define covariances between genes. Both GO distance and GO correlation can be defined in several ways, and the definitions used in any particular application could have a significant impact on the results. We therefore strongly suggest that alternative approaches to integrating GO data be considered when using it in practice.

CHAPTER 15

A misclassification model for inferring transcriptional regulatory networks

Ning Sun and Hongyu Zhao

15.1 Introduction

Understanding gene regulation through the underlying transcriptional regulatory networks (referred to as TRNs in the following) is a central topic in biology. A TRN can be thought of as consisting of a set of proteins (known as transcription factors), genes, small modules, and their mutual regulatory interactions. The potentially large number of components, the high connectivity among various components, and the transient stimulation in the network result in great complexity of TRNs. With the rapid advances of molecular technologies and enormous amounts of data being collected, intensive efforts have been made to dissect TRNs using data generated from the state-of-the-art technologies, including gene expression data and other data types (e.g., Chu et al. 1998; Ren et al. 2000; Davidson et al. 2002; Lee et al. 2002; Bar-Joseph et al. 2003; Zhang and Gerstein 2003). The computational methods include gene clustering (e.g., Eisen et al. 1998; Roberts et al. 2000), Boolean network modeling (e.g., Liang et al. 1998; Akutsu et al. 1999, 2000; Shmulevich et al. 2002), Bayesian network modeling (e.g., Friedman et al. 2000; Hartemink et al. 2001, 2002), differential equation systems (e.g., Gardner et al. 2003; Tegnér et al. 2003), information integration methods (e.g., Gao et al. 2004), and other approaches. For recent reviews, see de Jong (2002) and Sun and Zhao (2004). As discussed in our review (Sun and Zhao, 2004), although a large number of studies are devoted to infer TRNs from gene expression data alone, such data provide only a very limited amount of information. On the other hand, other data types, such as protein-DNA interaction data (which measure the binding targets of each transcription factor (TF) through direct biological experiments) may add more information and should be combined together for network inference.

In this chapter, we describe a Bayesian framework for TRN inference based on the combined analysis of gene expression data and protein-DNA interaction data. The statistical properties of our approach are investigated through extensive simulations, and our method is then applied to study TRNs in the yeast cell cycle.

15.2 Methods

In this chapter, we model a TRN as a bipartite graph: a one-layer network where a set of genes are regulated by a set of TFs. The TFs bind to the regulatory regions of their target genes to regulate (activate or inhibit) the transcription initiation of these genes. Transcription initiation is a principal mode of regulating the expression levels of many, if not most, genes (Carey and Smale, 1999). Because the number of genes largely exceeds the number of TFs in any organism (e.g., there are 374 TF entries in the updated TRANSFAC database (http://www.gene-regulation. com/pub/databases.html) and more than 6,000 genes in yeast), there is combinatorial control of the TFs on genes. That is, for a given gene, its expression level is controlled by the joint actions of its regulators. Two well-known concepts on the joint actions of TFs are *cooperativity*, which in the context of protein-DNA interaction refers to two or more TFs engaging in a protein-protein interaction stabilizing each other's binding to DNA sequences, and *transcriptional synergy*, which refers to the interacting effects among the Polymerase II general transcriptional machinery and the multiple TFs controlling transcription levels. In previous work (Zhao et al., 2003), we assumed that the expression level of a specific gene is controlled through the additive effects of its regulators. Liao et al. (2003) applied Hill's equation for the cooperative TF bindings on the regulatory regions of their target genes and the first-order kinetics for the rate of gene transcription. Under a quasi-steady-state assumption, they proved that the relative gene expression level has a linear relationship with the relative activities of the TFs that bind on the gene's regulatory region. In order to obtain a unique solution of the regulation matrix, they required full column-rank of the regulation and its reduced matrices. In this chapter, we extend our previous work (Zhao et al., 2003) to fully incorporate gene expression data and protein-DNA binding data to infer TRNs. Before the discussion of our model, we first give a brief overview of the protein-DNA binding data used in our method.

As the primary goal of TRN inference is to identify the regulation targets of each TF, the most direct biological approach for this goal is to experimentally identify the targets of various TFs. Many different biological methodologies are available to serve this purpose. The large-scale chromatin immunoprecipitation microarray data (ChIP-chip data) provide the *in vivo* measurements on TFs and DNA binding in yeast (Ren et al., 2000; Lee et al., 2002). In our study, the protein-DNA binding data thus collected are viewed as one measurement of the TRN with certain levels of measurement error due to biological and experimental variations, e.g., physical binding is not equivalent to regulation. We use the ChIP-chip data collected by Lee et al. (2002) as the data source for protein-DNA binding. These data represent a continuous measurement of the binding strength between each TF and its potential targets, and a p-value is derived based on replicated experiments to assess the statistical significance of binding. In our following work, the inferred binding p-values between a TF and its potential target genes are transformed into binary observations using a significance level cutoff of 0.05. That is, for all TF-gene pairs whose p-value is below 0.05, we denote the observation as 1, representing evidence for binding, and for those pairs whose p-value is larger than 0.05, we denote the observation as 0, representing insufficient evidence for binding. The reason that we utilize protein-DNA binding data is because we believe that the information from such data provides good predictions for the true underlying TRN.

In our previous work (Zhao et al., 2003), we treated protein-DNA binding data as representing the true underlying network, and used a simple linear model to describe the relationship between the transcript amounts of the genes considered and the activities of their regulators. In our current work, we extend this linear model to incorporate potential errors associated with protein-DNA binding data to integrate three components that are biologically important in transcription regulation, namely, the TRN as characterized by the covariate (or design) matrix in the linear model, protein regulation activities as defined by the predictors in the model, and gene expression levels as defined by the response variables. We propose a misclassification model to simultaneously extract information from protein-DNA binding data and gene expression data to reconstruct TRNs.

15.2.1 Model specification

Our model relating gene expression levels, TRNs, and TF activities can be described through three sub-models:

1. A linear regression model relating gene expression levels with the true underlying TRNs and regulators' activities;
2. A misclassification model relating the true underlying networks and the observed protein-DNA binding data;
3. A prior distribution on the TRNs.

The information on the measurement error can be incorporated in a flexible way into a graphical model (Richardson and Gilks, 1993; Richardson, 1999). The hierarchical structure of our graphical model is summarized in Figure 15.1. Each component is now described in detail.

Sub-model 1: Linear regression model   Let N denote the number of genes and M denote the number of TFs related to the regulation of these genes. We consider a total number T of gene expression experiments, where these experiments may represent a time course study, e.g., yeast cell cycle studies, or different knock out experiments. We focus on time course experiments in the following discussion. In this case, t represents a specific time point.

The observed gene expression levels at time t, $\mathbf{Y}_t$, are represented as a vector of N expression levels normalized over all time points for each gene i and serve as the

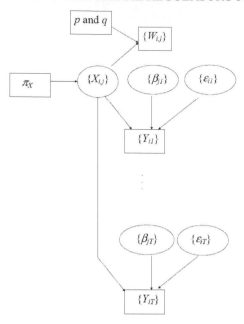

Figure 15.1 *The hierarchical structure of the misclassification model. Unknown parameters are in the ovals, and known parameters are in the rectangles.*

response in the linear model (15.1) with the following form:

$$\mathbf{Y}_t = \mathbf{X}\beta_t + \epsilon_t, \tag{15.1}$$

$$\epsilon_{it} \sim N(0, \sigma_t^2), \tag{15.2}$$

where $\mathbf{X}$ represents the true TRN, β represents the time-dependent regulator activities of the M TFs, and ϵ_t represents the errors that are associated with gene expression measurements. In the matrix $\mathbf{X}$, each row corresponds to a gene and each column corresponds to a TF. Therefore, the (i,j) entry in this matrix represents the regulation pattern of the jth TF to the ith gene. The value of this entry is 1 if the jth TF affects the transcription level of the ith gene, and the value is 0 otherwise. Therefore, when the primary interest is to infer the TRN, the overall objective is to infer the values in this matrix, either 0 or 1.

This model states that (1) the expression level of a gene is largely controlled by the additive regulation activities of its regulators, (2) the same regulator has the same relative effect on all its targets, (3) the TRN is identical across all time points, and (4) the errors associated with gene expression measurements have the same distribution across all the genes. We note that these assumptions are simplistic and may only provide a first-order approximation to reality. This model has nevertheless (implicitly or explicitly) been used by many research groups to successfully analyze TRNs.

The limitations and modifications of these assumptions are further discussed in the Summary (Section 15.5).

Because protein-DNA binding data are often obtained from a mixture of biological samples across all the time points (e.g., asynchronized cells), they measure an averaged protein-DNA binding over the whole cell cycle. Although we may use the time course gene expression data to investigate the fluctuation of the network over time, the information at one time point may not be sufficient for statistical inference (see results in the simulation study, Section 15.3). Therefore, we make the assumption that the network is time-independent and combine the information across time points. Consequently, the variation of the response variable, gene expression, across time points is accredited to the change in activities of the TFs, β_t. With the given activities of the predictors, the TRN of gene i ($\mathbf{X}_i$) is independent of the network of any other gene $\mathbf{X}_{i'}$, where $i' = 1, 2, ..., (i-1), (i+1), ..., N$.

Sub-model 2: Misclassification model

In our model setup, both the true and observed covariates are binary, where 0 corresponds to no regulation and 1 corresponds to regulation. We assume the following model (15.3-15.6):

$$
\begin{aligned}
P(W_{ij} &= 1|X_{ij} = 1) = 1 - p & (15.3)\\
P(W_{ij} &= 0|X_{ij} = 1) = p & (15.4)\\
P(W_{ij} &= 0|X_{ij} = 0) = 1 - q & (15.5)\\
P(W_{ij} &= 1|X_{ij} = 0) = q, & (15.6)
\end{aligned}
$$

where the values of p and q are the false negative and false positive rates, respectively, of the protein-DNA data. In practice, these values may be directly estimated from some control experiments, thus we treat these parameters as known or prior information in the misclassification model and specify their values. In the case where these values may not be precisely known, we also study the robustness of their misspecifications on statistical inference. Note that the false positive and false negative rates may be gene-TF specific; therefore, our assumption here represents a first-order approximation to reality that may need further extension in future studies. The binary binding matrix $\mathbf{W}$ serves as the measurement for the true TRN $\mathbf{X}$.

Sub-model 3: Exposure model

For this submodel, we need to specify the prior distribution of the regulatory matrix $\mathbf{X}$. The prior distribution of $\mathbf{X}$ (π_X) describes the probability of X_{ij} being 1, where X_{ij} represents the regulation between TF j and gene i. We assume that the X_{ij} are independent and have an identical distribution π_X. For a given true network $\mathbf{X}$, the value of π_X can be calculated from the data. When $\mathbf{X}$ is unknown and $\mathbf{W}$ serves as the surrogate of $\mathbf{X}$, π_X is a model parameter to be specified.

15.2.2 MCMC algorithm for statistical inference

In our model setup, a large number of unknown parameters $\{\mathbf{X}, \beta_t, \sigma_t^2\}$ need to be inferred based on the observations $\mathbf{Y}_t$, $t=1, , T$, and $\mathbf{W}$. We propose to use the Gibbs sampler for statistical inference (Gilks et al., 1995). The Gibbs sampler is alternated between two steps: (1) sample $\{\beta_t, \sigma_t^2\}$ conditional on $\mathbf{X}$, and (2) sample $\mathbf{X}$ conditional on $\{\beta_t, \sigma_t^2\}$. These two steps are described in detail in the following.

Step 1: Given a current estimate of $\mathbf{X}$, the model reduces to a standard linear regression model. The parameters $\{\beta_t, \sigma_t^2\}$ are sampled through the distributions

$$\sigma_t^2 \sim \text{ Inverse } \chi^2(df, s_t^2) \tag{15.7}$$

$$\beta_t \sim N(\widehat{\beta}_t, \mathbf{V}_\beta \sigma_t^2), \tag{15.8}$$

where $df = N - M$, $\mathbf{V}_\beta = (\widehat{\mathbf{X}}^T \widehat{\mathbf{X}})^{-1}$, $\widehat{\beta}_t = \mathbf{V}_\beta \widehat{\mathbf{X}}^T Y_t$, and s_t is the sample standard deviation. The matrix is the current estimate for the TRN.

Step 2: Given current estimates of $\{\beta_t, \sigma_t^2\}$, we update the TRN individually for each gene. If there are M TFs, a total of $K = 2^M$ combined patterns among the TFs to jointly regulate a specific gene are possible. The likelihood L_{ik} for each pattern k can be evaluated as

$$L_{ik} = L_{ik}^X + L_{ik}^Y, \tag{15.9}$$

where

$$L_{ik}^X = n_1 \log \pi_X \quad + \quad n_{11} \log(1 - p) + n_{10} \log p + n_0 \log(1 - \pi_X) \tag{15.10}$$
$$+ \quad n_{01} \log q + n_{00} \log(1 - q)$$

$$\text{and} \quad L_{ik}^Y = -\sum_{t=1}^T \frac{(Y_{it} - \widehat{Y_{ikt}})^2}{2\sigma_t^2}. \tag{15.11}$$

In the above expressions, L_{ik}^X and L_{ik}^Y represent the likelihood contributions from the protein-DNA binding data (X) and the expression data (Y), respectively. In the expression for L_{ik}^X, n_{so} represents the number of TF-gene pairs whose true regulation is s and the observed binding is o, where the values of s and o are 0 or 1. For example, n_{11} corresponds to the number of pairs whose true regulation and observed binding are both 1, $n_1 = n_{10} + n_{11}$, and $n_0 = n_{00} + n_{01}$. The expression for L_{ik}^Y represents the likelihood component derived from gene expression data across all time points. After evaluating the log-likelihood for all the patterns, we sample one pattern based on the following multinomial distribution:

$$L_{ik}^Y \sim \text{multinomial}\left(1, \frac{\exp(L_{ik})}{\sum_{k=1}^K \exp(L_{ik})}\right). \tag{15.12}$$

Therefore, in the updating of the TRN, our algorithm does an exhaustive search over all possible network patterns for each gene, sampling a specific network based on the relative likelihood of all possible networks. We repeat this search for each of the N genes to obtain the updated $\hat{\mathbf{X}}$ for the next iteration.

Based on the sampled parameter values, we can derive posterior distributions for all the unknown parameters. For example, we can obtain the inferred TRN describing the binding between the jth TF and the ith gene through the marginal posterior distribution, i.e. the proportion of samples that the value of X_{ij} is 1. These posterior probabilities can then be used to infer the presence or absence of regulation through specifying a cutoff value, e.g., 0.5, such that all the entries below this cutoff are inferred not to have regulation effect, whereas all the entries having values above this cutoff are inferred to have regulation.

15.2.3 Data analysis and simulation setup

Since our simulation model is based on the real data to be analyzed, we describe the data sources first. Based on the literature, we select eight important cell cycle TFs: Fkh1, Fkh2, Ndd1, Mcm1, Ace2, Swi5, Mbp1, and Swi4. Based on protein-DNA interaction data reported in Lee et al. (2002), we obtain a binary binding matrix for these regulators and all yeast genes. The binary observation is obtained by applying a 0.05 p-value cutoff to the p-values reported by Lee et al. (2002). We then remove those genes with no *in vivo* binding evidence with any of the eight TFs from the binding matrix, and further focus only on yeast cell cycle genes defined by Spellman et al. (1998). These steps result in a total of 295 genes to be analyzed, and the observed protein-DNA binding matrix has a dimension of 295 (genes) by 8 (TFs). For gene expression data, we use the α arrest cell cycle data with 18 time points collected by Spellman et al. (1998).

Now we describe the setup used to conduct simulation experiments to evaluate the performance of our proposed procedure. For the simulation model, we need to specify (1) the true TRN, (2) true protein regulation activities, (3) false positive and false negative rates in the observed binding matrix, and (4) measurement errors associated with microarray data. We consider all 295 genes used in the real data analysis, and select five TFs (Fkh2, Mcm1, Ace2, Mbp1, and Swi4, which are reported to control the gene expression at the four cell cycle stages) out of the total eight in our simulations to simplify the analysis and summary. For the specification of the "true" TRN in our simulations, we use the observed binding data to represent the true TRN. As for TF activity specifications, we estimate the activities of the chosen five TFs from the linear regression model using the above "true" TRN and the expression levels of all 295 genes at each time point. The activity levels of the five TFs over 18 time points are shown in Figure 15.2. We vary the false positive and false negative rates from 0.1 to 0.9 to examine their effects on statistical inference. Finally, we assess the effect of the measurement variation associated with microarray data on statistical inference. For the majority of simulations, we assume that the microarray data are collected from 18 time points as in Spellman et al. (1998). In one case, we vary the number of time points available to investigate the effect of the number of time points on statistical inference.

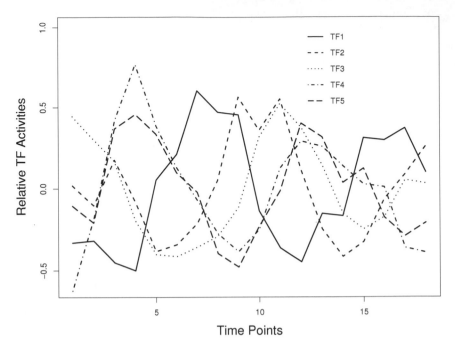

Figure 15.2 *The activities of five transcription factors vary over 18 time points. Two of the five transcription factors share similar variation, which may lead to an identifiability problem in the model. However, our results show that the slight difference between the TF activities is sufficient to avoid the problem.*

15.3 Simulation results

15.3.1 Convergence diagnosis of the MCMC procedure

Based on our simulation runs, we generally find good mixing of the proposed MCMC procedure. Both the traces of the parameter values and the autocorrelation of the parameter curves indicate that a burn-in run of 1,000 iterations out of 10,000 iterations is stable enough to obtain reliable posterior distributions. The posterior distributions of the five TF activities (β_t) and measure variations from microarrays σ_t^2 at a time point from a randomly chosen simulated data set are shown in Figure 15.3. We also investigate the effect of the initial network (covariate matrix) on MCMC results. When the measurement errors in gene expression data are low, the MCMC procedure has good convergence regardless of the initial network. In general, the observed protein-DNA binding data provide a good starting point for statistical inference.

In our model specification, there are two types of errors: the errors associated with the measured gene expression levels (responses, denoted by σ) and those associated with the observed protein-DNA binding data (denoted by p and q). In order to systematically investigate the effect of both types of errors, we consider seven pairs of p

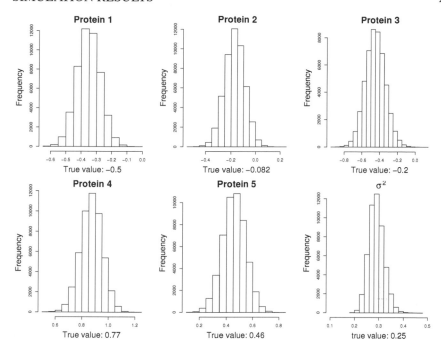

Figure 15.3 *Histograms of posterior distributions for the model parameters β_t and σ_t^2 at t = 4. The standard deviations of these posterior distributions are 0.075, 0.078, 0.092, 0.077, 0.091, and 0.027, respectively.*

and q: (0.1,0.1), (0.2,0.2), (0.2,0.4), (0.4,0.2), (0.3,0.3), (0.4,0.4), and (0.5,0.5). For each pair of p and q values, we simulate the observed protein-DNA binding data as well as gene expression data under 22 different σ values, ranging from 0.001 to 1.5. For each specification of the $22 \times 7 = 154$ sets of parameter values, we simulate data sets consisting of protein-DNA interaction data and gene expression data. Each data set is analyzed through our proposed MCMC approach with a burn-in of 1,000 iterations and a further run of 5,000 iterations. The posterior distribution for each unknown parameter is summarized and compared to the true underlying network. We use a cutoff of 0.5 to infer the presence or absence of interactions between TFs and genes. The inferred network is then compared to the true network to calculate the proportion of false positive and false negative inferences for each TF-gene pair. The overall false positive and false negative rates are then estimated through the average of all TF-gene pairs across all the simulated data sets. The results are summarized in Figure 15.4. In Figure 15.4(a), we plot the false positive rates for the inferred network. As can be seen from this figure, the false positive rates for the inferred network increase as σ, p, and q increase. The false negative rates for the inferred networks show a similar pattern. The major feature is that the information from gene expression data may significantly improve the estimation on **X**. When σ is small and p and q are not too high, there is a very good chance that the true network can be recovered from the joint analysis of gene expression data and protein-DNA binding data. For example, with 30% false positive and 30% false negative rates, when σ is less than 0.2, the whole network may be fully recovered. Even when σ is large, the false positive rates in the inferred network using both binding data and gene expression data still outperform the false positive rates in the observed protein-DNA expression data, i.e. gene expression data are not considered in the inference. The results for the false negative rates as shown in Figure 15.4(b) show similar patterns.

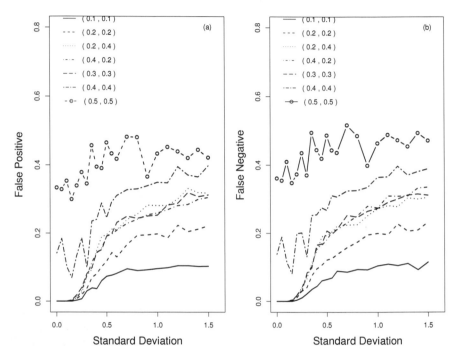

Figure 15.4 *False positive and false negative rates of the inferred network. The x-axis is the standard deviation in the gene expression data, while the y-axis is either the false positive rate or false negative of the posterior network with respect to the true regulatory network in the cell cycle. Different lines correspond to different levels of quality of the protein-DNA binding data.*

15.3.2 Misspecification of the model parameters p, q, and π_X

In the results summarized above, we assume that the true values of p and q are precisely known to us. However, their exact values may not be accurately inferred. Therefore, we conduct simulation experiments to examine the performance of the proposed procedure when the values of p and q are misspecified. In this set of simulations, we simulate data from three sets of p and q values: (0.1,0.1), (0.3,0.3), and (0.2,0.4). For each simulated data set under a given set of parameter values, we perform statistical analysis under different sets of specifications for p and q, including: (0.9,0.9), (0.8,0.8), (0.7,0.7), (0.6,0.6), (0.5,0.5), (0.4,0.4), (0.3,0.3), (0.2,0.2), (0.1,0.1), (0.05, 0.05), (0.01,0.01), and (0.05, 0.4). Throughout these simulations, we assume $\sigma = 0.2$. The performance of our procedure in terms of false positive and false negative rates is summarized in Figures 15.5(a)–(c). These results suggest that the statistical inference is robust to the misspecification of the parameters p and q when the specified values are not too distinct from the true parameter values. We observe similar patterns for other values of σ.

Another parameter that needs to be specified in our approach is π_X, the prior probability that there is an interaction between a TF and a gene. We further investigate the performance of our approach when π_X is misspecified. The true value of π_X is about 0.46 $(683/(295 \times 5))$, where there are 683 regulation pairs in the protein-DNA binding data) in the given true network $\mathbf{X}$, but we consider values 0.1, 0.2, 0.3, 0.4, 0.46, 0.5, 0.6, 0.7, 0.8, and 0.9 in the specification of π_X in our analysis. The results are summarized in Figure 15.5(d). Compared to the results for p and q, statistical inference is more sensitive to the value of π_X. However, when the specified parameter value is reasonably close to the true value, our approach yields generally robust estimates.

Overall, our simulation studies suggest that misspecification of model parameters p, q, and π_X within a reasonable range will not substantially affect the statistical inference of the true network.

15.3.3 Effect of the number of experiments used in the inference

In the above simulations, we simulate data from 18 time points and use all of them in the inference of the underlying network. In this subsection, we consider the effect of the number of time points on the inference. For this set of simulations, we simulate the protein-DNA binding data by fixing the values of p and q at 0.1, select the value of σ at 0.001, 0.2, and 0.5, and vary the number of time points used in the analysis from 1 to 18. When there is little error associated with gene expression data, i.e., $\sigma = 0.001$, the data at one time point can carry enough information to fully recover the true network. With increasing σ values, the number of time points affects the results on the inferred network (Figure 15.6). When σ is 0.2, our previous results show that there is a significant improvement of the inferred network from the binding data. As more time points are included in the analysis, we observe a more accurate inference of the underlying network. When σ is 0.5, the improvement of the inferred network from the binding data is still obvious but limited by too much noise in gene expression data.

15.4 Application to yeast cell cycle data

In this section, we apply our method to jointly analyze gene expression data from 295 genes over 18 time points (Spellman et al., 1998) and protein-DNA binding

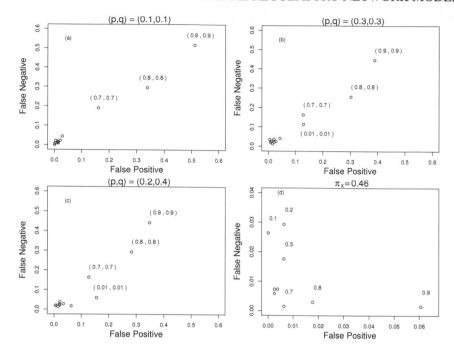

Figure 15.5 *The effects of the misspecification of the model parameters p, q, and π_X on the inferred network. The standard deviation of the simulated gene expression data is 0.2. The real values of parameters (p,q) or π_X are indicated in the title of each plot. In the first three plots, the true value of π_X is 0.46, but (p,q) are specified as (0.9,0.9), (0.8,0.8), (0.7,0.7), (0.6,0.6), (0.5,0.5), (0.4,0.4), (0.3,0.3), (0.2,0.2), (0.1,0.1), (0.05, 0.05), (0.01,0.01), and (0.05, 0.4). For the last plot, the values of (p, q) are (0.1,0.1), but π_X is specified at various levels: 0.1, 0.2, 0.3, 0.4, 0.46, 0.5, 0.6, 0.7, 0.8, and 0.9.*

data of Fkh1, Fkh2, Ndd1, Mcm1, Swi5, Ace2, Mbp1, and Swi4 (Lee et al., 2002). We consider eight sets of model parameters for $\{p, q, \pi_X\}$: {0.1, 0.1, 0.5}, {0.2, 0.2, 0.5}, {0.2, 0.1, 0.5}, {0.1, 0.2, 0.5}, {0.2, 0.2, 0.4}, {0.2, 0.2, 0.6}, {0.1, 0.1, 0.4}, and {0.1, 0.1, 0.6}. For each set of parameter specifications, we run MCMC with a burn-in of 1,000 runs and an additional 5,000 runs to obtain the posterior distributions for the parameters of interest. The overall inference is based on the average posterior probabilities over the eight model parameter settings, which yield similar results among different settings.

The posterior distributions of the protein activities for the eight TFs and the σ at every time point are summarized in Table 15.1. The average value of σ across 18 time points is about 0.55. Based on our simulation studies, at this level of expression errors, the incorporation of gene expression data should improve the inference of TRNs.

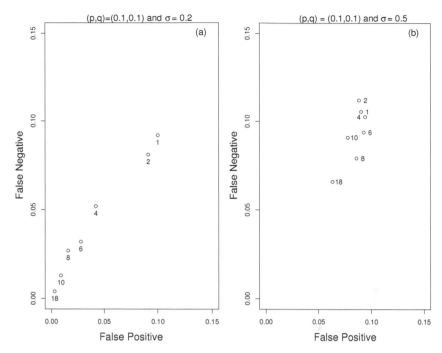

Figure 15.6 *The effect of sample size on the inferred network. The number beside each symbol indicates the number of the time points used in the simulated gene expression data. The value of π_X is 0.46, and the values of other parameters are indicated in the title of each plot.*

15.5 Summary

In this chapter, we have developed a misclassification model to integrate gene expression data and protein-DNA binding data to infer TRNs. Compared to other models, our model (1) integrates gene expression data and protein-DNA binding data through a consistent framework, (2) considers the misclassification associated with protein-DNA binding data explicitly, and (3) consists of a flexible model structure. The systematic simulation results indicate that this model performs well in the reconstruction of the underlying networks when the misclassification associated with gene expression data and (more importantly) protein-DNA binding data are within reasonable ranges. For example, in the case of less than 30% to 40% false positive and false negative rates in the observed binding data, our method may significantly reduce both types of error rates in the inferred network when the standard deviation in gene expression measurements is around 0.5 or less. In all the cases, the inclusion of gene expression data leads to improved inference of the underlying network compared to that solely based on the binding data even when the measurement error in gene expression data is very high.

In this chapter, we have considered five TFs in simulation studies and eight TFs in the application to the yeast cell cycle data. Because there are 133 TFs in yeast

Table 15.1 *Estimates (±) of transcription factor regulation activities and σ based on our model.*

Time Point	Fkh1	Fkh2	Ndd1	Mcm1	Ace2	Swi5	Mbp1	Swi4	σ
1	0.09	-0.81	-0.55	0.54	1.84	-0.29	-0.79	-0.27	0.88
	±0.13	±0.12	±0.13	±0.13	±0.14	±0.13	±0.12	±0.12	
2	-0.36	-1.00	0.24	0.28	1.18	-0.46	-0.18	-0.01	0.75
	±0.11	±0.11	±0.11	±0.11	±0.13	±0.12	±0.10	±0.11	
3	-0.53	-0.63	0.14	0.09	0.98	-0.35	1.43	0.06	0.66
	±0.10	±0.10	±0.10	±0.10	±0.14	±0.11	±0.09	±0.10	
4	-0.34	-0.31	-0.25	-0.29	0.17	-0.42	1.86	0.27	0.58
	±0.08	±0.09	±0.09	±0.08	±0.13	±0.10	±0.07	±0.08	
5	0.73	0.12	-0.62	-0.63	0.26	-0.67	0.79	0.13	0.54
	±0.07	±0.08	±0.08	±0.07	±0.09	±0.08	±0.07	±0.08	
6	0.72	0.20	-0.42	-0.49	-0.17	-0.49	0.28	-0.04	0.6
	±0.08	±0.08	±0.09	±0.08	±0.10	±0.09	±0.08	±0.08	
7	1.31	0.16	0.41	-0.61	-0.07	-0.55	-0.28	-0.28	0.53
	±0.08	±0.09	±0.09	±0.08	±0.10	±0.09	±0.08	±0.08	
8	0.44	0.18	0.61	0.01	-0.47	-0.31	-0.43	-0.57	0.44
	±0.06	±0.06	±0.06	±0.06	±0.08	±0.07	±0.06	±0.06	
9	0.17	0.09	1.03	0.58	-0.46	-0.00	-0.57	-0.74	0.5
	±0.07	±0.07	±0.07	±0.07	±0.09	±0.08	±0.07	±0.07	
10	-0.27	-0.48	0.81	0.47	-0.54	1.11	-0.39	-0.42	0.57
	±0.07	±0.08	±0.07	±0.07	±0.10	±0.08	±0.07	±0.07	
11	-0.90	0.02	-0.01	0.79	-0.32	1.23	0.13	0.08	0.75
	±0.10	±0.11	±0.11	±0.10	±0.13	±0.12	±0.10	±0.11	
12	-1.07	0.22	-0.29	0.14	-0.45	0.93	0.56	0.65	0.44
	±0.07	±0.06	±0.07	±0.06	±0.08	±0.07	±0.07	±0.06	
13	-0.20	0.44	-0.82	-0.28	-0.15	0.35	0.16	0.63	0.45
	±0.07	±0.07	±0.07	±0.06	±0.08	±0.07	±0.06	±0.06	
14	-0.35	0.42	-0.68	-0.37	-0.31	-0.08	-0.31	0.52	0.45
	±0.06	±0.07	±0.07	±0.07	±0.07	±0.07	±0.06	±0.06	
15	0.44	0.68	-0.61	-0.51	-0.08	-0.32	-0.44	0.38	0.44
	±0.06	±0.07	±0.07	±0.06	±0.08	±0.07	±0.06	±0.07	
16	0.09	0.59	-0.10	-0.16	-0.58	-0.04	-0.45	0.13	0.6
	±0.08	±0.08	±0.08	±0.08	±0.10	±0.09	±0.07	±0.08	
17	0.26	0.26	0.46	-0.02	-0.27	-0.08	-0.71	-0.26	0.62
	±0.08	±0.09	±0.09	±0.08	±0.10	±0.09	±0.07	±0.08	
18	-0.20	-0.15	0.66	0.48	-0.57	0.44	-0.63	-0.26	0.57
	±0.08	±0.09	±0.09	±0.08	±0.10	±0.10	±0.07	±0.08	

protein-DNA binding data, the inclusion of all TFs in the same model will create both statistical and computation challenges. In the context of yeast cell cycle data, protein-DNA binding data suggest that close to 20 TFs may be involved in the regulation of cell cycle genes (data not shown). From this study, we have found that (1) protein-DNA binding data can serve as a good starting point in the proposed MCMC procedure, and (2) the larger the number of gene expression data sets used, the more accurately we expect our procedure to perform, especially when the gene expression data have low to moderate measurement errors. Therefore, in general, when the number of TFs increases, we hope to collect more samples on relevant gene expressions. More samples can be achieved by increasing the number of experimental conditions or the number of replicates per experimental condition or both. The advantage of increasing the number of experimental conditions is to introduce more variation in TF activity profiles so as to better infer the underlying network. However, more parameters are needed to specify the model for the additional conditions. We also need to be cautious on how to pool the experiments to infer the TRN. In this work, we have assumed a time-independent TRN throughout the yeast cell cycle. This assumption may be true in this context and it allows us to pool information from across all time points. However, the TRN may differ under varying conditions, and the transient behavior of the TRN needs to be taken into account when using all the microarray data. The advantage of increasing the number of replicates per condition is to reduce errors associated with measured gene expression levels at each point without introducing more model parameters. In this study, the replicates were not included in the model setup, however, the flexible structure of our model allows an easy incorporation of such information into the model.

In our simulation studies, we have investigated the sensitivity of our method when some of the model parameters are misspecified, including the prior distribution on the network connections and our belief (measured by p and q) on the quality of protein-DNA binding data. We found that the method is not sensitive to the misspecifications of these model parameters unless the specified model parameters are drastically different from the true model parameters. In the analysis of yeast cell cycle data, we considered eight sets of model parameters and observed general agreements among results from different parameter specifications. In practice, we may take a full Bayesian approach to inferring the network through averaging inferred networks under certain prior distributions for the model parameters.

As discussed above, although we have treated the observed protein-DNA binding data as a 0-1 variable, the observed data are, in fact, continuous. In this case, our model can be modified within the measurement model framework so that the measured and true covariate values are continuous. To specify the prior distribution for the covariate values, we may use normal mixtures or more sophisticated models for the binding intensity. However, the interpretation of the model parameters will be somewhat different if the intensity levels are used because the parameter β_t cannot be simply interpreted as TF activities.

In our model setup, we assume that all the TFs act additively to affect the transcription levels of their target genes. This linear relationship between TF activities and the normalized expression levels is a key assumption for this model. Because of the complexity in transcription regulation, such as synergistic effects among TFs, a linear model can serve as an approximation at best. Nevertheless, linear models have been used in this context by several authors (Bussemaker et al., 2001; Liu et al., 2002; Wang et al., 2002; Liao et al., 2003; Gao et al., 2004). The potential departure from linearity may result from synergistic regulation effects of TFs bound to the upstream region of the same gene. We are therefore in the process of developing statistical approaches for analyzing nonlinear models.

To conclude, we note that our model can be extended in different ways to be more comprehensive and better represent the underlying biological mechanisms. For example, the linear form of the model can be extended to incorporate nonlinear inter-actions among different TFs as discussed above; the replicates per experiment can be incorporated into the model to improve the data quality; more prior information or more sophisticated statistical models can be used to construct the prior distribution of the network (π_X). In addition, our general framework has the potential to inte-grate more data types into the model, such as sequence data and mRNA decay data to further infer the transcriptional regulatory networks.

Acknowledgments

This work was supported in part by National Science Foundation grant DMS-0241160.

CHAPTER 16

Data integration for the study of protein interactions

Fengzhu Sun, Ting Chen, Minghua Deng, Hyunju Lee, and Zhidong Tu

16.1 Introduction

In recent years, an increasing number of genomes of model organisms have been sequenced. Using these genomic sequences, researchers have been able to make tremendous progress in the study of genomes, with numerous successes in the identification of genes, the detection of protein-binding DNA motifs, and the determination of gene regulation. Beyond these successes is the far more challenging and rewarding task of understanding proteomes by means of, e.g., (1) discovering signal transduction pathways, (2) determining protein structures, (3) detecting protein-protein, protein-DNA, and protein-metabolite interactions, (4) detecting post-translational modifications of proteins, and ultimately (5) elucidating the functions of genes and their protein products.

Unlike a genome, which is a stable feature of an organism, a proteome varies with the state of development, the tissue, and the environment. Among the many features of a protein, its interaction with other proteins is one of the most important aspects of its function. Traditionally, protein interactions have been studied individually by biochemical techniques. However, the speed of discovering new interactions increased dramatically in the last few years. Several high throughput techniques have produced a total of about 80,000 interactions between yeast proteins, which constitute a rough view of the actual protein-protein interaction network. The successful methods include yeast two-hybrid assays (Uetz et al., 2000; Ito et al., 2000, 2001), protein complex purification-mass spectrometry (Gavin et al., 2002; Ho et al., 2002), microarray gene expression profiles (Eisen et al., 1998), genetic interactions (Tong et al., 2002; Mewes et al., 2002), and computationally predicted protein associations (Enright et al., 1999; Marcotte et al., 1999a,b). These protein interactions will be very useful to study gene regulatory networks, pathways, as well as functions of proteins. To understand the interaction network and its applications for protein function prediction, it is essential to design a joint approach using tools from mathematics, statistics, computer science, and molecular biology. In recent years, several groups have developed computational tools to analyze and compare the different interaction data sets.

Two issues are important in assessing the usefulness of an experimentally observed protein-protein interaction data set. One is the *reliability*, defined as the fraction of real protein-protein interactions in the observed interactions. The other is the *coverage*, defined as the fraction of real interactions in the observed data over all the real interactions. A database of high coverage is not very useful if its reliability is low. Results of comparative analysis of multiple data sets have shown significantly different coverage and reliability for each technique (Deane et al., 2002; Mering et al., 2002; Mrowka et al., 2001). In this chapter, we review methods to study the following two problems:

1. Estimate the reliability of a putative observed interaction data set.
2. Give a score that a pair of proteins interact by combining different data sources.

Assigning functions to novel proteins is one of the most important problems in the post-genomic era. Many researchers have undertaken the task of functionally analyzing one of the most well-studied species, the yeast genome, comprising approximately 6,000 proteins, of which roughly 12% do not have known functions (Mewes et al., 2002), and those that do most likely have many other unknown functions. The annotation of the yeast genome will have a great impact on genomes of higher organisms such as the human genome: new genes can be annotated through their homologous yeast genes.

Several approaches have been applied to assign functions to genes, including analyzing gene expression patterns, phylogenetic profiles, protein fusions and protein-protein interactions. Gene expression analysis can cluster genes based on similar expression patterns. This makes it possible to assign a biological function to genes, depending on the knowledge of the functions of other genes in the cluster (Eisen et al., 1998). However, expression profiling gives an indirect measure of a gene product's biological and cellular function, because many cellular processes and biochemical events are ultimately achieved by interactions of proteins. A more complete study of protein functions can be achieved by looking at not only the mRNA levels but also the protein interaction network. We review the following methods for protein function prediction:

1. A Markov random field (MRF) model for assigning functions to proteins using highly reliable protein-protein interaction data and other data sources, including gene expression profiles, protein sequence similarities, and features of individual proteins, and correlations of protein functions.
2. The use of support vector machine (SVM) for protein function prediction combining different data sources.
3. A kernel-based MRF model for protein function prediction.

The remainder of the chapter is divided into two major sections: one on estimating the reliability of observed putative protein interactions, the other on predicting protein functions based on reliable protein interactions and other data sources. Finally, we discuss the connections of the two topics and future research questions.

16.2 Data sources

Protein interactions. These have traditionally been studied individually by genetic, biochemical, and biophysical techniques. However, these techniques are generally labor intensive and cannot keep up with the speed with which new proteins are discovered. Several high throughput methods for the detection of protein interactions have been developed. These include the yeast two-hybrid assays (Uetz et al., 2000; Ito et al., 2000, 2001), mass spectrometry (Gavin et al., 2002) and gene knockouts (Tong et al., 2002). *In silico* (computational) methods for interaction prediction include the chromosomal proximity method (Overbeek et al., 2000), the gene fusion method (Enright et al., 1999; Marcotte et al., 1999a), the phylogenetic method (Pellegrini et al., 1999), and the combined method (Marcotte et al., 1999b; Pavlidis and Weston, 2001; Zheng et al., 2003). Several databases have been developed to collect different sources of protein interaction data, including:

- MIPS: Munich Information Center for Protein Sequences (Mewes et al., 2002)
  `http://mips.gsf.de/`
- DIP: Database of Interacting Proteins (Xenarios et al., 2002)
  `http://dip.doe-mbi.ucla.edu/`
- BOND: Biomolecular Object Network Databank (Bader et al., 2003)
  `http://www.bind.ca/`
- BioGRID: General Repository for Interaction Datasets (Breitkreutz et al., 2003)
  `http://www.thebiogrid.org/`.

Gene expressions. These are widely used to study relationships between proteins. It is generally believed that the members of an interacting protein pair are more likely to be coexpressed than random protein pairs and thus that gene expression data can be useful for evaluating the reliability of protein interaction data as well as the probability that two proteins interact. It is also generally believed that if two proteins are highly correlated, they are more likely to have similar functions. Therefore, gene expression data can also be useful for protein function prediction. For this study, we use the yeast cell cycle gene expression data from Spellman et al. (1998). Other gene expression data can also be used.

Protein localizations. Proteins belong to different localizations in the cell, with proteins within the same locations being more likely to interact. Therefore, protein localization data can be useful for predicting protein interactions. We use the protein localization data of Huh et al. (2003) in this study.

Domains. The amino acid sequence of a protein is extremely important for the proteins function. The sequence of a protein determines its secondary and tertiary structure and thus determines its interaction partners and its biological functions. Protein domains are conserved regions of peptide sequences with relatively independent tertiary structures and represent important features for understanding protein function. We use Pfam domains as the source of domain information. `SwissPfam` (version 7.5) (`http://pfam.jouy.inra.fr/swisspfam.shtml`) defines the mapping between proteins Swiss-Prot/TrEMBL accession numbers and Pfam domains.

Gene Ontology (GO). The Gene Ontology (`http://www.geneontology.org`) describes gene products (proteins or RNA) based on three principles: Biological Process, Molecular Function, and Cellular Component. GO has a directed acyclic graph (DAG) structure. The high level categories are more general and contain many more genes than low level categories. We base protein function prediction on the known gene annotation given in GO. Shaw (Chapter 11 in this volume) provides a detailed overview of the Gene Ontology.

All the databases listed above are publicly available.

16.3 Assessing the reliability of protein interaction data

Many protein interaction data sets generated from different laboratories using a variety of techniques are available. It is difficult to compare interaction data from different sources because the varied conditions and experimental techniques may not detect the same type of interactions. Another difficulty comes from the fact that the true interactions are unknown. Without knowing the true interactions, it is difficult to study the coverage of a certain observed interaction data set. On the other hand, it is possible to study the reliability of an observed interaction data set using gene expressions and localizations.

Mrowka et al. (2001) first observed that the distribution of correlation coefficients of gene expressions for true interacting protein pairs is stochastically larger than that for random protein pairs. The distribution of gene expression correlation coefficients for observed interacting protein pairs from high throughput yeast two-hybrid assays is between that for random protein pairs and that for true interaction pairs. The observations indicate that the set of observed protein interactions from high throughput experiment is a mixture of random protein pairs and true interaction pairs. Several problems are of interest:

1. How do we choose the true interaction set (the gold standard)?
2. How do we estimate the fraction of true interactions among a set of observed interactions?
3. Is it possible to give a reliability score for an individual observed interaction?

16.3.1 Estimating the reliability of putative protein interactions based on gene expression

There is no consensus on the gold standard set of true protein interactions. Mrowka et al. (2001) used MIPS physical interactions (excluding those from high through-put experiments) as the gold standard. They used a bootstrap method to count how many random pairs need to be added to the reference data such that it has the same statistical behavior of gene expression correlation coefficients as that of the observed protein-protein interaction data, and then estimate the reliability using the sampling data. On the other hand, Deane et al. (2002) used INT, a subset of DIP interactions which are derived from small-scale experiments, as the gold standard for real inter-actions. They formalized the above idea assuming that the distribution of the square of Euclidian distance between expression profiles of putative interacting pairs is a mixture of that for the real interacting pairs and that of random pairs. They then used a least squares approach to estimate the reliability of the putative protein interaction data. Deng et al. (2003) further extended the idea in Deane et al. (2002) and used a maximum likelihood estimation (MLE) approach to estimate the reliability of a putative interaction data set. Like Mrowka et al. (2001), they used MIPS physical interactions as a reference set for true interactions. The same approach can be ap-plied to estimate the fraction of protein pairs that belong to the same complex in an observed complex data set. The method can be briefly described as follows.

Let α be the reliability of a given set of putative protein interactions. Let $O_e(\cdot)$, $T_e(\cdot)$, and $R_e(\cdot)$ be the distribution of the correlation coefficients for gene pairs based on gene expressions for the given set of putative protein interactions, the true protein interaction set, and the random protein pairs, respectively. Then

$$O_e(\cdot) = \alpha T_e(\cdot) + (1 - \alpha)R_e(\cdot). \tag{16.1}$$

$T_e(\cdot)$ and $R_e(\cdot)$ can be approximated based on the correlation coefficients for pairs of proteins within the gold standard set of protein interactions and the correlation coefficients of all the protein pairs, respectively.

Deng et al. (2003) split the values of correlation coefficients into $K = 20$ bins. Let n_k be the number of observed interaction pairs in the bin k. Let p_k and q_k be the fractions of real interactions and random pairs in bin k, respectively. Then the likelihood function can be defined as:

$$L(\alpha) = \prod_{k=1}^{K} (\alpha p_k + (1 - \alpha)q_k)^{n_k}. \tag{16.2}$$

$L(\alpha)$ is a convex function and a classical gradient algorithm can be used to estimate the parameter α by maximizing $L(\alpha)$. The variance of the resulting estimate $\widehat{\alpha}$ is obtained as

$$\text{Var}(\widehat{\alpha}) = \left(\sum_{k=1}^{K} n_k \frac{(p_k - q_k)^2}{(\widehat{\alpha}p_k + (1 - \widehat{\alpha})q_k)^2} \right)^{-1}. \tag{16.3}$$

*16.3.2 Estimating the reliability of putative protein interactions based on gene
expression and protein localization*

Huh et al. (2003) generated a large-scale protein localization map of yeast, showing that protein interactions are strongly enriched among co-localized proteins and
proteins between specific cellular locations. Therefore, both gene expressions and
localizations can be used for reliability estimation (Lee et al., 2005a). Again, we
model the putative interaction data set as a mixture of true interactions and random
pairs. Let $\theta_{ll'}$ and $\delta_{ll'}$ be the probability that a true interacting pair and random protein pair belong to locations (l, l'), respectively. Let $n_{kll'}$ be the number of observed
protein pairs within the putative interaction data set with correlation coefficient in the
k-th bin and with localizations (l, l'). Combining gene expression data and protein
localization data results in the following likelihood function

$$L(\alpha) = \prod_{k=1}^{K} \prod_{l,l'=1}^{L_0} (\alpha p_k \theta_{ll'} + (1 - \alpha) q_k \delta_{ll'})^{n_{kll'}}, \qquad (16.4)$$

where L_0 is the number of locations being considered. The parameter α can again be
estimated by maximizing $L(\alpha)$. The variance of $\widehat{\alpha}$ is computed as

$$\mathrm{Var}(\widehat{\alpha}) = \left(\sum_{k=1}^{K} \sum_{l,l'=1}^{L_0} n_{kll'} \frac{(p_k \theta_{ll'} - q_k \delta_{ll'})^2}{(\widehat{\alpha} p_k \theta_{ll'} + (1 - \widehat{\alpha}) q_k \delta_{ll'})^2} \right)^{-1}. \qquad (16.5)$$

16.3.3 Applications to protein interactions from high throughput experiments

We applied the above methods to protein interaction data sets from several high
throughput experiments. Two groups of interaction data sets were studied. The first
group includes pairwise physical interactions including the MIPS, DIP, Uetz et al.
(2000), and Ito et al. (2000, 2001) interaction data sets. The notation Ito$i$IST indicates the set of protein pairs observed to interact i times. The MIPS physical interactions are used as a true interaction data set. Table 16.1 gives the estimated reliability
together with their standard deviations of the estimates using gene expression and
protein localization alone or combined.

The second group includes the protein complexes such as the MIPS complex data,
the TAP complex data, and the HMS-PCI complex data. Any pair of proteins within
the same complex are considered interacting. We treat the MIPS complex data as
a true protein complex data set. Table 16.1 also gives the estimated reliability and
the corresponding standard deviation for the various protein complex data sets. The
standard deviation of the estimate using gene expression alone is very large, with
the estimated reliability showing irregular patterns. For example, the estimated reliability for Ito4IST (0.895) is much higher than the estimated reliability of Ito6IST
(0.676) contradicting our intuition. Consistent with intuition, the standard deviation
of the estimated reliability using localization alone is much smaller and the estimated
reliability for Ito$i$IST increases as i increases. Finally, the standard deviation of the

Table 16.1 *Reliability of the protein physical interaction data (DIP, Uetz, and Ito with different IST hits), and the protein complex data (TAP and HMS-PCI) using the protein localization data, the gene expression data, and both data sets.*

Data	Localization		Gene Expression		Both	
	Reliability	SE	Reliability	SE	Reliability	SE
Physical Interactions						
DIP	0.587	0.0082	0.815	0.0244	0.619	0.0076
Uetz	0.685	0.0273	0.529	0.0843	0.699	0.0257
Ito1IST	0.268	0.0140	0.167	0.0383	0.293	0.0133
Ito2IST	0.411	0.0259	0.558	0.0831	0.470	0.0253
Ito3IST	0.532	0.0345	0.753	0.1144	0.611	0.0321
Ito4IST	0.552	0.0397	0.895	0.1436	0.640	0.0366
Ito5IST	0.547	0.0429	0.964	0.1567	0.640	0.0394
Ito6IST	0.556	0.0491	0.676	0.1768	0.641	0.0451
Ito7IST	0.608	0.0544	0.791	0.1942	0.682	0.0492
Ito8IST	0.614	0.0572	0.878	0.2054	0.684	0.0514
Complexes						
TAP	0.4544	0.0063	0.585	0.0081	0.516	0.0056
HMS-PCI	0.1975	0.0042	0.248	0.0053	0.205	0.0037

estimate based on the combined data is smaller than that using gene expressions or protein localizations alone.

16.3.4 Estimating the probability of interaction for individual protein pairs

The above approach can only estimate the fraction of true interactions in a putative interaction data set. However, it does not give a reliability score for a particular observed interaction. Saito et al. (2003) proposed the criterion "interaction generality" to assess the reliability of a particular interaction protein pair based on the idea that a protein cannot interact with too many interacting partners. If a protein interacts with a large number of proteins, it is most likely a "stick" protein and the observed interactions associated with this protein do not have real functional associations. Recently, Troyanskaya et al. (2003) and Jansen et al. (2003) developed Bayesian approaches to give a reliability measure for a particular putative interaction based on the observations that interacting proteins are more likely to have similar functions, to have similar gene expression patterns, and to be in the same location. Troyanskaya et al. (2003) gave a reliability score for two proteins to be functionally related and Jansen et al. (2003) gave a reliability score for two proteins to be in the same complex. Methods have also been developed to evaluate the contributions of individual features as well as combined features for predicting protein interaction (Lin et al., 2004; Lu et al., 2005). It is found that only a relatively small number of features, for example, protein function, is adequate for predicting protein interactions. Jaimovich et al. (2005)

proposed a Markov random field (MRF) model for predicting protein interactions. They assumed an MRF model for the interaction network based on the theory of random graphs (Frank and Strauss, 1986). Conditional on the true interaction network, they assumed probability models for the observed data. Machine learning approaches were used to estimate the parameters as well as to predict the posterior probability of interactions for protein pairs conditional on the observations from different data sources. More details can be found in Jaimovich et al. (2005).

16.4 Protein function prediction using protein interaction data

It has been observed that interacting proteins are more likely to have similar functions (Mering et al., 2002). Therefore, protein interaction networks can be useful for protein function prediction. The *neighbors* of a given protein are all the proteins interacting with it. Fellenberg et al. (2000) and Schwikowski et al. (2000) developed a neighbor counting method for protein function prediction. For an unknown protein, they counted the number of known proteins of its neighbors for each function of interest and assigned the unknown protein with the function category having the highest frequency. One problem with this approach is that it does not consider the frequency of the proteins having certain functions of interest. Hishigaki et al. (2001) developed a χ^2-statistic based approach for protein function prediction. For an unknown protein and a function of interest, a χ^2-statistic is calculated by comparing the observed frequency with the expected frequency of neighbors having the function of interest. The unknown protein is assigned the function with the highest χ^2 statistic. Both the counting method and the χ^2 method do not consider unknown protein neighbors. Several novel methods have been developed for protein function prediction based on interaction networks and other data sources. We review these approaches here.

Suppose a genome has N proteins $P_1, \ldots, P_N$. Let $P_1, \ldots, P_n$ be the unknown proteins and $P_{n+1}, \ldots, P_{n+m}$ be the known proteins, $N = n + m$. A protein may have several different functions. To simplify the problem, we study each functional category separately. For a function of interest, let $X_i = 1$ if the protein i has the function and 0 otherwise. The problem is to assign values to $X = (X_1, \ldots, X_n)$ conditional on the protein interaction networks, other pairwise relationships, features of individual proteins, and the functions of the known proteins.

16.4.1 A Markov random field (MRF) model for protein function prediction

Based on the principle of guilt by association, Deng et al. (2002) first developed an MRF model for protein function prediction. The basic idea is to assign a prior probability for $X = (X_1, \ldots, X_{n+m})$, the configuration of function labeling based on the protein interaction network. Under this model, they calculated the posterior probability distribution for $(X_1, \ldots, X_n)$ conditional on the network and $(X_{n+1}, \ldots, X_{n+m})$. The key is how to assign the prior probability distribution. Different priors give different accuracy for protein function prediction.

An MRF model based on one network

Deng et al. (2002) assigned the prior as follows. Let π be the probability of a protein having the function of interest. Without considering the interaction network, the probability of a configuration of X is proportional to

$$\prod_{i=1}^{N} \pi^{x_i}(1-\pi)^{1-x_i} = \left(\frac{\pi}{1-\pi}\right)^{N_1}(1-\pi)^N, \qquad (16.6)$$

where $N_1 = \sum_{i=1}^{N} x_i$.

Deng et al. (2002) then considered one interaction network. Let S denote all the interacting protein pairs. The probability of the functional labeling conditional on the network is proportional to

$$\exp(\beta N_{01} + \gamma N_{11} + \kappa N_{00}), \qquad (16.7)$$

where $N_{ll'}$ is the number of (l, l')-interacting pairs in S, and

$$N_{11} = \sum_{(i,j)\in S} x_i x_j$$
$$= \#\{(1 \leftrightarrow 1) \text{ pairs in S}\},$$
$$N_{10} = \sum_{(i,j)\in S} (1-x_i)x_j + (1-x_j)x_i \qquad (16.8)$$
$$= \#\{(1 \leftrightarrow 0) \text{ pairs in S}\}, \text{ and}$$
$$N_{00} = \sum_{(i,j)\in S} (1-x_i)(1-x_j)$$
$$= \#\{(0 \leftrightarrow 0) \text{ pairs in S}\}.$$

Thus, the total probability of the functional labeling is proportional to $\exp(-U(x))$, where

$$U(x) = -\alpha N_1 - \beta N_{10} - \gamma N_{11} - \kappa N_{00}$$
$$= -\alpha \sum_{i=1}^{N} x_i - \beta \sum_{(i,j)\in S} x_i x_j$$
$$\quad - \gamma \sum_{(i,j)\in S} (1-x_i)x_j + (1-x_j)x_i \qquad (16.9)$$
$$\quad - \kappa \sum_{(i,j)\in S} (1-x_i)(1-x_j),$$

and $\alpha = \log(\frac{\pi}{1-\pi})$.

$U(x)$ is referred as the *potential function* in the field of MRF and defines a global Gibbs distribution of the entire network,

$$\Pr(X \mid \theta) = \frac{1}{Z(\theta)} \exp(-U(x)), \qquad (16.10)$$

where $\theta = (\alpha, \beta, \gamma, \kappa)$ are parameters and $Z(\theta)$ is a normalized constant calculated by summing over all the configurations:

$$Z(\theta) = \sum_x \exp(-U(x)).$$

$Z(\theta)$ is called the *partition function*.

Several other approaches for protein function prediction based on one interaction network have been developed. In particular, Vazquez et al. (2003) considered multiple function categories and proposed to maximize the number of interactions within the same function categories. For a single function of interest, it is equivalent to maximize $N_{00} + N_{11}$, where N_{00} and N_{11} are defined as above. The Deng et al. (2002) model differs from the Vazquez et al. (2003) model in two significant ways. (1) Vazquez et al. (2003) used only the interaction network and did not consider the fraction of proteins having the function of interest in the known proteins. (2) Vazquez et al. (2003) gave an equal weight to intra-function class interactions. Letovsky and Kasif (2003) proposed a model to assign functions to proteins based on a probabilistic analysis of graph neighborhoods in a protein-protein interaction network, which is fundamentally an MRF model, and the belief propagation algorithm was used to assign function probabilities for proteins in the network.

An MRF model for multiple networks

Deng et al. (2004a) further extended the above model to multiple networks and to include features of individual proteins. Assume that L sources of protein pairwise relationships that may be useful for protein function prediction are available. A network can be built based on each pairwise relationship denoted as $\text{Net}_1, \text{Net}_2, \ldots, \text{Net}_L$, respectively. The entire network we consider is the union of all the networks, denoted as S.

Similar to Equation 16.7, our belief for the functional labeling of all the proteins based on network Net_l is proportional to

$$P\{\text{ labeling} \mid \text{Net}_l\} \propto \exp(\beta_l N_{10}^{(l)} + \gamma_l N_{11}^{(l)} + \kappa_l N_{00}^{(l)}), \qquad (16.11)$$

where $(N_{10}^{(l)}, N_{11}^{(l)}, N_{00}^{(l)})$ are defined similarly as in Equation 16.8.

Multiplying over all the networks, our belief for the functional labeling of all the proteins is proportional to

$$P\{\text{ labeling} \mid \text{networks }\} \propto \prod_{l=1}^{L} \exp(\beta_l N_{10}^{(l)} + \gamma_l N_{11}^{(l)} + \kappa_l N_{00}^{(l)})$$

$$= \exp \sum_{l=1}^{L} \left(\beta_l N_{10}^{(l)} + \gamma_l N_{11}^{(l)} + \kappa_l N_{00}^{(l)} \right). \qquad (16.12)$$

Our total belief for the functional labeling of all the proteins is proportional to the product of Equations 16.6 and 16.12.

Then an MRF over all the functional labeling is defined by

$$P\{\text{labeling, networks}\} = \exp(-U(x))/Z(\theta), \tag{16.13}$$

where

$$U(x) = -\sum_{i=1}^{n+m} x_i \alpha - \sum_{l=1}^{L} \left(\beta_l N_{10}^{(l)} + \gamma_l N_{11}^{(l)} + \kappa_l N_{00}^{(l)} \right), \tag{16.14}$$

θ indicates the vector of parameters, and $Z(\theta)$ is the summation of $\exp(-U(x))$ over all the functional labeling. Under the above model, the parameters $(\kappa_1, \kappa_2, \ldots, \kappa_L)$ are redundant and are set to 1. In the terminology of MRF, $U(x)$ is called the potential function.

Incorporating features of individual proteins

In addition to protein pairwise relationships, features of individual proteins can be very important for protein function prediction. A feature refers to an observation about a protein. It can be the presence or absence of a motif signal, the protein's conservation and localization, its isoelectric point, its absolute mRNA expression level, or mutant phenotypes from experiments about the sensitivity or resistance of disruption mutants under various growth conditions. Several investigators have developed protein function prediction methods based on features of individual proteins (Clare and King, 2002; Gupta and Brunak, 2002; Hegyi and Gerstein, 1999; Jensen et al., 2002; Kell and King, 2000; King et al., 2001; Stawiki et al., 2002; Drawid and Gerstein, 2000). Deng et al. (2004a) integrated features into the MRF models for protein function prediction.

Suppose we have M features of interest, $F_1, F_2, \ldots, F_M$. The mth feature can take values $0, 1, 2, \ldots k_m - 1$ where k_m is the number of categories for the mth feature. Let the feature vector corresponding to protein P_i be $f_i = (f_{i1}, f_{i2}, \ldots, f_{iM})$, where f_{im} is the index for the mth feature of protein i. For the mth feature, let $p_{1m}(k)$ $(p_{0m}(k))$ be the conditional probability that a protein has feature index k given that a protein has (does not have) the function of interest. For simplicity, we assume that all the features contribute independently to the functions of proteins.

For a given feature vector $f = (f_1, f_2, \ldots, f_M)$, define

$$P_1(f) = \prod_{m=1}^{M} p_{1m}(f_m),$$

$$P_0(f) = \prod_{m=1}^{M} p_{0m}(f_m).$$

The probability of the features of all the proteins given the functional labeling is

$$P\{\text{features} \mid \text{labeling}\} = \prod_{i:X_i=1} P_1(f_i) \times \prod_{i:X_i=0} P_0(f_i). \tag{16.15}$$

Multiplying Equations 16.13 and 16.15, we have the following probability model

$$P\{\text{labeling, networks, domain features}\} =$$
$$P\{\text{labeling, networks}\} \times P\{\text{domain features} \mid \text{labeling}\}. \qquad (16.16)$$

Deng et al. (2004a) described methods to estimate the posterior distribution of the functions of the unknown proteins given the features of all the proteins, the different sources of protein pairwise relationship, and the annotations of the known proteins.

Computational issues

Given the above models, the problem is to estimate the posterior probability distribution given the annotation of the known proteins, the features of all the proteins, and the network. The parameters are also unknown. Using Equation 16.16, it can be shown that

$$\log \frac{Pr(X_i = 1 \mid F, X_{[-i]}, \theta)}{1 - Pr(X_i = 1 \mid F, X_{[-i]}, \theta)}$$

$$= \alpha_i + \sum_{l=1}^{L} (\beta_l - 1) M_0^{(i)}(l) + (\gamma_l - \beta_l) M_1^{(i)}(l), \qquad (16.17)$$

where F is the feature information, $X_{[-i]} = (X_1, \ldots, X_{i-1}, X_{i+1}, \ldots, X_{n+m})$, $\alpha_i = \log \frac{\pi P_1(f_i)}{(1-\pi) P_0(f_i)}$, $M_0^{(i)}(l)$ and $M_1^{(i)}(l)$ are the numbers of neighbors of protein P_i labeled with 0 and 1 according to the lth network, respectively. The parameters can be estimated based on the network consisting of the known proteins by an S-PLUS routine (Venables and Ripley, 2002).

Once all the parameters have been defined, a Gibbs sampler (Liu, 2001) can be used to estimate the posterior probability distribution of $(X_1, \ldots, X_n)$. The algorithm can be described as follows:

1. Randomly set the value of missing data $X_i = \lambda_i, i = 1, \ldots, n$ with probability π.
2. For each protein P_i, update the value of X_i using Equation 16.17.
3. Repeat Step 2 a total of T times until all the posterior probabilities $Pr(X_i \mid D, X_{[-i]}, \theta)$ are stabilized.

16.4.2 Kernel-based methods for protein function prediction

In the MRF formulation, we consider only immediate neighbors for proteins. The protein interaction network can be used to define similarity between any pair of proteins using the diffusion kernel (Kondor and Lafferty, 2002). We first briefly describe kernel-based methods of Lanckriet et al. (2004a,b,c) to combine different data sources for protein function prediction. Then we describe our effort to combine the idea of kernel-based method with the MRF model.

Support vector machine (SVM) and semidefinite programming (SDP)

Lanckriet et al. (2004a,b,c) developed kernel-based methods for protein function prediction using SVM. Suppose that there are L data sources such as protein interactions, gene expressions, domains, localizations, etc. For the lth data source, a kernel matrix K_l (semi-positive definite) is defined. For continuous data such as gene expressions, the Gaussian diffusion kernel can be used. For protein interactions, diffusion kernel on graphs can be used (Kondor and Lafferty, 2002). Several other kernel matrixes have been developed for different sources of data structures in Lanckriet et al. (2004a,b,c). To integrate the different data sources, Lanckriet and colleagues considered the linear combinations of the kernel matrixes

$$K = \sum_{l=1}^{L} \mu_l K_l,$$

where $\mu_l \geq 0, l = 1, \ldots, L$ are parameters to be determined.

They used the standard 1-norm soft margin SVM to build a classifier. We now have the following constrained maximization problem:

$$\max_{\alpha,t} 2\alpha^T e - ct$$
$$\text{subject to } t \geq \frac{1}{r_i}\alpha^T \text{diag}(y) K_l \text{diag}(y)\alpha, \quad l = 1, 2, \ldots, L \tag{16.18}$$
$$\alpha^T y = 0,$$
$$C \geq \alpha \geq 0,$$

where $r_i = \text{trace}(K_i)$, $c = \mu^T r$ and y is the annotation of the known proteins. This problem is a quadratically-constrained quadratic program (QCQP) problem (Boyd and Vandenberghe, 2004) and can be solved using standard software such as SeDuMi (Sturm, 1999). The computational time is $O(n^3)$, where n is the number of proteins in the training set.

Combining kernel with the MRF model for protein function prediction

Lanckriet et al. (2004b) showed that the SVM described above outperformed the MRF approach in almost all the function categories considered. One of the main reasons is likely due to the inclusion of multiple level neighbors in the kernel-based methods. Note that $K_l(i,j)$ defines a similarity between protein P_i and protein P_j based on the lth data source. Similar to Equation 16.11, the probability of the labeling based on the lth network N_l can be modeled as

$$\exp(\beta_l D_{10}(l) + \gamma_l D_{11}(l) + \kappa_l D_{00}(l)), \tag{16.19}$$

where β_l, γ_l, and κ_l are constants, and

$$D_{11}(l) = \sum_{i<j} K_l(i,j)I\{x_i = 1, x_j = 1\},$$

$$D_{10}(l) = \sum_{i<j} K_l(i,j)I\{(x_i = 1, x_j = 0) \text{ or } (x_i = 0, x_j = 1)\}, \qquad (16.20)$$

$$D_{00}(l) = \sum_{i<j} K_l(i,j)I\{x_i = 0, x_j = 0\},$$

where the summations are over all the protein pairs. Multiplying Equations 16.6 and 16.19 for $l = 1, 2, \dots, L$, we obtain the total probability proportional to

$$\exp\left(\alpha N_1 + \sum_{l=1}^{L}\left(\beta_l D_{10}(l) + \gamma_l D_{11}(l) + \kappa_l D_{00}(l)\right)\right). \qquad (16.21)$$

From Equation 16.21, it can be shown that

$$\log \frac{P(X_i = 1 \mid X_{[-i]}, \theta)}{1 - P(X_i = 1 \mid X_{[-i]}, \theta)} \qquad (16.22)$$

$$= \alpha + (\beta_l - \kappa_l)K_0^{(i)}(l) + (\gamma_l - \beta_l)K_1^{(i)}(l),$$

where

$$K_0^{(i)}(l) = \sum_{j \neq i} K_l(i,j)I\{x_j = 0\},$$

$$K_1^{(i)}(l) = \sum_{j \neq i} K_l(i,j)I\{x_j = 1\}.$$

Note that if we let $K_l(i,j) = 1$ when protein i interacts with protein j in network l, and $K_l(i,j) = 0$ otherwise, this new model is the same as the MRF model of Deng et al. (2002). The probability that an unknown protein has the function of interest can be approximated using MCMC (Gilks et al., 1995). We refer to the above approach as kernel-based MRF (KMRF).

16.4.3 Applications to real data

All the methods described above have been applied to predict protein functions. The MRF model has been used for protein function prediction first based on the MIPS function classification (Deng et al., 2002, 2004a) and later extended to functions defined in GO (Deng et al., 2004b). The SVM approach has been used to predict protein functions based on MIPS (Lanckriet et al., 2004b) and to predict ribosomal proteins and membrane proteins (Lanckriet et al., 2004a). Lanckriet et al. (2004c) summarize protein function prediction based on SVM. The new KMRF method has been applied for protein function prediction based on GO (Lee et al., 2006) and for prediction of protein essentiality (Tu et al., 2005). The KMRF approach can be easily extended to incorporate correlated functions. For most functions that have been considered so far, the SVM approach outperformed the MRF approach. The KMRF approach has similar performance to that of the SVM approach. For example, for predicting protein essentiality, the receiver operating characteristic (ROC) scores for the MRF, SVM, and the KMRF approaches are 0.804. 0.812, and 0.831, respectively, based on the core interaction data set. Integrating protein function based on cellular processes, conservation, and localizations into the model increased the ROC score of the KMRF model to 0.869.

16.5 Discussion

Enormous amounts of biological data have been generated and stored in public and private databases. These data sources are extremely important for biological studies. However, the data are generally noisy and contain many false positive and false negative errors. There are no systematic statistical tools to choose the most reliable data from the noisy data. The various data sources can most likely contribute to our understanding of the biological problems of interest. The data sources are usually correlated though, so their contributions to our understanding of the underlying biology are not independent. An important issue is how to integrate noisy and correlated data sources to analyze biological problems.

In this chapter, we have reviewed some of our recent efforts in integrating different data sources for biological studies. We described likelihood-based methods for estimating the reliability of putative interaction data sets and showed that the localization data give more accurate estimation of the reliability than the gene expression data. Integrating the localization and gene expression data can give even more accurate estimates of the reliability of the different data sets. Other statistical methods integrating different data sources for estimating the probability of two proteins interacting have also been developed.

We have also described methods for protein function prediction based on heterogeneous information on interaction networks, genetic interactions, other pairwise relationships, as well as features of individual proteins. These approaches include MRF, SVM, and KMRF. The combination of kernels with MRF appears to be novel in protein function prediction. The simplicity of KMRF and its high accuracy in protein function prediction warrant further studies of the approach in other fields. Alternative approaches to data integration for protein prediction exist, with many more to be expected as different data sources continue to be generated. A recent assessment of nine different methods is given by Peña-Castillo et al. (2008).

In protein function prediction, we implicitly assume that the networks under consideration, such as the protein interaction network and genetic interaction network, are highly reliable. Therefore we used the core interaction data in DIP in all our studies on protein function prediction. We also tried using all the interactions, including the unreliable ones, in DIP for protein function prediction. As expected, prediction accuracy is lowered. A problem is how best to use all the interactions for protein function prediction. The effect of incompleteness of the interaction data on protein function prediction is also unknown.

In summary, we show the power of integrating multiple data sources for biological studies. Significant questions remain as to how to integrate noisy and incomplete data in biological studies. It is also important to develop methods to evaluate the dependence among the different data sources, with the aim of integrating correlated data sources for biological studies.

Acknowledgments

The research was partly supported by NIH/NSF joint mathematical biology initiative DMS-0241102.

CHAPTER 17

Gene trees, species trees, and species networks

Luay Nakhleh, Derek Ruths, and Hideki Innan

17.1 Introduction

The availability of whole-genome sequence data from multiple organisms has provided a rich resource for investigating several biological, medical and pharmaceutical problems and applications. Yet, along with the insights and promises these data are providing, genome-wide data have given rise to more complex problems and have challenged *traditional* biological paradigms. One of these paradigms is the inference of a *species phylogeny*. Traditionally, a biologist would proceed by obtaining the molecular sequence of a single locus, or gene, in a set of species, inferring the phylogeny (or evolutionary history) of this locus, and taking it to be an accurate representation of the species pattern of divergence. Although this approach may work for several groups of organisms, particularly when taking extra caution in selecting the locus, the availability of sequence data for multiple loci from a variety of organisms and populations has highlighted the deficiencies and inaccuracies of this traditional approach. Different loci in a group of organisms may have different gene tree topologies. In this case, there is no single gene tree topology to declare as the species tree. Further, in the presence of *reticulate* evolutionary events, such as horizontal gene transfer, the species phylogeny may not be a tree; instead, a *network* of relationships is the more appropriate model.

In a seminal paper, Maddison (1997) discussed the issue of species/gene tree incongruence, the implications it has on the inference of a species tree, and the processes that can cause such incongruence and for which explicit modeling is necessary for accurate inferences. The three main processes discussed were *lineage sorting*, *gene duplication and loss*, and *reticulate evolution*.

Lineage sorting occurs because of random contribution of genetic material from each individual in a population to the next generation. Some fail to have offspring while some happen to have multiple offspring. In population genetics, this process was first modeled by R. A. Fisher and S. Wright, in which each gene of the population at a particular generation is chosen independently from the gene pool of the previous generation, regardless of whether the genes are in the same individual or in different individuals. Under the Wright-Fisher model, *the coalescent* considers the process backward in time (Kingman, 1982; Hudson, 1983a; Tajima, 1983). That is, the ancestral lineages of genes of interest are traced from offspring to parents. A coalescent event occurs when two (or sometimes more) genes "merge" at the same parent, called the *most recent common ancestor* (MRCA) of the two genes. In certain cases, two genes coalesce at a branch in the species tree that is deeper than their MRCA. When this happens, it may be that coalescence patterns result in trees that do not reflect the divergence patterns of the species. Evidence of extensive lineage sorting has been reported in several groups of organisms; (see e.g., Rokas et al. 2003; Syring et al. 2005; Pollard et al. 2006; Than et al. 2008c; Kuo et al. 2008).

Gene duplication is considered a major mechanism of evolution, particularly in generating new genes and biological functions (Ohno, 1970; Graur and Li, 2000). Duplication events result in multiple gene copies, which when transmitted to descendant organisms, produce complex gene genealogies. As some of these genes may go extinct (Olson, 1999), inferring the gene tree from only those copies present in the organisms can result in a topology that may disagree with that of the species tree. A similar effect can be obtained as an artifact of sampling some, but not all, of the gene copies in an organism's genome.

The third process discussed by Maddison (1997) is reticulate evolution. For example, evidence shows that bacteria may obtain a large proportion of their genetic diversity through the acquisition of sequences from distantly related organisms, via horizontal gene transfer (HGT; Ochman et al., 2000; Doolittle, 1999b,a; Kurland et al., 2003; Hao and Golding, 2004; Nakamura et al., 2004). There is also recent evidence of widespread HGT in plants (Bergthorsson et al., 2003, 2004; Mower et al., 2004). Interspecific recombination is believed to be ubiquitous among viruses (Posada et al., 2002; Posada and Crandall, 2002), and hybrid speciation is a major evolutionary mechanism in plants, and groups of fish and frogs (Ellstrand et al., 1996; Rieseberg and Carney, 1998; Rieseberg et al., 2000; Linder and Rieseberg, 2004; Mallet, 2005; Noor and Feder, 2006; Mallet, 2007).

There is a major difference between lineage sorting and gene duplication/loss on the one hand and reticulate evolutionary events on the other, in terms of the reconciliation outcome. Gene trees may disagree with each other, as well as with the species phylogenies, due to lineage sorting or gene duplication/loss events. In this case, their reconciliation yields a tree topology, with the deep coalescences, duplications, and losses taking place within the species tree branches. However, when horizontal gene transfer or hybrid speciation occur, the evolutionary history of the genomes can no longer be adequately modeled by a tree; instead, a *phylogenetic network* is a more appropriate model.

Incorporating these processes into computational methods for inferring accurate evolutionary histories will have significant implications on reconstructing accurate evolutionary histories of genomes and better understanding of their diversification. Biologists have long acknowledged the presence of these processes, their significance, and their effects. The computational research community has responded in recent years, proposing a plethora of methods for reconstructing complex evolutionary histories by reconciling incongruent gene trees.

In this chapter, we review the processes that cause gene tree incongruence, issues that must be accounted for when dealing with these processes, and methods for reconciling gene trees and inferring species phylogenies despite gene tree incongruence. Inferring species trees from gene trees can be seen as an information combination problem, where the combination can take place at different levels: combining the sequence data followed by tree reconstruction or separate analyses which are combined by reconciliation. In Section 17.2 we discuss in more detail the processes that result in discord. In Section 17.3, we discuss approaches for reconciling gene trees when incongruence is solely due to lineage sorting. In Section 17.4, we discuss incongruence due to gene duplication and loss, and review some of the methods for reconciliation under this scenario. In Section 17.5, we describe the *phylogenetic network* model and discuss the problem of reconciling gene trees into species networks, assuming that incongruence is due to reticulate evolutionary events. In Section 17.6 we discuss preliminary attempts at establishing unified frameworks for distinguishing among the various processes. Such frameworks are crucial to accurate reconstruction of species phylogenies, since in general the cause of incongruence may not be known, and assuming one of the three processes arbitrarily may result in an incorrect reconciliation. We close with our conclusions (Section 17.7).

17.2 Gene tree incongruence

A *gene tree* is a model of how a gene evolves through not only nucleotide substitution, but also other mechanisms that act on a larger scale, such as duplication, loss, and horizontal gene transfer. As a gene at a locus in the genome replicates and its copies are passed on to more than one offspring, branching points are generated in the gene tree. Because the gene has a single ancestral copy, barring recombination, the resulting history is a branching tree (Maddison, 1997). Sexual reproduction and meiotic recombination within populations break up the genomic history into many small pieces, each of which has a strictly treelike pattern of descent (Hudson, 1983a; Hein, 1990; Maddison, 1995). Thus, within a species, many tangled gene trees can be found, one for each nonrecombined locus in the genome. A *species tree* depicts the pattern of branching of species lineages via the process of speciation. When reproductive communities are split by speciation, the gene copies within these communities likewise are split into separate bundles of descent. Within each bundle, the gene trees continue branching and descending through time. Thus, the gene trees are contained within the branches of the species phylogeny (Maddison, 1997).

Gene trees can differ from one another as well as from the species tree. Disagreements (incongruence) among gene trees may be an artifact of the data and/or methods used (statistical error). A number of studies show the effects of statistical error on the performance of phylogenetic tree reconstruction methods (e.g., Hillis et al.

1993; Hillis and Huelsenbeck 1994, 1995; Nakhleh et al. 2001a,b, 2002; Moret et al. 2002). Statistical errors confound the accurate reconstruction of evolutionary relationships, and must be handled in the first stage of a phylogenetic analysis. Incongruence among gene trees due to the three aforementioned processes, on the other hand, is a reflection of true biological events.

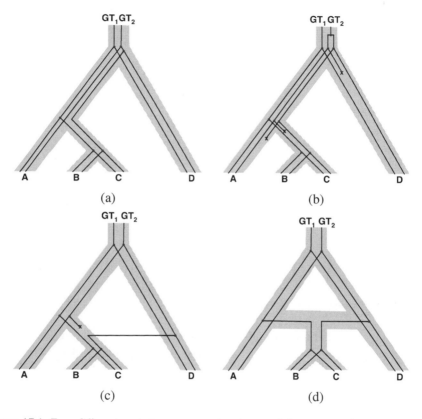

Figure 17.1 *Four different evolutionary scenarios that result in gene tree incongruence. (a) Gene tree GT_1 has a topology that is identical to that of the species phylogeny, whereas gene tree GT_2 is incongruent with both the species phylogeny and GT_1, due to lineage sorting. (b) Gene tree GT_1 has a topology that is identical to that of the species phylogeny, whereas gene tree GT_2 is incongruent with both the species phylogeny and GT_1, due to multiple duplication/loss events. (c) Gene tree GT_1 has a topology that is identical to that of the species phylogeny, whereas gene tree GT_2 is incongruent with both the species phylogeny and GT_1, due to horizontal gene transfer. (d) The species phylogeny is a network, since the clade (B, C) is a hybrid; the two gene trees GT_1 and GT_2 are incongruent due to hybrid speciation. The two different gene trees that arise from each of the four scenarios are shown in Figure 17.2.*

Figure 17.1 illustrates (a) the effect of lineage sorting, (b) gene duplication and loss, and reticulate evolution ((c) horizontal gene transfer and (d) hybrid speciation) on gene tree incongruence. The species phylogeny is represented by the shaded "tubes";

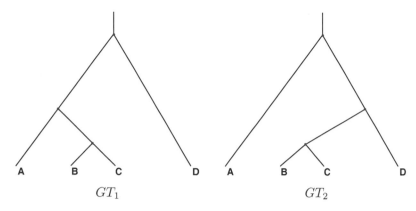

Figure 17.2 *The two gene trees GT_1 and GT_2 that arise in each of the four evolutionary scenarios depicted in Figure 17.1.*

it has B and C as sister taxa whose most recent common ancestor (MRCA) is a sister taxon of A, and the MRCA of all three taxa is a sister taxon of D. In the case of hybrid speciation (the scenario in Figure 17.1(d)), the MRCA of taxa B and C is a sister taxon of both A and D, since it is the outcome of hybridization. Each of the four scenarios shows the trees of two genes evolving within the branches of the species phylogeny. For clarity, the topologies of the two gene trees are shown separately in Figure 17.2. How such a scenario arises is further elaborated in Section 17.3.

Figure 17.1(b) shows an evolutionary scenario involving only gene duplication and loss events that result in identical gene tree topologies to that of the lineage sorting scenario. In this scenario, the gene whose tree is depicted by GT_1 has a divergence pattern identical to that of the species. The second gene, on the other hand, has undergone a duplication event prior to the splitting of D from the MRCA of the other three taxa, and then copies of the gene went extinct in the D lineage, the A lineage, and the lineage of the MRCA of B and C. This combination of duplication and loss events results in gene tree GT_2, whose topology disagrees with that of GT_1 as well as the species tree. The topologies of the two gene trees in this scenario are also identical to the ones shown in Figure 17.2.

As can be seen in Figures 17.1(a) and 17.1(b), even though the two gene trees are incongruent, they are reconciled within the branches of a species tree that is identical in both cases. In other words, the incongruence in these two scenarios does not necessitate deviating from a tree-like pattern of divergence at the species level. This stands in contrast to the two scenarios illustrated in Figures 17.1(c) and 17.1(d), where reticulate evolutionary events (in this case, horizontal gene transfer (HGT) and hybrid speciation, respectively) do necessitate the adoption of a phylogenetic network as a more appropriate model of the evolutionary history of the genomes.

Views as to the extent of HGT in bacteria vary between two extremes (Doolittle, 1999b,a; Kurland et al., 2003; Welch, 2002; Hao and Golding, 2004; McClelland et al., 2004; Nakamura et al., 2004). There is a sizeable "ideological and rhetorical" gap between the researchers who believe that HGT is so rampant that a prokaryotic phylogenetic tree is useless, as opposed to those who believe HGT is mere "background noise" that does not affect the reconstructibility of a phylogenetic tree for bacterial genomes. Supporting arguments for each of these two views have been published. For example, the heterogeneity of genome composition between closely related strains (only 40% of the genes in common with three *E. coli* strains (Welch, 2002)) supports the former view, whereas the well-supported phylogeny reconstructed by Lerat et al. (2003) from about 100 "core" genes in γ-Proteobacteria gives evidence in favor of the latter view. Nonetheless, regardless of the views and the accuracy of the various analyses, the occurrence of HGT may result in networks, rather than trees, of evolutionary relationships of the genomes.

Three mechanisms of HGT are

1. *transformation*, which is the uptake of naked DNA from the environment (see e.g., Stewart and Sinigalliano 1991; Woegerbauer et al. 2002);

2. *conjugation*, which is the transfer of DNA by direct physical interaction between a donor and a recipient, usually mediated by conjugal plasmids or conjugal transposons (see e.g., Nakamura et al. 2004); and

3. *transduction*, which is the transfer of DNA by phage infection (Brüssow et al., 2004).

Transducing phages have been observed in many bacteria, including *Salmonella* (Schicklmaier and Schmieger, 1995), *Streptomyces* (Burke et al., 2001), and *Listeria* (Hodgson, 2000). Transduction typically occurs among closely related bacteria, whereas conjugation may occur among more distant organisms.

In the case of HGT, shown in Figure 17.1(c), genetic material is transferred from one lineage to another. Sites that are not involved in a horizontal transfer are inherited from the parent (as in GT_1), while other sites are horizontally transferred from another species (as in GT_2).

In the case of hybrid speciation, as illustrated in Figure 17.1(d), two lineages recombine to create a new species. We can distinguish *diploid hybridization*, in which the new species inherits one of the two homologs for each chromosome from each of its two parents (so that the new species has the same number of chromosomes as its parents), and *polyploid hybridization*, in which the new species inherits the two homologs of each chromosome from both parents (so that the new species has the sum of the numbers of chromosomes of its parents). Under this last heading, we can further distinguish *allopolyploidization*, in which two lineages hybridize to create a new species whose ploidy level is the sum of the ploidy levels of its two parents (the expected result), and *auto-polyploidization*, a regular speciation event that does not involve hybridization, but which doubles the ploidy level of the newly created lineage. Prior to hybridization, each site on each homolog has evolved in a tree-like fashion, although, due to meiotic recombination, different strings of sites may have different histories. Thus, each site in the homologs of the parents of the hybrid evolved in a tree-like fashion on one of the trees induced by (contained inside) the network representing the hybridization event. Figure 17.1(d) shows a network with one hybrid speciation event, followed by speciation that results in the clade (B, D). Assuming no (incomplete) lineage sorting (i.e., that all alleles from B and C coalesce at the MRCA of these two taxa), each site evolves down exactly one of the two trees shown in Figures 17.2(a) and 17.2(b).

In the next three sections, we briefly address issues and methodologies for reconciling incongruence gene trees due to lineage sorting, gene duplication and loss, and reticulate evolutionary events, respectively.

17.3 Lineage sorting

The basic coalescent process can be treated as follows. Consider a pair of genes at time τ_1 in a random mating haploid population. The population size at time τ is denoted by $N(\tau)$. The probability that the pair are from the same parental gene at the previous generation (time $\tau_1 + 1$) is $1/N(\tau_1 + 1)$. Therefore, starting at τ_1, the probability that the coalescence between the pair occurs at τ_2 is given by

$$P(\tau_2) = \frac{1}{N(\tau_2)} \prod_{\tau=\tau_1+1}^{\tau_2-1} \left(1 - \frac{1}{N(\tau)} \right). \tag{17.1}$$

When $N(\tau)$ is constant, the probability density function (pdf) of the coalescent time (i.e., $t = \tau_2 - \tau_1$) is given by a geometric distribution, and can be approximated by an exponential distribution for large N:

$$P(t) = \frac{1}{N} e^{-t/N}. \tag{17.2}$$

The coalescent process is usually ignored in phylogenetic analysis, but has a significant effect (causing lineage sorting) when closely related species are considered (Hudson, 1983b; Takahata, 1989; Rosenberg, 2002). The situation of Figure 17.1(b) is reconsidered under the framework of the coalescent in Figure 17.3. Here, it is assumed that species A and B split $T_1 = 5$ generations ago, and the ancestral species of A and B and species C split $T_2 = 19$ generation ago. The ancestral lineage of a gene from species A and that from B meet in their ancestral population at time $\tau = 6$, and they coalesce at $\tau = 33$, which predates T_2, the speciation time between (A, B) and C. The ancestral lineage of B enters in the ancestral population of the three species at time $\tau = 20$, and first coalesces with the lineage of C. Therefore, the gene tree is represented by $A(BC)$ while the species tree is $(AB)C$. That is, the gene tree and species tree are "incongruent." Under the model in Figure 17.3, the probability that the gene tree is congruent with the species tree is 0.863, which is one minus the product of the probability that the ancestral lineages of A and B do not coalesce between $\tau = 6$ and $\tau = 9$, and the probability that the first coalescence in the ancestral population of the three species occurs between (A and C) or (B and C). The former probability is $\frac{15}{16} \cdot \frac{14}{15} \cdot \frac{12}{13} \cdot \frac{11}{12} \cdot \frac{10}{11} \cdot \frac{8}{9} \cdots (1 - \frac{7}{8})^8 = 0.26$ and the latter is $\frac{2}{3}$.

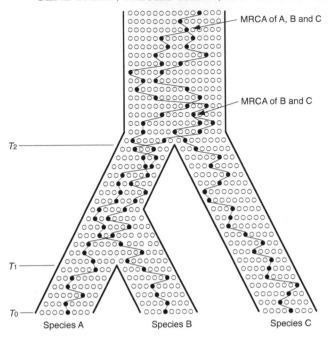

Figure 17.3 *An illustration of the coalescent process in a three species model with discrete generations. The process is considered backward in time from present, T_0, to the past. Circles represent haploid individuals. We are interested in the gene tree of the three genes (haploids) from the three species. Their ancestral lineages are represented by closed circles connected by lines. A coalescent event occurs when a pair of lineages happen to share a single parental gene (haploid).*

Under the three-species model (Figure 17.3), there are three possible types of gene tree, $(AB)C$, $(AC)B$ and $A(BC)$. Let $P[(AB)C]$, $P[(AC)B]$, and $P[A(BC)]$ be the probabilities of the three types of gene tree. These three probabilities are simply expressed with a continuous time approximation when all populations have equal and constant population sizes, N, where N is large:

$$P[(AB)C] = 1 - \frac{2}{3}e^{-(T_2-T_1)/N}, \qquad (17.3)$$

and

$$P[(AC)B] = P[A(BC)] = \frac{1}{3}e^{-(T_2-T_1)/N}. \qquad (17.4)$$

An interesting application of this three species problem is in hominoids; A: human, B: chimpanzee and C: gorilla. It is believed that the species tree is $(AB)C$. Chen and Li (2001) investigated DNA sequences from 88 autosomal intergenic regions, and estimated the gene tree for each region. They found that 36 regions support the species tree, $(AB)C$, while 10 estimated trees are $(AC)B$ and 6 are $A(BC)$. No resolution is obtained for the remaining 36 regions (see below). It is possible to estimate the time between two speciation events, $T_2 - T_1$, assuming all populations have equal and constant diploid population sizes N (Wu, 1991). Since 36 out of 52 gene trees are congruent with the species tree, $T_2 - T_1$ is estimated to be $-\ln[(3/2)(36/52)] = 0.77$ times $2N$ generations. It should be noted that $2N$ is used for the coalescent time scale instead of N because hominoids are diploids. If we assume N to be between 5×10^4 and 1×10^5 (Takahata et al., 1995; Takahata and Satta, 1997), the time between two speciation events is between 7.7×10^4–15.5×10^4 generations, which is roughly 1–3 million years, assuming a generation time of 15–20 years.

It is important to notice that the estimation of the gene tree from DNA sequence data is based on the nucleotide differences between sequences, and that the gene tree is sometimes unresolved. One of the reasons for that is a lack of nucleotide differences such that DNA sequence data are not informative enough to resolve the gene tree. This possibility strongly depends on the mutation rate. Let μ be the mutation rate per region per generation, and consider the effect of mutation on the estimation of the gene tree. We consider the simplest model of mutations on DNA sequences, the infinite site model (Kimura, 1969), in which mutation rate per site is so small that no multiple mutations at a single site are allowed. Consider a gene tree, $(AB)C$, and suppose that there is a reasonable outgroup sequence such that we know the sequence of the MRCA of the three sequences. It is obvious that mutations on the internal branch between the MRCA of the three and the MRCA of A and B are informative. If at least one mutation occurred on this branch, the gene tree can be resolved from the DNA sequence alignment. This effect is investigated by assuming that the number of mutations on a branch with length t follows a Poisson distribution with mean μt.

A few methods have been introduced recently for analyzing gene trees, reconciling their incongruities, and inferring species trees despite these incongruities, when these incongruities are assumed to be caused solely by lineage sorting. Generally speaking, each of these methods follows one of two approaches: the *combined analysis* approach or the *separate analysis* approach; see Figure 17.4. In the combined analysis approach, the sequences from multiple loci are concatenated, and the resulting "supergene" data set is analyzed using traditional phylogenetic methods, such as maximum parsimony and maximum likelihood (Rokas et al., 2003). In the separate analysis approach, the sequence data from each locus are first analyzed individually, then combined via reconciliation of the gene trees. One way to reconcile the gene trees is by taking their majority consensus (Kuo et al., 2008). Another is the "democratic vote" method, which entails taking the tree topology occurring with the highest frequency among all gene trees as the species tree. Shortcomings of these methods have been analyzed (Degnan and Rosenberg, 2006; Kubatko and Degnan, 2007). Recently, Bayesian methods following the separate analysis approach were developed (Edwards et al., 2007; Liu and Pearl, 2007). Although these methods are accurate, they are very time-consuming, taking hours or days even on moderately sized data sets, which limits their scalability. For example, the analysis of the yeast data set of Rokas et al. (2003) using the Bayesian approach of Edwards et al. (2007) took about 800 hours. Than et al. (2008c) and Than and Nakhleh (2009) have recently introduced

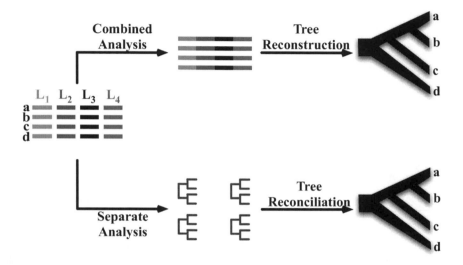

Figure 17.4 *Approaches for inferring species trees. In the combined analysis approach (top), gene sequences are concatenated, to create a "super gene," and a gene tree is inferred from this super gene. In the separate analysis approach (bottom), each gene is analyzed individually, and the analyses are then integrated in the form of gene tree reconciliation.*

very efficient *integer linear programming* (ILP) formulations of the "minimize deep coalescences" approach proposed by Maddison (1997) and Maddison and Knowles (2006). The resulting methods are very efficient, inferring accurate species trees in mere seconds, from data sets containing thousands of loci.

17.4 Gene duplication and loss

Various reports of instances and effects of gene loss and duplication exist in the literature (see e.g., Moore 1995; Nichols 2001; Ruvolo 1997). When losses and duplications are the only processes acting on the genes, a mathematical formulation of the gene tree reconciliation problem can be given as follows:

Definition 17.1 *The Gene Tree Reconciliation Problem.*

> **Input:** *A set* T *of rooted gene trees, a cost* w_D *for duplications, and a cost* w_L *for losses.*
>
> **Output:** *Rooted tree* T *with each gene tree* $t \in T$ *mapped onto* T, *so as to minimize the sum over all genes of* $w_D n_D + w_L n_L$, *where* n_D *is the total number of duplications and* n_L *is the total number of losses.*

This problem was shown to be NP-hard by Fellows et al. (1998) and Ma et al. (1998). Heuristics for the problem exist, but these do not solve the optimization problem (see Ma et al. 1998; Page and Charleston 1997). Various fixed-parameter approaches have been proposed by Stege (1999) and Hallett and Lagergren (2000); some variants can be approximated to within a factor of 2 (Ma et al., 1998).

When loss and duplication are the only processes acting on the genes, two different questions can be posed, depending on the input data:

1. *Gene tree reconciliation problem*: when the gene trees are known and the species tree is known, what is the best set of duplication and loss events that reconcile each gene tree with the species tree?
2. *Species tree construction problem*: when the gene trees are known, but the evolutionary relationships among the species involved is not known, can the gene trees provide the information necessary to derive an estimate of the species tree?

Both of these questions require the assumption of a model of gene duplication and loss. The complexity of the gene-tree reconciliation problem is determined by the model chosen, whereas the general species tree construction problem is NP-hard under all commonly used models of gene duplication and loss.

The simplest version of either problem uses a duplication-only model (i.e., losses do not occur) . During the period from 1995 and 2000, this was a commonly used model (Eulenstein et al., 1997; Page and Charleston, 1997; Page, 1998; Eulenstein, 1997; Stege, 1999; Ma et al., 1998; Zhang, 1997; Ma et al., 2000). Under the duplication-only model, the gene tree reconciliation problem has linear-time solutions (Zhang, 1997; Eulenstein, 1997), as well as other polynomial-time algorithms that report better performance on real biological datasets (Zmasek and Eddy, 2001). The species tree construction problem is NP-hard (Ma et al., 1998). Approaches that have been taken to solving the species tree construction problem include heuristics (Page and Charleston, 1997), approximation algorithms (Ma et al., 2000), and fixed parameter tractable algorithms obtained by parameterizing by the number of gene duplications separating a gene tree from the species tree (Stege, 1999).

The other commonly used model is the more general duplication-loss model, which admits both duplication and loss events within gene trees. The gene tree reconciliation problem has been shown to be polynomial-time under conditions where the evolution of the sequences themselves is not considered (Arvestad et al., 2004; Chen et al., 2000; Durand et al., 2005); if this is taken into account, the problem becomes

NP-hard (Fellows et al., 1998; Ma et al., 1998). Some efficient heuristics for the problem are currently available (Arvestad et al., 2003, 2004). Early work on the gene tree reconciliation problem under this model borrowed techniques from biogeography and host/parasite evolution (Charleston, 1998; Page and Charleston, 1998).

17.5 Reticulate evolution

As described in Section 17.2, when events such as horizontal gene transfer or hybrid speciation occur, the evolutionary history can no longer be adequately modeled by a tree; rather, *phylogenetic networks* are more appropriate in this case. In this section, we describe a phylogenetic network model that models reticulate evolution explicitly, and discuss approaches for reconstructing networks from gene trees.

17.5.1 Terminology and notation

Given a (directed) graph G, let $E(G)$ denote the set of (directed) edges of G and $V(G)$ denote the set of nodes of G. Let (u, v) denote a directed edge from node u to node v; u is the *tail* and v the *head* of the edge and u is a *parent* of v. The *indegree* of a node v is the number of edges whose head is v, while the *outdegree* of v is the number of edges whose tail is v. A node of outdegree 0 is a *leaf* (often called a *tip* by systematists). A directed path of length k from u to v in G is a sequence $u_0 u_1 \cdots u_k$ of nodes with $u = u_0$, $v = u_k$, and $\forall i$, $1 \leq i \leq k$, $(u_{i-1}, u_i) \in E(G)$; we say that u is the tail of p and v is the head of p. Node v is *reachable* from u in G, denoted $u \rightsquigarrow v$, if there is a directed path in G from u to v; we then also say that u is an *ancestor* of v. A *cycle* in a graph is a directed path from a vertex back to itself; trees never contain cycles: in a tree, there is always a unique path between two distinct vertices. Directed acyclic graphs (or DAGs) play an important role in our model; note that every DAG contains at least one vertex of indegree 0. A *rooted directed acyclic graph*, in the context of this paper, is then a DAG with a single node of indegree 0, the *root*; note that all other nodes are reachable from the root by a (directed) path of graph edges. We denote by $r(T)$ the root of tree T and by $L(T)$ the leaf set of T.

17.5.2 Evolutionary phylogenetic networks

Moret et al. (2004) modeled phylogenetic networks using directed acyclic graphs (DAGs), and differentiated between "model" networks and "reconstructible" ones.

Model networks   A phylogenetic network $N = (V, E)$ is a rooted DAG obeying certain constraints. We begin with a few definitions.

Definition 17.2 *A node $v \in V$ is a* tree node *if one of these three conditions holds:*

- $indegree(v) = 0$ and $outdegree(v) = 2$: root;
- $indegree(v) = 1$ and $outdegree(v) = 0$: leaf; or
- $indegree(v) = 1$ and $outdegree(v) = 2$: internal tree node.

A node v is a network node if $indegree(v) = 2$ and $outdegree(v) = 1$.

Tree nodes correspond to regular speciation or extinction events, whereas network nodes correspond to reticulation events (such as hybrid speciation and horizontal gene transfer). Clearly $V_T \cap V_N = \emptyset$ and we can easily verify that $V_T \cup V_N = V$.

Definition 17.3 *An edge $e = (u,v) \in E$ is a* tree edge *if v is a tree node; it is a* network edge *if v is a network node.*

The tree edges are directed from the root of the network toward the leaves and the network edges are directed from their tree-node endpoint toward their network-node endpoint.

A phylogenetic network $N = (V,E)$ defines a partial order on the set V of nodes. We can also assign times to the nodes of N, associating time $t(u)$ with node u; such an assignment, however, must be consistent with the partial order. Call a directed path p from node u to node v that contains at least one tree edge a *positive-time directed path*. If there exists a positive-time directed path from u to v, then it must be that $t(u) < t(v)$. Moreover, if $e = (u,v)$ is a network edge then $t(u) = t(v)$, because a reticulation event is effectively instantaneous at the scale of evolution; thus, reticulation events act as synchronization points between lineages.

Definition 17.4 *Given a network N, two nodes u and v cannot* co-exist *(in time) if there exists a sequence $P = \langle p_1, p_2, \ldots, p_k \rangle$ of paths such that:*

- p_i *is a positive-time directed path, for every $1 \leq i \leq k$;*
- u *is the tail of p_1, and v is the head of p_k; and*
- *for every $1 \leq i \leq k-1$, there exists a network node whose two parents are the head of p_i and the tail of p_{i+1}.*

Obviously, if two nodes x and y cannot co-exist in time, then a reticulation event between them cannot occur.

Definition 17.5 *A* model phylogenetic network *is a rooted DAG obeying the following constraints:*

1. *Every node has indegree and outdegree defined by one of the four combinations $(0,2)$, $(1,0)$, $(1,2)$, or $(2,1)$—corresponding to, respectively, root, leaves, internal tree nodes, and network nodes.*
2. *If two nodes u and v cannot co-exist in time, then there does not exist a network node w with edges (u,w) and (v,w).*
3. *Given any edge of the network, at least one of its endpoints must be a tree node.*

Reconstructible networks

Definition 17.5 of model phylogenetic networks assumes that complete information about every step in the evolutionary history is available. Such is the case in simulations and in artificial phylogenies evolved in a laboratory setting; hence, our use of the term *model*. When attempting to reconstruct a phylogenetic network from sample data, however, normally a researcher will have only incomplete information, due to a combination of extinctions, incomplete sampling, and abnormal model conditions. Extinctions and incomplete sampling have the same consequences: the data do not reflect all of the various lineages that have contributed to the current situation. Abnormal conditions include insufficient differentiation along edges, in which case some of the edges may not be reconstructible, leading to polytomies and thus to nodes of outdegree larger than 2. All three types of problem may lead to the reconstruction of networks that violate the constraints of Definition 17.5. (The distinction between a model phylogeny and a reconstructible phylogeny is common with trees as well: for instance, model trees are always rooted, whereas reconstructed trees are usually unrooted. In networks, both the model network and the reconstructed network must be rooted: reticulations only make sense with directed edges.) Clearly, then, a reconstructible network will require changes from the definition of a model network. In particular, the degree constraints must be relaxed to allow arbitrary outdegrees for both network nodes and internal tree nodes. In addition, the time coexistence property must be reconsidered.

There are at least two types of problems in reconstructing phylogenetic networks. First, slow evolution may give rise to edges so short that they cannot be reconstructed, leading to polytomies. This problem cannot be resolved within the DAG framework, so we must relax the constraints on the outdegree of tree nodes. Secondly, missing data may lead methods to reconstruct networks that violate indegree constraints or time coexistence. In such cases, we can postprocess the reconstructed network to restore compliance with most of the constraints in the following three steps:

1. For each network node w with outdegree larger than 1 and with edges (w, v_1), $\ldots$, (w, v_k), add a new tree node u with edge (w, u) and, for each i, $1 \le i \le k$, replace edge (w, v_i) by edge (u, v_i).
2. For each network node w whose parents u and v violate time coexistence, add two tree nodes w_u and w_v and replace the two network edges (u, w) and (v, w) by four edges: the two tree edges (u, w_u) and (v, w_v) and the two network edges (w_u, w) and (w_v, w).
3. For each edge (u, v) where both u and v are network nodes, add a new tree node w and replace the edge (u, v) by the two edges (u, w) and (w, v).

The resulting network is consistent with the original reconstruction, but now satisfies the outdegree requirement for network nodes, obeys time coexistence (the introduction of tree edges on the paths to the network node allows arbitrary time delays), and no longer violates the requirement that at least one endpoint of each edge be a tree node. Moreover, this postprocessing is unique and quite simple. We can thus define a reconstructible network in terms similar to a model network.

Definition 17.6 *A* reconstructible phylogenetic network *is a rooted DAG obeying the following constraints:*

1. *Every node has indegree and outdegree defined by one of the three combinations of (indegree,outdegree):* $(0, x)$, $(1, y)$, *or* $(z, 1)$, *for* $x \geq 1$, $y \geq 0$, *and* $z \geq 2$, *corresponding to, respectively, root, other tree nodes (internal nodes and leaves), and network nodes.*
2. *If two nodes u and v cannot co-exist in time, then there does not exist a network node w with edges (u, w) and (v, w).*
3. *Given any edge of the network, at least one of its endpoints must be a tree node.*

Definition 17.7 *A network N induces a tree T' if T' can be obtained from N by the following two steps:*

1. *For each network node in N, remove all but one of the edges incident into it; and*
2. *for every node v such that $indegree(v) = outdegree(v) = 1$, the parent of v is u, and the child of v is w, remove v and the two edges (u, v) and (v, w), and add new edge (u, w) (this is referred to in the literature as the* forced contraction *operation).*

For example, the network N shown in Figure 17.1(d) induces both trees shown in Figure 17.2(a) and Figure 17.2(b).

17.5.3 Reconstructing networks from gene trees

From a graph-theoretic point of view, the problem can be formulated as pure phylogenetic network reconstruction (Moret et al., 2004; Nakhleh et al., 2004, 2005b). In the case of HGT, and despite the fact the evolutionary history of the set of organisms is a network, Lerat et al. (2003) showed that an underlying species tree can still be inferred. In this case, a phylogenetic network is a pair (T, Ξ), where T is the species (organismal) tree, and Ξ is a set of HGT edges whose addition to T results in a phylogenetic network N that induces all the gene trees. The problem can be formulated as follows.

Definition 17.8 *The HGT Reconstruction Problem.*

Input: A species tree ST and a set G of gene trees.

Output: Set Ξ of minimum cardinality whose addition to ST yields a phylogenetic network N that induces each of the gene trees in G.

However, in the case of hybrid speciation, there is no underlying species tree; instead the problem is one of reconstructing a phylogenetic network N that induces a given set of gene trees.

Definition 17.9 *The Hybrid Speciation Reconstruction Problem.*

Input: A set G of gene trees.

Output: A phylogenetic network N with minimum number of network nodes that induces each of the gene trees in G.

The minimization criterion reflects the fact that the simplest solution is sought; in this case, the simplest solution is one with the minimum number of HGT or hybrid speciation events. A major issue that impacts the identifiability of reticulate evolution is that of extinction and incomplete taxon sampling. Moret et al. (2004) illustrated some of the scenarios that lead to non-identifiability of reticulation events from a set of gene trees.

Hallett and Lagergren (2001) gave an efficient algorithm for solving the HGT Reconstruction Problem; however, their algorithm handles limited special cases of the problem in which the number of HGT events is very small, and the number of times a gene is transferred is very low. Also, their tool handles only binary trees (Addario-Berry et al., 2002). Nakhleh et al. (2004) gave efficient algorithms for solving the Hybrid Speciation Reconstruction Problem, but for constrained phylogenetic networks, referred to as *gt-networks*; further, they handled only binary trees. Nakhleh et al. (2005b) introduced RIATA-HGT for solving the general case of the HGT Reconstruction Problem, with later extensions to handle non-binary trees as well as to identify multiple minimal solutions by Than and Nakhleh (2008). The method is implemented in the PhyloNet software package by Than et al. (2008b). Other methods for reconciling species and gene trees, assuming reticulate evolution, include tree and reticulogram reconstruction (T-REX; Makarenkov, 2001), a recursive procedure of consolidation and rearrangement implemented in HorizStory (MacLeod et al., 2005), and Efficient Evaluation of Edit Paths (EEEP; Beiko and Hamilton, 2006).

Recently, Nakhleh and colleagues have extended the maximum parsimony and maximum likelihood criteria to the domain of phylogenetic network reconstruction and evaluation (Nakhleh et al., 2005a; Jin et al., 2006, 2007a,b,c; Than et al., 2008a).

17.6　Distinguishing lineage sorting from HGT

In Section 17.3 we showed that the gene tree is not always identical to the species tree. Than et al. (2007) considered the effect of horizontal gene transfer (HGT) on gene tree under the framework of the coalescent, which we review here. The application of the coalescent theory to bacteria is straightforward. Bacterial evolution is better described by the Moran model (Moran, 1958) rather than the Wright-Fisher model because bacteria do not fit a discrete generation model. Suppose that each haploid individual in a bacterial population with size N has a lifespan that follows an exponential distribution with mean l. When an individual dies, another individual randomly chosen from the population replaces it to keep the population size constant. In other words, one of the $N - 1$ alive lineages is duplicated to replace the dead one. Under the Moran model, the ancestral lineages of individuals of interest can be traced backward in time, and the coalescent time between a pair of individuals follows an exponential distribution with mean $lN/2$ (Ewens, 1979; Rosenberg,

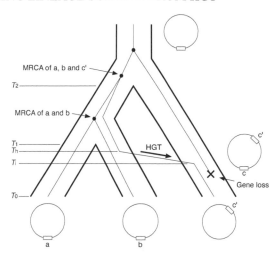

Figure 17.5 *A three species model with a HGT event, demonstrating that a congruent tree can be observed even with HGT.*

2005). This means that one half of the mean lifetime in the Moran model corresponds to one generation in the Wright-Fisher model.

It may usually be thought that HGT can be detected when the gene tree and species tree are incongruent (see Section 17.5). However, the situation is complicated when lineage sorting is also involved. Consider a model with three species, A, B, and C, in which an HGT event occurs from species B to C. Suppose the ancient circular genome has a single copy of a gene, as illustrated in Figure 17.5. Let a, b and c be the focal orthologous genes in the three species, respectively. At time T_h, a gene escaped from species B and was inserted in a genome in species C at T_i, which is denoted by c'. Following the HGT event, c was physically deleted from the genome, so that each of the three species currently has a single copy of the focal gene.

If there is no lineage sorting, the gene tree should be $a(bc')$. Since this tree is incongruent with the species tree $(AB)C$, we could consider it as an evidence for HGT. However, as demonstrated in Section 17.2, lineage sorting could also produce the incongruence between the gene tree and species tree without HGT. It is also important to note that lineage sorting, coupled with HGT, could produce a congruent gene tree, as illustrated in Figure 17.5. Although b and c' have more chance to coalesce first, the probability that the first coalescence occurs between a and b or between a and c' may not be negligible, especially when $T_1 - T_h$ is short.

The probabilities of the three types of gene tree can be formulated under this tri-species model with HGT as illustrated in Figure 17.5. Here, T_h could exceed T_1, in such a case it can be considered that HGT occurred before the speciation between A and B. Assuming that all populations have equal and constant population sizes, N, the three probabilities can be obtained modifying Equations 17.3 and 17.4:

$$P[(AB)C] = \begin{cases} \frac{1}{3}e^{-(T_1-T_h)/N} & \text{if } T_h \leq T_1 \\ 1 - \frac{2}{3}e^{-(T_h-T_1)/N} & \text{if } T_h > T_1 \end{cases}, \qquad (17.5)$$

$$P[(AC)B] = \begin{cases} \frac{1}{3}e^{-(T_1-T_h)/N} & \text{if } T_h \leq T_1 \\ \frac{1}{3}e^{-(T_h-T_1)/N} & \text{if } T_h > T_1 \end{cases}, \qquad (17.6)$$

and

$$P[A(BC)] = \begin{cases} 1 - \frac{2}{3}e^{-(T_1-T_h)/N} & \text{if } T_h \leq T_1 \\ \frac{1}{3}e^{-(T_h-T_1)/N} & \text{if } T_h > T_1 \end{cases}. \qquad (17.7)$$

Thus, lineage sorting due to the coalescent process works as a noise component for detecting and reconstructing HGT based on a gene tree, sometimes mimicking the evidence for HGT and sometimes "canceling" such evidence. Therefore, to distinguish HGT and lineage sorting, statistical methods based on the theory introduced in this chapter are needed. We have only considered very simple cases with three species here, but it is straightforward to extend the theory to more complicated models. Recently, Meng and Kubatko (2009) considered a similar scenario involving lineage sorting and hybrid speciation.

17.7 Summary

In the post-genomic era, evidence of massive gene tree incongruence is accumulating in large and diverse groups of organisms. In this chapter, we have discussed three major processes that may lead to such incongruence, namely lineage sorting, gene duplication and loss, and reticulate evolution. In the case of the first two processes, the evolutionary history of the set of species still takes the shape of a tree, with gene trees reconciled within the branches of such a tree. However, in the case of reticulate evolution, the evolutionary history of the set of the species' genomes may be more appropriately modeled by a phylogenetic network. We discussed general approaches for reconciliations under each of the three processes. We also briefly reviewed preliminary work that is being done on extending the coalescent framework in order to enable distinction between lineage sorting as a cause of gene tree incongruence and reticulate evolution as an alternative explanation.

The development of computational tools for identifying gene tree incongruence and inferring species trees despite such incongruence is still in its infancy. There are several major directions for future research in this area, which include, but are not limited to:

1. Developing computational tools for simulating evolution of whole-genomes, or multi-locus data, while incorporating all processes that cause gene tree incongruence. Such tools would play a crucial role in understanding these processes, as well as in enabling the study of the performance of existing and newly developed computational tools.

2. Developing computational tools for reconciling gene trees that can scale to genome-wide data.

3. Methods for addressing lineage sorting already consider multiple loci, but most computational methods for addressing gene duplication/loss and reticulate evolution work with only a pair of trees. It is important that methods are developed for handling multiple trees.

4. We expect that development of sound, unified frameworks for simultaneously analyzing all three processes and distinguishing among them will be central to progress in this area.

Acknowledgments

This work was supported in part by DOE grant DE-FG02-06ER25734, NSF grant CCF-0622037, and grant R01LM009494 from the National Library of Medicine. The contents are solely the responsibility of the author and do not necessarily represent the official views of the DOE, NSF, National Library of Medicine or the National Institutes of Health.

References

Abecasis, G.R. et al., Merlin – rapid analysis of dense genetic maps using sparse gene flow trees, *Nat. Genet.*, 30:97–101, 2002.

Addario-Berry, L., M.T. Hallett, and J. Lagergren, Towards identifying lateral gene transfer events, in Altman, R.B. et al., eds., *Proceedings of the Seventh Pacific Symposium on Biocomputing (PSB02)*, Proceedings of the Pacific Symposium on Biocomputing, pp. 279–290, River Edge, NJ: World Scientific, 2002.

Affymetrix, *Statistical Algorithms Description Document*, Santa Clara, CA, 2002.

Akutsu, T., S. Miyano, and S. Kuhara, Identification of genetic networks from a small number of gene expression patterns under the boolean network model, in Altman, R., ed., *Pacific Symposium on Biocomputing 1999*, pp. 17–28, River Edge, NJ: World Scientific, 1999.

Akutsu, T., S. Miyano, and S. Kuhara, Inferring qualitative relations in genetic networks and metabolic pathways, *Bioinformatics*, 16:727–734, 2000.

Al-Shahrour, F., R. Diaz-Uriarte, and J. Dopazo, FatiGO: a web tool for finding significant associations of Gene Ontology terms with groups of genes, *Bioinformatics*, 20:578–580, 2004.

Allison, D.B. et al., Testing the robustness of the new Haseman-Elston quantitative-trait loci-mapping procedure, *Am. J. Hum. Genet.*, 67:249–252, 2000.

Allison, D.B. et al., A mixture model approach for the analysis of microarray gene expression data, *Comput. Stat. Data An.*, 39:1–20, 2002.

Allison, D.B. and M. Heo, Meta-analysis of linkage data under worst-case conditions: a demonstration using the human OB region, *Genetics*, 148:859–865, 1998.

Almasy, L. and J. Blangero, Multipoint quantitative-trait linkage analysis in general pedigrees, *Am. J. Hum. Genet.*, 62:1198–1211, 1998.

Alter, O., P.O. Brown, and D. Botstein, Singular value decomposition for genome-wide expression data processing and modeling, *Proc. Natl. Acad. Sci. USA*, 97:10101–10106, 2000.

Altmüller, J. et al., Genomewide scans of complex human diseases: True linkage is hard to find, *Am. J. Hum. Genet.*, 69:936–950, 2001.

Altschul, S.F. et al., Gapped BLAST and PSI-BLAST: a new generation of protein database search programs, *Nucleic Acids Res.*, 25:3389–3402, 1997.

Amos, C.I., Robust variance-components approach for assessing genetic linkage in pedigrees, *Am. J. Hum. Genet.*, 54:535–543, 1994.

Amos, C.I., Successful design and conduct of genome-wide association studies, *Hum. Mol. Genet.*, 16:R220–R225, 2007.

Amos, C.I., M. de Andrade, and D.K. Zhu, Comparison of multivariate tests for genetic linkage, *Hum. Hered.*, 51:133–144, 2001.

Amos, C.I. and R.C. Elston, Robust methods for the detection of genetic linkage for quantitative data from pedigrees, *Genet. Epidemiol.*, 6:349–360, 1989.

Amos, C.I. et al., A more powerful robust sib-pair test of linkage for quantitative traits, *Genet. Epidemiol.*, 6:435–449, 1989.

An, P. et al., Genome-wide linkage scans for fasting glucose, insulin, and insulin resistance in the National Heart, Lung, and Blood Institute Family Blood Pressure Program: evidence of linkages to chromosome 7q36 and 19q13 from meta-analysis, *Diabetes*, 54:909–914, 2005.

Andersen, P.K., J.P. Klein, and M.J. Zhang, Testing for centre effects in multi-centre survival studies: A Monte Carlo comparison of fixed and random effects tests, *Stat. Med.*, 18:1489–1500, 1999.

Armitage, P., Tests for linear trends in proportions and frequencies, *Biometrics*, 11:375–386, 1955.

Arvestad, L. et al., Bayesian gene/species tree reconciliation and orthology analysis using MCMC, *Bioinformatics*, 19 Suppl 1:i7–i15, 2003.

Arvestad, L. et al., Gene tree reconstruction and orthology analysis based on an integrated model for duplications and sequence evolution, in Gusfield, D. et al., eds., *RECOMB 2004: Proceedings of the Eighth Annual International Conference on Research in Computational Molecular Biology*, pp. 326–335, New York: Association for Computing Machinery, 2004.

Ashburner, M. et al., Gene ontology: tool for the unification of biology. The Gene Ontology Consortium, *Nat. Genet.*, 25:25–29, 2000.

Ashburner, M. et al., Creating the Gene Ontology resource: Design and implementation, *Genome Res.*, 11:1425–1433, 2001.

Babron, M.C. et al., Meta and pooled analysis of European coeliac disease data, *Eur. J. Hum. Genet.*, 11:828–834, 2003.

Bader, G.D., D. Betel, and C.W. Hogue, BIND: the Biomolecular Interaction Network Database, *Nucleic Acids Res.*, 31:248–250, 2003.

Badner, J.A. and E.S. Gershon, Meta-analysis of whole-genome linkage scans of bipolar disorder and schizophrenia, *Mol. Psychiatry*, 7:405–411, 2002a.

Badner, J.A. and E.S. Gershon, Regional meta-analysis of published data supports linkage of autism with markers on chromosome 7, *Mol. Psychiatry*, 7:56–66, 2002b.

Bailey, K.R., Inter-study differences: how should they influence the interpretation and analysis of results?, *Stat. Med.*, 6:351–358, 1987.

Balasubramanian, R. et al., A graph-theoretic approach to testing associations between disparate sources of functional genomics data, *Bioinformatics*, 20:3353–3362, 2004.

Baldi, P. and A.D. Long, A Bayesian framework for the analysis of microarray expression data: regularized t-test and statistical inferences of gene changes, *Bioinformatics*, 17:509–519, 2001.

Bar-Joseph, Z. et al., Computational discovery of gene modules and regulatory networks, *Nat. Biotechnol.*, 21:1337–1342, 2003.

Barrett, T. et al., NCBI GEO: mining millions of expression profiles-database and tools, *Nucleic Acids Res.*, 33:D562–D566, 2005.

Beasley, T.M. et al., Empirical Bayes method for incorporating data from multiple genome scans, *Hum. Hered.*, 60:36–42, 2005.

Beer, D.G. et al., Gene-expression profiles predict survival of patients with lung adenocarcinoma, *Nat. Med.*, 9:816–824, 2002.

Beiko, R. and N. Hamilton, Phylogenetic identification of lateral genetic transfer events, *BMC Evol. Biol.*, 6:15, 2006.

Beissbarth, T. and T.P. Speed, GOstat: find statistically overrepresented Gene Ontologies within a group of genes, *Bioinformatics*, 20:1464–14655, 2004.

Benito, M. et al., Adjustment of systematic microarray data biases, *Bioinformatics*, 20:105–114, 2004.

Benjamini, Y. and Y. Hochberg, Controlling the false discovery rate: A practical and powerful approach to multiple testing, *J. Roy. Stat. Soc. B*, 57:289–300, 1995.

Benjamini, Y. and D. Yekutieli, The control of the false discovery rate in multiple testing under dependency, *Ann. Stat.*, 29:1165–1188, 2001.

Berger, R.L., Multiparameter hypothesis testing and acceptance sampling, *Technometrics*, 24:295–300, 1982.

Berger, R.L., Bioequivalence trials, intersection-union tests, and equivalence confidence sets, *Stat. Sci.*, 11:283–319, 1996.

Bergthorsson, U. et al., Widespread horizontal transfer of mitochondrial genes in flowering plants, *Nature*, 424:197–201, 2003.

Bergthorsson, U. et al., Massive horizontal transfer of mitochondrial genes from diverse land plant donors to the basal angiosperm Amborella, *Proc. Natl. Acad. Sci. USA*, 101:17747–17752, 2004.

Berners-Lee, T., J. Hendler, and O. Lassila, The semantic web, *Sci. Am.*, 284:29–37, 2001.

Bhattacharjee, A. et al., Classification of human lung carcinomas by mRNA expression profiling reveals distinct adenocarcinoma subclasses, *Proc. Natl. Acad. Sci. USA*, 98:13790–13795, 2001.

Birnbaum, A., Combining independent tests of significance, *J. Am. Stat. Assoc.*, 49:559–574, 1954.

Blacker, D. et al., Results of a high-resolution genome screen of 437 Alzheimer's disease families, *Hum. Mol. Genet.*, 12:23–32, 2003.

Blanchard, A.P., R.J. Kaiser, and L.E. Hood, High-density oligonucleotide arrays, *Biosens. Bioelectron.*, 11:687–690, 1996.

Böhning, D., E. Dietz, and P. Schlattmann, Recent developments in computer-assisted analysis of mixtures, *Biometrics*, 54:525–536, 1998.

Bolstad, B.M., *affyPLM: Probe Level Models*, 2008, URL `http://www.bioconductor.org/download/`.

Bolstad, B.M. et al., A comparison of normalization methods for high density oligonucleotide array data based on bias and variance, *Bioinformatics*, 19:185–193, 2003.

Bonney, G.E. et al., An application of empirical Bayes methods to updating linkage information on chromosome 21, *Cytogenet. Cell Genet.*, 59:112–113, 1992.

Borozan, I. et al., MAID: An effect size based model for microarray data integration across laboratories and platforms, *BMC Bioinformatics*, 9:305, 2008.

Botstein, D. et al., Construction of a genetic map in man using restriction fragment length polymorphisms, *Am. J. Hum. Genet.*, 32:314–331, 1980.

Boyd, S. and L. Vandenberghe, *Convex Optimization*, Cambridge, U. K.: Cambridge University Press, 2004.

Brazma, A. et al., Minimum information about a microarray experiment (MIAME) – toward standards for microarray data, *Nat. Genet.*, 29:373, 2001.

Breitkreutz, B.J., C. Stark, and M. Tyers, The GRID: The General Repository for Interaction Datasets, *Genome Biol.*, 4:R23, 2003.

Breitling, R. et al., Rank products: a simple, yet powerful, new method to detect differentially regulated genes in replicated microarray experiments, *FEBS Lett.*, 573:83–92, 2004.

Brem, R.B. et al., Genetic dissection of transcriptional regulation in budding yeast, *Science*, 296:752–755, 2002.

Brenner, S. et al., Gene expression analysis by massively parallel signature sequencing (MPSS) on microbead arrays, *Nat. Biotechnol.*, 18:630–634, 2000.

Brettschneider, J. et al., Quality assessment for short oligonucleotide microarray data, *Technometrics*, 50:241–264, 2008.

Broman, K.W., Mapping expression in randomized rodent genomes, *Nat. Genet.*, 37:209–210, 2005.

Brüssow, H., C. Canchaya, and W.D. Hardt, Phages and the evolution of bacterial pathogens: from genomic rearrangements to lysogenic conversion, *Microbiol. Mol. Biol. Rev.*, 68:560–602, 2004.

Burke, J., D. Schneider, and J. Westpheling, Generalized transduction in *Streptomyces coelicolor*, *Proc. Natl. Acad. Sci. USA*, 98:6289–6294, 2001.

Bush, C.R. et al., Functional genomic analysis reveals cross-talk between peroxisome proliferator-activated receptor gamma and calcium signaling in human colorectal cancer cells, *J. Biol. Chem.*, 282:23387–23401, 2007.

Bussemaker, H.J., H. Li, and E.D. Siggia, Regulatory element detection using correlation with expression, *Nat. Genet.*, 27:167–171, 2001.

Bystrykh, L. et al., Uncovering regulatory pathways that affect hematopoietic stem cell function using "genetical genomics", *Nat. Genet.*, 37:225–232, 2005.

Cardon, L.R. and L.J. Palmer, Population stratification and spurious allelic association, *Lancet*, 361:598–604, 2003.

Carey, M. and S.T. Smale, *Transcriptional Regulation in Eukaryotes*, Cold Spring Harbor, NY: Cold Spring Harbor Laboratory Press, 1999.

Carlin, B.P. and T.A. Louis, *Bayes and Empirical Bayes Methods for Data Analysis*, Boca Raton, FL: Chapman and Hall, 2nd ed., 2000a.

Carlin, B.P. and T.A. Louis, Empirical Bayes: Past and present and future, *J. Am. Stat. Assoc.*, 95:1286–1289, 2000b.

Chan, E.Y. et al., Increased huntingtin protein length reduces the number of polyglutamine-induced gene expression changes in mouse models of Huntington's disease, *Hum. Mol. Genet.*, 11:1939–1951, 2002.

Charleston, M.A., Jungles: a new solution to the host/parasite phylogeny reconciliation problem, *Math. Biosci.*, 149:191–223, 1998.

Chen, F.C. and W.H. Li, Genomic divergences between humans and other hominoids and the effective population size of the common ancestor of humans and chimpanzees, *Am. J. Hum. Genet.*, 68:444–456, 2001.

Chen, H., J. Chen, and J. Kalbfleisch, A modified likelihood ratio test for homogeneity in finite mixture models, *J. Roy. Stat. Soc. B*, 63:19–29, 2001.

Chen, K., D. Durand, and M. Farach-Colton, NOTUNG: A program for dating gene duplications and optimizing gene family trees, *J. Comput. Biol.*, 7:429–447, 2000.

Chesler, E.J. et al., Complex trait analysis of gene expression uncovers polygenic and pleiotropic networks that modulate nervous system function, *Nat. Genet.*, 37:233–242, 2005.

Cheung, V.G. et al., Natural variation in human gene expression assessed in lymphoblastoid cells, *Nat. Genet.*, 33:422–425, 2003.

Chiodini, B.D. and C.M. Lewis, Meta-analysis of 4 coronary heart disease genome-wide linkage studies confirms a susceptibility locus on chromosome 3q, *Arterioscler. Thromb. Vasc. Biol.*, 23:1863–1868, 2003.

Chipping Forecast, The Chipping Forecast, *Nat. Genet.*, 21:Special Supplement, 1999, Special Supplement.

Chipping Forecast, The Chipping Forecast II, *Nat. Genet.*, 32:Special Supplement, 2002.

Cho, H. and J.K. Lee, Bayesian hierarchical error model for analysis of gene expression data, *Bioinformatics*, 20:2016–2025, 2004.

Cho, H. and J.K. Lee, Error-pooling empirical Bayes model for enhanced statistical discovery of differential expression in microarray data, *IEEE Trans. Syst. Man Cybern. A, Syst. Humans*, 38:425–436, 2008.

Choi, J.K. et al., Combining multiple microarray studies and modeling interstudy variation, *Bioinformatics*, 19 Suppl 1:i84–i90, 2003.

Christian, J.B., Incorporating annotation data in quantitative trait loci mapping with mRNA transcripts, Ph.D. thesis, Department of Statistics, Rice University, 2009.

Chu, S. et al., The transcriptional program of sporulation in budding yeast, *Science*, 282:699–705, 1998.

Churchill, G.A. and R.W. Doerge, Empirical threshold values for quantitative trait mapping, *Genetics*, 138:963–971, 1994.

Clare, A. and R.D. King, Machine learning of functional class from phenotype data, *Bioinformatics*, 18:160–166, 2002.

Clayton, D.G. et al., Population structure, differential bias and genomic control in a large-scale, case-control association study, *Nat. Genet.*, 37:1243–1246, 2005.

Cochran, W.G., The combination of estimates from different experiments, *Biometrics*, 10:101–129, 1954a.

Cochran, W.G., Some methods of strengthening the common χ^2 tests, *Biometrics*, 10:417–451, 1954b.

Cole, D.A. et al., How the power of MANOVA can both increase and decrease as a function of the intercorrelations among the dependent variables, *Psychol. Bull.*, 115:465–474, 1994.

Coleman, D.L. and K.P. Hummel, The influence of genetic background on the expression of the obese (ob) gene in the mouse, *Diabetologia*, 9:287–293, 1973.

Colinayo, V.V. et al., Genetic loci for diet-induced atherosclerotic lesions and plasma lipids in mice, *Mamm. Genome*, 14:464–471, 2003.

Collin, F., Analysis of oligonucleotide data with a view to data quality assessment, Ph.D. thesis, Department of Statistics, University of California, Berkeley, 2004.

Conlon, E.M. et al., Integrating regulatory motif discovery and genome-wide expression analysis, *Proc. Natl. Acad. Sci. USA*, 100:3339–3344, 2003.

Cooper, H.M. and L.V. Hedges, *The Handbook of Research Synthesis*, New York: Russell Sage Foundation, 1994.

Copas, J. and J.Q. Shi, Meta-analysis, funnel plots and sensitivity analysis, *Biostatistics*, 1:247–262, 2000.

Cope, L.M. et al., A benchmark for Affymetrix GeneChip expression measures, *Bioinformatics*, 20:323–331, 2004.

Cordell, H.J., Sample size requirements to control for stochastic variation in magnitude and location of allele-sharing linkage statistics in affected sibling pairs, *Ann. Hum. Genet.*, 65:491–502, 2001.

Cox, N.J., An expression of interest, *Nature*, 12:733–734, 2004.

Cui, X. and G.A. Churchill, Statistical tests for differential expression in cDNA microarray experiments, *Genome Biol.*, 4:210, 2003.

Cui, X. et al., Improved statistical tests for differential gene expression by shrinking variance components estimates, *Biostatistics*, 6:59–75, 2005.

Dabney, A. and J.D. Storey, *qvalue: Q-value estimation for false discovery rate control*, 2008, URL http://cran.r-project.org/.

Davidson, E.H. et al., A genomic regulatory network for development, *Science*, 295:1669–1678, 2002.

de Jong, H., Modeling and simulation of genetic regulatory systems: a literature review, *J. Comput. Biol.*, 9:67–103, 2002.

Deane, C.M. et al., Protein interactions: Two methods for assessment of the reliability of high-throughput observation, *Mol. Cell. Proteomics*, 1:349–356, 2002.

Degnan, J.H. and N.A. Rosenberg, Discordance of species trees with their most likely gene trees, *PLoS Genet.*, 2:e68, 2006.

Demenais, F. et al., A meta-analysis of four European genome screens (GIFT consortium) shows evidence for a novel region on chromosome 17p11.2-q22 linked to type 2 diabetes, *Hum. Mol. Genet.*, 12:1865–1873, 2003.

Dempfle, A. and S. Loesgen, Meta-analysis of linkage studies for complex diseases: An overview of methods and a simulation study, *Ann. Hum. Genet.*, 68:69–83, 2004.

Dempster, A.P., N.M. Laird, and D.B. Rubin, Maximum likelihood from incomplete data via the EM algorithm (with discussion), *J. Roy. Stat. Soc. B*, 39:1–38, 1977.

Deng, M.H., T. Chen, and F.Z. Sun, An integrated probabilistic model for functional prediction of proteins, *J. Comput. Biol.*, 11:463–475, 2004a.

Deng, M.H., F.Z. Sun, and T. Chen, Assessment of the reliability of protein-protein interactions and protein function prediction, in Altman, R.B. et al., eds., *Biocomputing 2003: Proceedings of the Pacific Symposium*, pp. 140–151, River Edge, NJ: World Scientific, 2003.

Deng, M.H. et al., Mapping gene ontology to proteins based on protein-protein interaction data, *Bioinformatics*, 20:895–902, 2004b.

Deng, M.H. et al., Prediction of protein function using protein-protein interaction data, in *Proceedings of the First IEEE Computer Society Bioinformatics Conference (CSB2002)*, pp. 197–206, Los Alamitos, CA: IEEE Computer Society Press, 2002.

Dennis, G.J. et al., DAVID: Database for annotation, visualization, and integrated discovery, *Genome Biol.*, 4:P3, 2003.

DeRisi, J.L., V.R. Iyer, and P.O. Brown, Exploring the metabolic and genetic control of gene expression on a genomic scale, *Science*, 278:680–685, 1997.

DerSimonian, R. and N.M. Laird, Meta-analysis in clinical trials, *Control. Clin. Trials*, 7:177–188, 1986.

Devlin, B. and K. Roeder, Genomic control for association studies, *Biometrics*, 55:997–1004, 1999.

Diehn, M. et al., SOURCE: a unified genomic resource of functional annotations, ontologies, and gene expression data, *Nucleic Acids Res.*, 31:219–223, 2003.

Dobbin, K.K. et al., Interlaboratory comparability study of cancer gene expression analysis using oligonucleotide microarrays, *Clin. Cancer Res.*, 11:565–572, 2005.

Doolittle, W.F., Lateral genomics, *Trends Biochem. Sci.*, 24:M5–M8, 1999a.

Doolittle, W.F., Phylogenetic classification and the universal tree, *Science*, 284:2124–2129, 1999b.

Draghici, S. et al., Onto-Tools, the toolkit of the modern biologist: Onto-Express, Onto-Compare, Onto-Design and Onto-Translate, *Nucleic Acids Res.*, 31:3775–3781, 2003.

Drawid, A. and M. Gerstein, A Bayesian system integrating expression data with squence patterns for localizing proteins: Comprehensive application to the yeast genome, *J. Mol. Biol.*, 301:1059–1075, 2000.

Drigalenko, E., How sib pairs reveal linkage, *Am. J. Hum. Genet.*, 63:1242–1245, 1998.

Dudoit, S., J.P. Shaffer, and J.C. Boldrick, Multiple hypothesis testing in microarray experiments, *Stat. Sci.*, 18:71–103, 2003.

Dudoit, S. and T.P. Speed, Triangle constraints for sib-pair identity by descent probabilities under a general multilocus model for disease susceptibility, in Halloran, M.E. and S. Geisser, eds., *Statistics in Genetics*, vol. 112 of *IMA Volumes in Mathematics and its Applications*, pp. 181–221, New York: Springer-Verlag, 1999.

Dudoit, S., M.J. van der Laan, and K.S. Pollard, Multiple testing, Part I. Single-step procedures for control of general Type I error rates, *Stat. Appl. Genet. Mol. Biol.*, 3:Article 13, 2004.

Dudoit, S. et al., Statistical methods for identifying differentially expressed genes in replicated cDNA microarray experiments, *Stat. Sinica*, 12:111–139, 2002.

Dupuis, J. and D. Siegmund, Statistical methods for mapping quantitative trait loci from a dense set of markers, *Genetics*, 151:373–386, 1999.

Durand, D., B. Halldorsson, and B. Vernot, A hybrid micro-macroevolutionary approach to gene tree reconstruction, in Miyano, S. et al., eds., *Proceedings of the Ninth International Conference on Computational Molecular Biology (RECOMB2005)*, pp. 250–264, New York: Springer, 2005.

Durbin, B. et al., A variance-stabilizing transformation for gene-expression microarray data, *Bioinformatics*, 18 Suppl 1:S105–S110, 2002.

Edgar, R., M. Domrachev, and A.E. Lash, Gene Expression Omnibus: NCBI gene expression and hybridization array data repository, *Nucleic Acids Res.*, 30:207–210, 2002.

Edwards, S.V., L. Liu, and D.K. Pearl, High-resolution species trees without concatenation, *Proc. Natl. Acad. Sci. USA*, 104:5936–5941, 2007.

Efron, B. and C. Morris, Stein's estimator and its competitors – an empirical Bayes approach, *J. Am. Stat. Assoc.*, 68:117–130, 1973.

Efron, B. and C. Morris, Data analysis using Stein's estimator and its generalizations, *J. Am. Stat. Assoc.*, 70:379–421, 1975.

Efron, B. et al., Empirical Bayes analysis of a microarray experiment, *J. Am. Stat. Assoc.*, 96:1151–1160, 2001.

Egger, M. and G.D. Smith, Bias in location and selection of studies, *BMJ*, 316:61–66, 1998.

Egger, M. et al., Bias in meta-analysis detected by a simple, graphical test, *BMJ*, 315:629–634, 1997.

Eisen, M.B. et al., Cluster analysis and display of genome-wide expression patterns, *Proc. Natl. Acad. Sci. USA*, 95:14863–14868, 1998.

Ekstrom, C.T. and P. Dalgaard, Linkage analysis of quantitative trait loci in the presence of heterogeneity, *Hum. Hered.*, 55:16–26, 2003.

Elkon, R. et al., Genome-wide in silico identification of transcriptional regulators controlling the cell cycle in human cells, *Genome Res.*, 13:773–780, 2003.

Ellstrand, N.C., R. Whitkus, and L.H. Rieseberg, Distribution of spontaneous plant hybrids, *Proc. Natl. Acad. Sci. U S A.*, 93:5090–5093, 1996.

Elston, R.C. et al., Haseman and Elston revisited, *Genet. Epidemiol.*, 19:1–17, 2000.

Enright, A.J. et al., Protein interaction maps for complete genomes based on gene fusion events, *Nature*, 402:86–90, 1999.

Etzel, C.J. and T.J. Costello, Assessing linkage of immunoglobulin E using a meta-analysis approach, *Genet. Epidemiol.*, 21 Suppl 1:S97–S102, 2001.

Etzel, C.J. and R. Guerra, Meta-analysis of genetic-linkage analysis of quantitative-trait loci, *Am. J. Hum. Genet.*, 71:56–65, 2002.

Etzel, C.J., M. Liu, and T.J. Costello, An updated meta-analysis approach for genetic linkage, *BMC Genet.*, 6 Suppl 1:S43, 2005.

Eulenstein, O., A linear time algorithm for tree mapping, Tech. rep., Arbeitspapiere der GMD 1046, St Augustine, Germany, 1997, URL http://taxonomy.zoology.gla.ac.uk/rod/genetree/maths/Linear.pdf.

Eulenstein, O., B. Mirkin, and M. Vingron, Comparison of an annotatng duplication, tree mapping, and copying as methods to compare gene trees within species trees, in Mirkin, B. et al., eds., *Mathematical Hierarchies and Biology*, vol. 37 of *DIMACS series in Discrete Mathematics and Theoretical Computer Science*, pp. 71–93, Providence, RI: American Mathematical Society, 1997.

Everett, K. et al., Genome-wide high-density SNP-based linkage analysis of infantile hypertrophic pyloric stenosis identifies loci on chromosomes 11q14-q22 and Xq23, *Am. J. Hum. Genet.*, 82:756–762, 2008.

Everitt, B.S., L. Landau, and M. Leese, *Cluster Analysis*, New York: Oxford University Press, 4th ed., 2001.

Ewens, W.J., *Mathematical Population Genetics*, Berlin: Springer-Verlag, 1979.

Faraway, J.J., Distribution of the admixture test for the detection of linkage under heterogeneity, *Genet. Epidemiol.*, 10:75–83, 1993.

Feingold, E., Regression-based quantitative-trait-locus mapping in the 21st century, *Am. J. Hum. Genet.*, 71:217–222, 2002.

Feingold, E., P.O. Brown, and D. Siegmund, Gaussian models for genetic linkage analysis using complete high-resolution maps of identity by descent, *Am. J. Hum. Genet.*, 53:234–251, 1993.

Fellenberg, M. et al., An integrated probabilistic model for functional prediction of proteins, in Altman, R., ed., *Proceedings, Eighth International Conference on Intelligent Systems for Molecular Biology*, pp. 152–161, Menlo Park, CA: AAAI Press, 2000.

Fellows, M.R. et al., Analogs and duals of the MAST problem for sequences and trees, in Bilardi, G., ed., *Algorithms–ESA 1998: Proceedings of the 6th Annual European Symposium*, vol. 1461 of *Lecture Notes in Computer Science*, pp. 103–114, New York: Springer-Verlag, 1998.

Ferreira, M.A.R. et al., Collaborative genome-wide association analysis supports a role for ANK3 and CACNA1C in bipolar disorder, *Nat. Genet.*, 40:1056–1058, 2008.

Fisher, R.A., *Statistical Methods for Research Workers*, London: Oliver and Lloyd, 4th ed., 1932.

Fisher, S.A. et al., Meta-analysis of genome scans of age-related macular degeneration, *Hum. Mol. Genet.*, 14:2257–2264, 2005.

Fisher, S.A., J.S. Lanchbury, and C.M. Lewis, Meta-analysis of four rheumatoid arthritis genome-wide linkage studies – confirmation of a susceptibility locus on chromosome 16, *Arthritis Rheum.*, 48:1200–1206, 2003.

Folks, J.L., Combination of independent tests, in Krishnaiah, P.R. and P.K. Sen, eds., *Handbook of Statistics*, vol. 4, pp. 113–121, New York: North-Holland, 1984.

Fox, J., *car: Companion to Applied Regression*, 2008, URL http://socserv.socsci.mcmaster.ca/jfox/.

Frank, O. and D. Strauss, Markov graphs, *J. Am. Stat. Assoc.*, 81:832–842, 1986.

Friedman, N. et al., Using Bayesian networks to analyze expression data, *J. Comput. Biol.*, 7:601–620, 2000.

Fulker, D.W., S.S. Cherny, and L.R. Cardon, Multipoint interval mapping of quantitative trait loci, using sib pairs, *Am. J. Hum. Genet.*, 56:1224–1233, 1995.

Gadbury, G.L. et al., Evaluating statistical methods using plasmode data sets in the age of massive public databases: an illustration using false discovery rates, *PLoS Genet.*, 20:e1000098, 2008.

GAMES and The Transatlantic Multiple Sclerosis Genetics Cooperative, A meta-analysis of whole genome linkage screens in multiple sclerosis, *J. Neuroimmunol.*, 143:39–46, 2003.

Gao, F., B.C. Foat, and H.J. Bussemaker, Defining transcriptional networks through integrative modeling of mRNA expression and transcription factor binding data,

BMC Bioinformatics, 5:31, 2004.

Gardner, T.S. et al., Inferring genetic networks and identifying compound mode of action via expression profiling, *Science*, 301:102–105, 2003.

Gavin, A. et al., Functional organization of the yeast proteome by systematic analysis of protein complexes, *Nature*, 415:141–147, 2002.

Ge, Y., S. Dudoit, and T.P. Speed, Resampling-based multiple testing for microarray data analysis (with discussion), *Test*, 12:1–77, 2003.

Gentleman, R., Using GO for statistical analyses, in Antoch, J., ed., *COMPSTAT 2004 – Proceedings in Computational Statistics: 16th Symposium Held in Prague, Czech Republic, 2004*, pp. 171–180, 2004.

Gentleman, R., *annotate: Annotation for microarrays*, 2008, URL `http://www.bioconductor.org/download/`.

Gentleman, R.C. et al., BioConductor: Open software development for computational biology and bioinformatics, *Genome Biol.*, 5:R80, 2004, URL `http://genomebiology.com/2004/5/10/R80`.

Ghosh, D., Mixture models for assessing differential expression in complex tissues using microarray data, *Bioinformatics*, 20:1663–1669, 2004.

Ghosh, D. et al., Statistical issues and methods for meta-analysis of microarray data: a case study in prostate cancer, *Funct. Integr. Genomics*, 3:180–188, 2003.

Gilks, W.R., S. Richardson, and D.J. Spiegelhalter, *Markov Chain Monte Carlo in Practice*, Boca Raton, Florida: Princeton University Press, 1995.

Goldstein, D.R. and M. Delorenzi, Statistical design and data analysis for microarray experiments, in Berger, A. and M.A. Roberts, eds., *Unravelling Lipid Metabolism with Microarrays*, New York: Dekker, 2004.

Goldstein, D.R. et al., Meta-analysis by combining parameter estimates: simulated linkage studies, *Genet. Epidemiol.*, 17 Suppl 1:S581–S586, 1999.

Graur, D. and W.H. Li, *Fundamentals of Molecular Evolution*, Sunderland, MA: Sinauer, 2nd ed., 2000.

Gruber, T.R., A translational approach to portable ontologies, *Knowl. Acquis.*, 5:199–220, 1993.

Gruber, T.R., Toward principles for the design of ontologies used for knowledge sharing, *Int. J. Hum.-Comput. St.*, 43:907–928, 1995.

Gu, C. et al., Meta-analysis methodology for combining non-parametric sibpair linkage results: genetic homogeneity and identical markers, *Genet. Epidemiol.*, 15:609–626, 1998.

Guerra, R., Meta-analysis in human genetics studies, in Elston, R.C., J. Olson, and L. Palmer, eds., *Biostatistcal Genetics and Genetic Epidemiology*, New York: Wiley Reference in Biostatistics, 2002.

Guerra, R. et al., Meta-analysis by combining p-values: simulated linkage studies, *Genet. Epidemiol.*, 17 Suppl 1:S605–S609, 1999.

Gupta, R. and S. Brunak, Prediction of glycosylation across the human proteome and the correlation to protein function, in Altman, R.B. et al., eds., *Proceedings of the Seventh Pacific Symposium on Biocomputing (PSB02)*, pp. 310–322, River Edge, NJ: World Scientific, 2002.

Hallett, M.T. and J. Lagergren, New algorithms for the duplication-loss model, in Shamir, R., ed., *Proceedings of the Fourth International Conference on Computational Molecular Biology (RECOMB2000)*, pp. 138–146, New York: Association for Computing Machinery, 2000.

Hallett, M.T. and J. Lagergren, Efficient algorithms for lateral gene transfer problems, in Lengauer, T., ed., *Proceedings of the Fifth International Conference on Computational Molecular Biology (RECOMB2001)*, pp. 149–156, New York: Association for Computing Machinery, 2001.

Hanahan, D. and R.A. Weinberg, The hallmarks of cancer, *Cell*, 100:57–70, 2000.

Hao, W. and G.B. Golding, Patterns of bacterial gene movement, *Mol. Biol. Evol.*, 21:1294–1307, 2004.

Harbord, R.M., M. Egger, and J.A.C. Sterne, A modified test for small-study effects in meta-analyses of controlled trials with binary endpoints, *Stat. Med.*, 25:3443–3457, 2006.

Hartemink, A.J. et al., Using graphical models and genomic expression data to statistically validate models of genetic regulatory networks, in Altman, R., ed., *Pacific Symposium on Biocomputing 2001*, pp. 422–433, River Edge, NJ: World Scientific, 2001.

Hartemink, A.J. et al., Combining location and expression data for principled discovery of genetic regulatory network models, in Altman, R., ed., *Proceedings of the Seventh Pacific Symposium on Biocomputing (PSB02)*, pp. 437–449, River Edge, NJ: World Scientific, 2002.

Haseman, J.K. and R.C. Elston, The investigation of linkage between a quantitative trait and a marker locus, *Behav. Genet.*, 2:3–19, 1972.

Hauser, E.R. and M. Boehnke, Genetic linkage analysis of complex genetic traits by using affected sibling pairs, *Biometrics*, 54:1238–1246, 1998.

Hedges, L.V. and I. Olkin, *Statistical Methods for Meta-Analysis*, New York: Academic Press, 1985.

Hegyi, H. and M. Gerstein, The relationship between protein structure and function: a comprehensive survey with application to yeast genome, *J. Mol. Biol.*, 288:147–164, 1999.

Heijmans, B.T. et al., Meta-analysis of four new genome scans for lipid parameters and analysis of positional candidates in positive linkage regions, *Eur. J. Hum. Genet.*, 13:1143–1153, 2005.

Hein, J., Reconstructing evolution of sequences subject to recombination using parsimony, *Math. Biosci.*, 98:185–200, 1990.

Heo, M. et al., A meta-analytic investigation of linkage and association of common leptin receptor (LEPR) polymorphisms with body mass index and waist circumference, *Int. J. Obes. Relat. Metab. Disord.*, 26:640–646, 2002.

Higami, Y. et al., Adipose tissue energy metabolism: altered gene expression profile of mice subjected to long-term caloric restriction, *FASEB J.*, 18:415–417, 2004.

Hillis, D.M. et al., Experimental approaches to phylogenetic analysis, *Syst. Biol.*, 42:90–92, 1993.

Hillis, D.M. and J.P. Huelsenbeck, To tree the truth: Biological and numerical simulations of phylogeny, in Fambrough, D.M., ed., *Molecular Evolution of Physiological Processes*, vol. 49 of *Society of General Physiologists Series*, pp. 55–67, New York: Rockefeller University Press, 1994.

Hillis, D.M. and J.P. Huelsenbeck, Assessing molecular phylogenies, *Science*, 267:255–256, 1995.

Hishigaki, H. et al., Assessment of prediction accuracy of protein function from protein-protein interaction data, *Yeast*, 18:523–531, 2001.

Ho, Y. et al., Systematic identification of protein complexes in *Saccharomyces cerevisiae* by mass spectrometry, *Nature*, 415:180–183, 2002.

Hoaglin, D.C., F. Mosteller, and J.W. Tukey, *Understanding Robust and Exploratory Data Analysis*, New York: Wiley, 1983.

Hodgson, D.A., Generalized transduction of serotype 1/2 and serotype 4b strains of *Listeria monocytogenes*, *Mol. Microbiol.*, 35:312–323, 2000.

Holden, M. et al., GSEA-SNP: applying gene set enrichment analysis to SNP data from genome-wide association studies, *Bioinformatics*, 24:2784–2785, 2008.

Holliday, E. et al., The importance of modelling heterogenity in complex disease: application to NIMH schizophrenia genetics initiative data, *Hum. Genet.*, 117:160–167, 2005.

Huang, D. and W. Pan, Incorporating biological knowledge into distance-based clustering analysis of microarray gene expression data, *Bioinformatics*, 22:1259, 2006.

Huang, D., P. Wei, and W. Pan, Combining gene annotations and gene expression data in model-based clustering: Weighted method, *OMICS A Journal of Integrative Biology*, 10, 2006.

Huang, E. et al., Gene expression phenotypic models that predict the activity of oncogenic pathways, *Nat. Genet.*, 34:226–230, 2003.

Huang, X. and W. Pan, Comparing three methods for variance estimation with duplicated high density oligonucleotide arrays, *Funct. Integr. Genomics*, 2:126–183, 2002.

Hubner, N. et al., Integrated transcriptional profiling and linkage analysis for identification of genes underlying disease, *Nat. Genet.*, 37:243–253, 2005.

Hudson, R.R., Properties of the neutral allele model with intergenic recombination, *Theor. Popul. Biol.*, 23:183–201, 1983a.

Hudson, R.R., Testing the constant-rate neutral allele model with protein sequence data, *Evolution*, 37:203–217, 1983b.

Hughes, T.R. et al., Expression profiling using microarrays fabricated by an ink-jet oligonucleotide synthesizer, *Nat. Biotechnol.*, 19:342–347, 2001.

Huh, W.K. et al., Global analysis of protein localization in budding yeast, *Nature*, 425:686–691, 2003.

Ibrahim, J.G., M.H. Chen, and R.J. Gray, Bayesian models for gene expression with DNA microarray data, *J. Am. Stat. Assoc.*, 97:88–99, 2002.

Ideker, T., T. Galitski, and L. Hood, A new approach to decoding life: systems biology, *Annu. Rev. Genom. Hum. G.*, 2:343–372, 2001.

Ihaka, R. and R. Gentleman, R: A language for data analysis and graphics, *J. Comput. Graph. Stat.*, 5:299–314, 1996.

International Human Genome Sequencing Consortium, Initial sequencing and analysis of the human genome, *Nature*, 409:860–921, 2001.

Ionita-Laza, I. et al., Genetic association analysis of copy-number variation (CNV) in human disease pathogenesis, *Genomics*, 93:22–26, 2009.

Irizarry, R.A. et al., Summaries of Affymetrix GeneChip probe level data, *Nucleic Acids Res.*, 31:e15, 2003a.

Irizarry, R.A. et al., *affy: Methods for Affymetrix Oligonucleotide Arrays*, 2008, URL `http://www.bioconductor.org/download/`.

Irizarry, R.A. et al., Exploration, normalization, and summaries of high density oligonucleotide array probe level data, *Biostatistics*, 4:249–264, 2003b.

Irizarry, R.A. et al., Multiple-laboratory comparison of microarray platforms, *Nat. Methods*, 2:345–50, 2005.

Irizarry, R.A., Z. Wu, and H.A. Jaffee, Comparison of Affymetrix GeneChip expression measures, *Bioinformatics*, 22:789–794, 2006.

Ito, T. et al., A comprehensive two hybrid analysis to explore the yeast protein interactome, *Proc. Natl. Acad. Sci. USA*, 98:4569–4574, 2001.

Ito, T. et al., Toward a protein-protein interaction map of the budding yeast: a comprehensive system to examine two-hybrid interactions in all possible combinations between the yeast proteins, *Proc. Natl. Acad. Sci. USA*, 97:1143–1147, 2000.

Ivanova, N.B. et al., A stem cell molecular signature, *Science*, 298:601–604, 2002.

Iyengar, S. and J. Greenhouse, Selection models and the file drawer problem, *Stat. Sci.*, 3:109–117, 1988.

Iyengar, S.K. et al., Improved evidence for linkage on 6p and 5p with retrospective pooling of data from three asthma genome screens, *Genet. Epidemiol.*, 21 Suppl 1:S130–S135, 2001.

Jaimovich, A. et al., Towards an integrated protein-protein interaction network, in Miyano, S. et al., eds., *Proceedings of the Ninth International Conference on Computational Molecular Biology (RECOMB2005)*, pp. 14–30, New York: Association for Computing Machinery, 2005.

Jain, N. et al., Local pooled error test for identifying differentially expressed genes with a small number of replicated microarrays, *Bioinformatics*, 19:1945–1951, 2003.

Jansen, R. et al., A Bayesian networks approach for predicting protein-protein interactions from genomic data, *Science*, 302:449–453, 2003.

Jensen, F.V., *Bayesian Networks and Decision Graphs*, New York: Springer-Verlag, 2001.

Jensen, L.J. et al., Prediction of human protein function from post-translational modifications and localization features, *J. Mol. Biol.*, 319:1257–1265, 2002.

Ji, Y. et al., RefSeq refinements of UniGene-based gene matching improves the correlation between microarray platforms, Tech. rep., MD Anderson Cancer Center, Department of Biostatistics and Applied Mathematics, 2005.

Jiang, H. et al., Joint analysis of two microarray gene-expression data sets to select lung adenocarcinoma marker genes, *BMC Bioinformatics*, 5:81, 2004.

Jin, C. et al., Selective phenotyping for increased efficiency in genetic mapping studies, *Genetics*, 168:2285–2293, 2004.

Jin, G. et al., Maximum likelihood of phylogenetic networks, *Bioinformatics*, 22:2604–2611, 2006.

Jin, G. et al., Efficient parsimony-based methods for phylogenetic network reconstruction, *Bioinformatics*, 23:e123–e128, 2007a.

Jin, G. et al., Inferring phylogenetic networks by the maximum parsimony criterion: a case study, *Mol. Biol. Evol.*, 24:324–337, 2007b.

Jin, G. et al., A new linear-time heuristic algorithm for computing the parsimony score of phylogenetic networks: Theoretical bounds and empirical performance, in Mandoiu, I. and A. Zelikovsky, eds., *Proceedings of the International Symposium on Bioinformatics Research and Applications*, vol. 4463 of *Lecture Notes in Bioinformatics*, pp. 61–72, 2007c.

Jin, W. et al., The contributions of sex, genotype and age to transcriptional variance in *Drosophila melanogaster*, *Nat. Genet.*, 29:389–395, 2001.

Juo, S.H.H. et al., Mild association between the A/G polymorphism in the promoter of the apolipoprotein A-I gene and apolipoprotein A-I levels: A meta-analysis, *Am. J. Med. Genet.*, 82:235–241, 1978.

Kacser, H. and J. Burns, The control of flux, *Symp. Soc. Exp. Biol.*, 27:65–104, 1973.

Kamb, A. and A. Ramaswami, A simple method for statistical analysis of intensity differences in microarray-derived gene expression data, *BMC Biotechnology*, 1:8, 2001.

Kanehisa, M. and S. Goto, KEGG: Kyoto Encyclopedia of Genes and Genomes, *Nucleic Acids Res.*, 28:27–30, 2000.

Kaufman, L. and P. Rousseeuw, *Finding Groups in Data: An Introduction to Cluster Analysis*, New York: Wiley, 1990.

Kell, D.B. and R.D. King, On the optimization of classes for the assignment of unidentified reading frames in functional genomics programmes: the need for machine learning, *Trends Biotechnol.*, 18:93–98, 2000.

Kendziorski, C. et al., Statistical methods for expression trait loci (ETL) mapping, Tech. rep., University of Wisconsin, Dept. of Biostatistics and Medical Informatics, Technical Report 184, Madison, WI, 2004.

Kendziorski, C. et al., Statistical methods for expression quantitative trait loci (eQTL) mapping, *Biometrics*, 62:19–27, 2006.

Kendziorski, C.M. et al., On parametric empirical Bayes methods for comparing multiple groups using replicated gene expression profiles, *Stat. Med.*, 22:3899–3914, 2003.

Kerr, M.K. and G.A. Churchill, Statistical design and the analysis of gene expression microarray data, *Genet. Res.*, 77:123–128, 2001.

Kim, K. et al., Picking the most likely candidates for further development: Novel intersection-union tests for addressing multi-component hypotheses in comparative genomics, in *Proceedings of the American Statistical Association, ENAR Section*, Alexandria, VA: American Statistical Association, 2004, CD-ROM.

Kimura, M., The number of heterozygous nucleotide sites maintained in a finite population due to steady flux of mutations, *Genetics*, 61:893–903, 1969.

King, R.D. et al., The utility of different representations of protein sequence for predicting functional class, *Bioinformatics*, 17:445–454, 2001.

Kingman, J.F.C., The coalescent, *Stoch. Proc. Appl.*, 13:235–248, 1982.

Koivukoski, L. et al., Meta-analysis of genome-wide scans for hypertension and blood pressure in Caucasians shows evidence of susceptibility regions on chromosomes 2 and 3, *Hum. Mol. Genet.*, 13:2325–2332, 2004.

Kondor, R. and J. Lafferty, Diffusion kernels on graphs and other discrete input spaces, in Sammut, C. and A. Hoffmann, eds., *Proceedings of the International Conference on Machine Learning (ICML)*, pp. 315–322, San Francisco, CA: Morgan Kaufmann, 2002.

Kooperberg, C. et al., Significance testing for small sample microarray experiments, *Stat. Med.*, 24:2281–2298, 2005.

Korn, J.M. et al., Integrated genotype calling and association analysis of SNPs, common copy number polymorphisms and rare CNVs, *Nat. Genet.*, 40:1253–1260, 2008.

Koziol, J.A. and M.D. Perlman, Combining independent chi-squared tests, *J. Am. Stat. Assoc.*, 73:753–763, 1978.

Kruglyak, L., The road to genome-wide association studies, *Nat. Rev. Genet.*, 9:314–318, 2008.

Kruglyak, L. et al., Parametric and nonparametric linkage analysis: a unified multipoint approach, *Am. J. Hum. Genet.*, 58:1347–1363, 1996.

Kruglyak, L. and E.S. Lander, Complete multipoint sib-pair analysis of qualitative and quantitative traits, *Am. J. Hum. Genet.*, 57:439–454, 1995.

Kubatko, L.S. and J.H. Degnan, Inconsistency of phylogenetic estimates from concatenated data under coalescence, *Syst. Biol.*, 56:17–24, 2007.

Kuhn, K. et al., A novel, high-performance random array platform for quantitative gene expression profiling, *Genome Res.*, 14:2347–2356, 2004.

Kulinskaya, E., S. Morgenthaler, and R.G. Staudte, *A Guide to Calibrating and Combining Statistical Evidence*, New York: Wiley-Interscience, 2008.

Kuo, C.H., J.P. Wares, and J.C. Kissinger, The Apicomplexan whole-genome phylogeny: an analysis of incongruence among gene trees, *Mol. Biol. Evol.*, 25:2689–2698, 2008.

Kuo, W.P. et al., Analysis of matched mRNA measurements from two different microarray technologies, *Bioinformatics*, 18:405–412, 2002.

Kurland, C.G., B. Canback, and O.G. Berg, Horizontal gene transfer: A critical view, *Proc. Natl. Acad. Sci. USA*, 100:9658–9662, 2003.

Lamb, J. et al., A mechanism of cyclin D1 action encoded in the patterns of gene expression in human cancer, *Cell*, 114:323–334, 2003.

Lan, H. et al., Dimension reduction for mapping mRNA abundance as quantitative traits, *Genetics*, 164:1607–1614, 2003.

Lanckriet, G.R.G. et al., A statistical framework for genomic data fusion, *Bioinformatics*, 20:2626–2635, 2004a.

Lanckriet, G.R.G. et al., Kernel-based data fusion and its application to protein function prediction in yeast, in Altman, R., ed., *Pacific Symposium on Biocomputing 2004*, pp. 300–311, River Edge, NJ: World Scientific, 2004b.

Lanckriet, G.R.G. et al., Kernel-based integration of genomic data using semidefinite programming, in Schölkopf, B., K. Tsuda, and J.P. Vert, eds., *Kernel Methods in Computational Biology*, pp. 71–92, Cambridge, MA: MIT Press, 2004c.

Lander, E. and L. Kruglyak, Genetic dissection of complex traits – guidelines for interpreting and reporting linkage results, *Nat. Genet.*, 11:241–247, 1995.

Lander, E.S. and D. Botstein, Mapping Mendelian factors underlying quantitative traits using RFLP linkage maps, *Genetics*, 121:185–199, 1989.

Lange, K. et al., Mendel version 4.0: A complete package for the exact genetic analysis of discrete traits in pedigree and population data sets, *Am. J. Hum. Genet.*, 69 (supplement):504, 2001.

Lauritzen, S., *Graphical Models*, Oxford: Oxford University Press, 1996.

Lebrec, J., H. Putter, and J.C. van Houwelingen, Score test for detecting linkage to complex traits in selected samples, *Genet. Epidemiol.*, 27:97–108, 2004.

Lee, H.J. et al., Assessment of the reliability of protein-protein interactions using protein localization and gene expression data, in *Proceedings of the International Joint Conference of InCoB, AASBi and KSBI (BIOINFO 2005)*, 2005a, URL http://www-hto.usc.edu/people/tingchen/MyPublication/Bioinfo-Localization-2005.pdf.

Lee, H.J. et al., Diffusion kernel based logistic regression models for protein function prediction, *OMICS*, 10:40–55, 2006.

Lee, H.K. et al., Coexpression analysis of human genes across many microarray data sets, *Genome Res.*, 14:1085–1094, 2004.

Lee, J.K., Discovery and validation of microarray gene expression patterns, *LabMedica Intern.*, 19:8–10, 2002.

Lee, J.K. and M. O'Connell, An S-PLUS library for the analysis of differential expression, in Parmigiani, G. et al., eds., *The Analysis of Gene Expression Data: Methods and Software*, pp. 163–184, New York: Springer-Verlag, 2003.

Lee, T.I. et al., Transcriptional regulatory networks in *Saccharomyces cerevisiae*, *Science*, 298:799–804, 2002.

Lee, Y.H. et al., Microarray profiling of isolated abdominal subcutaneous adipocytes from obese vs non-obese Pima Indians: increased expression of inflammation-related genes, *Diabetologia*, 48:1776–1783, 2005b.

Lerat, E., V. Daubin, and N.A. Moran, From gene trees to organismal phylogeny in prokaryotes: The case of the γ-proteobacteria, *PLoS Biol.*, 1:1–9, 2003.

Letovsky, S. and S. Kasif, Predicting protein function from protein/protein interaction data: a probabilistic approach, *Bioinformatics*, 19 (Suppl. 1):197–204, 2003.

Levinson, D.F. et al., No major schizophrenia locus detected on chromosome 1q in a large multicenter sample, *Science*, 296:739–741, 2002.

Levinson, D.F. et al., Genome scan meta-analysis of schizophrenia and bipolar disorder, part I: Methods and power analysis, *Am. J. Hum. Genet.*, 73:17–33, 2003.

Lewin, B., ed., *Genes IX*, Boston, MA: Jones and Bartlett Boston, 9th ed., 2007.

Lewis, C.M. et al., Genome scan meta-analysis of schizophrenia and bipolar disorder, part II: Schizophrenia, *Am. J. Hum. Genet.*, 73:34–48, 2003.

Li, C. and W.H. Wong, Model-based analysis of oligonucleotide arrays: Expression index computation and outlier detection, *Proc. Natl. Acad. Sci. USA*, 98:31–36, 2001.

Li, J. and M. Burmeister, Genetical genomics: combining genetics with gene expression analysis, *Hum. Mol. Genet.*, 14:R163–R169, 2005.

Li, Z. and D.C. Rao, Random effects model for meta-analysis of multiple quantitative sibpair linkage studies, *Genet. Epidemiol.*, 13:377–383, 1996.

Liang, S., S. Fuhrman, and R. Somogyi, REVEAL, a general reverse engineering algorithm for inference of genetic network architectures, in Altman, R., ed., *Pacific Symposium on Biocomputing 1998*, pp. 18–29, River Edge, NJ: World Scientific, 1998.

Liao, J.C. et al., Network component analysis: reconstruction of regulatory signals in biological systems, *Proc. Natl. Acad. Sci. USA*, 100:15522–15527, 2003.

Lin, N. et al., Information assessment on predicting protein-protein interactions, *BMC Bioinformatics*, 5:154, 2004.

Linder, C.R. and L.H. Rieseberg, Reconstructing patterns of reticulate evolution in plants, *Am. J. Bot.*, 91:1700–1708, 2004.

Linn, S.C. et al., Gene expression patterns and gene copy number changes in dermatofibrosarcoma protuberans, *American Journal of Pathology*, 163:2383–2395, 2003.

Liu, J.S., *Monte Carlo Strategies in Scientific Computing*, New York: Springer-Verlag, 2001.

Liu, L. and D.K. Pearl, Species trees from gene trees: reconstructing Bayesian posterior distributions of a species phylogeny using estimated gene tree distributions, *Syst. Biol.*, 56:504–514, 2007.

Liu, X.S., D.L. Brutlag, and J.S. Liu, An algorithm for finding protein-DNA binding sites with applications to chromatin-immunoprecipitation microarray experiments, *Nat. Biotechnol.*, 20:835–839, 2002.

Lockhart, D.J. et al., Expression monitoring by hybridization to high-density oligonucleotide arrays, *Nat. Biotechnol.*, 14:1675–1680, 1996.

Lockwood, J.R., K. Roeder, and B. Devlin, A Bayesian hierarchical model for allele frequencies, *Genet. Epidemiol.*, 20:17–33, 2001.

Loesgen, S. et al., Weighting schemes in pooled linkage analysis, *Genet. Epidemiol.*, 21 Suppl 1:S142–S147, 2001.

Lönnstedt, I. et al., Microarray analysis of two interacting treatments: a linear model, Tech. rep., Uppsala University, Sweden, Department of Mathematics, 2001.

Lönnstedt, I. and T.P. Speed, Replicated microarray data, *Stat. Sinica*, 12:31–46, 2002.

Lu, L.J. et al., Assessing the limits of genomic data integration for predicting protein networks, *Genome Res.*, 15:945–953, 2005.

Lumley, T., *rmeta: Meta-analysis*, 2008, URL `http://cran.r-project.org/`.

Luthi-Carter, R. et al., Decreased expression of striatal signaling genes in a mouse model of Huntington's disease, *Hum. Mol. Genet.*, 9:1259–1271, 2000.

Luthi-Carter, R. et al., Dysregulation of gene expression in the R6/2 model of polyglutamine disease: parallel changes in muscle and brain, *Hum. Mol. Genet.*, 11:1911–1926, 2002a.

Luthi-Carter, R. et al., Polyglutamine and transcription: gene expression changes shared by DRPLA and Huntington's disease mouse models reveal context-independent effects, *Hum. Mol. Genet.*, 11:1927–1937, 2002b.

Ma, B., M. Li, and L. Zhang, On reconstructing species trees from gene trees in terms of duplications and losses, in Istrail, S., P. Pevzner, and M. Waterman, eds., *Proceedings of the Second Annual International Conference on Computational Molecular Biology (RECOMB98)*, pp. 182–191, New York: Association for Computing Machinery, 1998.

Ma, B., M. Li, and L. Zhang, From gene trees to species trees, *SIAM J. Comput.*, 30:729–752, 2000.

Ma, S. and J. Huang, Regularized gene selection in cancer microarray meta-analysis,

BMC Bioinformatics, 10:1, 2009.

MacLeod, D. et al., Deduction of probable events of lateral gene transfer through comparison of phylogenetic trees by recursive consolidation and rearrangement, *BMC Evol. Biol.*, 8:27, 2005.

Maddison, W.P., Phylogenetic histories within and among species, in Hoch, P.C. and A.G. Stephenson, eds., *Experimental and Molecular Approaches to Plant Biosystematics*, vol. 53 of *Monographs in Systematics*, pp. 273–287, St. Louis, MO: Missouri Botanical Garden, 1995.

Maddison, W.P., Gene trees in species trees, *Syst. Biol.*, 46:523–536, 1997.

Maddison, W.P. and L.L. Knowles, Inferring phylogeny despite incomplete lineage sorting, *Syst. Biol.*, 55:21–30, 2006.

Mah, N. et al., A comparison of oligonucleotide and cDNA-based microarray systems, *Physiol. Genomics*, 16:361–370, 2004.

Makarenkov, V., T-REX: reconstructing and visualizing phylogenetic trees and reticulation networks, *Bioinformatics*, 17:664–668, 2001.

Mallet, J., Hybridization as an invasion of the genome, *Trends Ecol. Evol.*, 20:229–237, 2005.

Mallet, J., Hybrid speciation, *Nature*, 446:279–283, 2007.

Mangiarini, L. et al., Exon 1 of the HD gene with an expanded CAG repeat is sufficient to cause a progressive neurological phenotype in transgenic mice, *Cell*, 87:493–506, 1996.

Marazita, M.L. et al., Meta-analysis of 13 genome scans reveals multiple cleft lip/palate genes with novel loci on 9q21 and 2q32-35, *Am. J. Hum. Genet.*, 75:161–173, 2004.

Marcotte, E.M. et al., Detecting protein function and protein-protein interactions from genome sequences, *Science*, 285:751–753, 1999a.

Marcotte, E.M. et al., A combined algorithm for genome-wide prediction of protein function, *Nature*, 402:83–86, 1999b.

Marshall, E., Getting the noise out of gene arrays, *Science*, 306:630–631, 2004.

McClelland, M. et al., Comparison of genome degradation in Paratyphi A and Typhi, human-restricted serovars of *Salmonella enterica* that cause typhoid, *Nat. Genet.*, 36:1268–1274, 2004.

McQueen, M.B. et al., Combined analysis from eleven linkage studies of bipolar disorder provides strong evidence of susceptibility loci on chromosomes 6q and 8q, *Am. J. Hum. Genet.*, 77:582–595, 2005.

Mecham, B.H. et al., Sequence-matched probes produce increased cross-platform consistency and more reproducible biological results in microarray-based gene expression measurements, *Nucleic Acids Res.*, 32:e74, 2004a.

Mecham, B.H. et al., Increased measurement accuracy for sequence-verified microarray probes, *Physiol. Genomics*, 18:308–315, 2004b.

Mehta, C., N. Patel, and R. Gray, Computing an exact confidence interval for the common odds ratio in several 2x2 contingency tables, *J. Am. Stat. Assoc.*, 80:969–973, 1985.

Meng, C. and L.S. Kubatko, Detecting hybrid speciation in the presence of incomplete lineage sorting using gene tree incongruence: A model, *Theor. Popul. Biol.*, 2009, in press.

Mering, C.V. et al., Comparative assessment of large scale data sets of protein-protein interactions, *Nature*, 417:399–403, 2002.

Mewes, H. et al., MIPS: analysis and annotation of proteins from whole genomes, *Nucleic Acids Res.*, 32:D41–D44, 2004.

Mewes, H.W. et al., MIPS: a database for genomes and protein sequences, *Nucleic Acids Res.*, 30:31–34, 2002.

Mok, S.C. et al., Prostasin, a potential serum marker for ovarian cancer: identification through microarray technology, *J. Natl. Cancer Inst.*, 93:1458–1464, 2001.

Moore, W.S., Inferring phylogenies from mtDNA variation: mitochondrial gene trees versus nuclear gene trees, *Evolution*, 49:718–726, 1995.

Mootha, V.K. et al., PGC-1alpha-responsive genes involved in oxidative phosphorylation are coordinately downregulated in human diabetes, *Nat. Genet.*, 34:267–273, 2003.

Moran, P., Random processes in genetics, *Proc. Cambridge Philos. Soc.*, 54:60–71, 1958.

Moreau, Y. et al., Comparison and meta-analysis of microarray data: from the bench to the computer desk, *Trends Genet.*, 19:570–577, 2003.

Moret, B.M.E. et al., Phylogenetic networks: modeling, reconstructibility, and accuracy, *IEEE/ACM Trans. Comput .Biol. Bioinform.*, 1:13–23, 2004.

Moret, B.M.E., U. Roshan, and T. Warnow, Sequence length requirements for phylogenetic methods, in Guigó, R. and D. Gusfield, eds., *Proceedings of the Second International Workshop Algorithms in Bioinformatics (WABI02)*, vol. 2452 of *Lecture Notes in Computer Science*, pp. 343–356, New York: Springer-Verlag, 2002.

Morley, M. et al., Genetic analysis of genome-wide variation in human gene expression, *Nature*, 430:743–747, 2004.

Morris, J.S. et al., Pooling information across different studies and oligonucleotide microarray chip types to identify prognostic genes for lung cancer, in Shoemaker, J. and S.M. Lin, eds., *Methods of Microarray Data Analysis IV*, pp. 51–66, New York: Springer-Verlag, 2005.

Morris, R.J. et al., Capturing and profiling adult hair follicle stem cells, *Nat. Biotechnol.*, 22:411–417, 2004.

Morton, N.E., Sequential tests for the detection of linkage, *Am. J. Hum. Genet.*, 7:277–318, 1955.

Mower, J.P. et al., Plant genetics: gene transfer from parasitic to host plants, *Nature*, 432:165–166, 2004.

Mrowka, R., A. A. Patzak, and H. Herzel, Is there a bias in proteome research?, *Genome Res.*, 11:1971–1973, 2001.

Naggert, J.K. et al., Hyperproinsulinaemia in obese fat/fat mice associated with a carboxypeptidase E mutation which reduces enzyme activity, *Nat. Genet.*, 10:135–142, 1995.

Nakamura, Y. et al., Biased biological functions of horizontally transferred genes in prokaryotic genomes, *Nat. Genet.*, 36:760–766, 2004.

Nakhleh, L. et al., Reconstructing phylogenetic networks using maximum parsimony, in *Proceedings of the 2005 IEEE Computational Systems Bioinformatics Conference (CSB2005)*, pp. 93–102, Los Alamitos, California: IEEE Computer Society Press, 2005a, doi:10.1109/CSB.2005.47.

Nakhleh, L. et al., The accuracy of phylogenetic methods for large datasets, in Altman, R., ed., *Proceedings of the Seventh Pacific Symposium on Biocomputing (PSB02)*, pp. 211–222, River Edge, NJ: World Scientific, 2002.

Nakhleh, L. et al., Designing fast converging phylogenetic methods, *Bioinformatics*, 17 Suppl 1:S190–S198, 2001a.

Nakhleh, L. et al., The performance of phylogenetic methods on trees of bounded diameter, in Gascuel, O. and B.M.E. Moret, eds., *Proceedings of the First International Workshop Algorithms in Bioinformatics (WABI01)*, vol. 2149 of *Lecture Notes in Computer Science*, pp. 214–226, New York: Springer-Verlag, 2001b.

Nakhleh, L., D. Ruths, and L.S. Wang, RIATA-HGT: A fast and accurate heuristic for reconstrucing horizontal gene transfer, in Wang, L., ed., *Proceedings of the 11th International Conference Computing and Combinatorics (COCOON05)*, vol. 3595 of *Lecture Notes in Computer Science*, New York: Springer-Verlag, 2005b.

Nakhleh, L., T. Warnow, and C.R. Linder, Reconstructing reticulate evolution in species – theory and practice, in Gusfield, D. et al., eds., *Proceedings of the Eighth International Conference on Computational Molecular Biology (RECOMB2004)*, pp. 337–346, New York: Association for Computing Machinery, 2004.

National Research Council, *Combining Information: Statistical Issues and Opportunities for Research*, Washington, D.C.: National Academy Press, 1992.

NCI-NHGRI Working Group on Replication in Association Studies, Replicating genotype-phenotype associations, *Nature*, 447:655–660, 2007.

Neale, M.C. et al., *Mx: Statistical Modeling*, Department of Psychiatry, Virginia Commonwealth University; Richmond, VA, 5th ed., 1999.

Newton, M. and C. Kendziorski, Parametric empirical Bayes methods for microarrays, in Parmigiani, G. et al., eds., *The Analysis of Gene Expression Data: Methods and Software*, pp. 254–271, New York: Springer-Verlag, 2003.

Newton, M.A. et al., On differential variability of expression ratios: Improving statistical inference about gene expression changes from microarray data, *J. Comput. Biol.*, 8:37–52, 2001.

Newton, M.A. et al., Detecting differential gene expression with a semiparametric hierarchical mixture method, *Biostatistics*, 5:155–176, 2004.

Nica, A.C. and E.T. Dermitzakis, Using gene expression to investigate the genetic basis of complex disorders, *Hum. Mol. Genet.*, 17:R129–R134, 2008.

Nichols, R., Gene trees and species trees are not the same, *Trends Ecol. Evol.*, 16:358–364, 2001.

Nielsen, T.O. et al., Molecular characterization of soft tissue tumours: a gene expression study, *Lancet*, 359:1301–1307, 2002.

Noor, M.A. and J.L. Feder, Speciation genetics: evolving approaches, *Nat. Rev. Genet.*, 7:851–861, 2006.

Normand, S.L., Meta-analysis: formulating, evaluating, combining and reporting, *Stat. Med.*, 18:321–359, 1999.

North, B.V., D. Curtis, and P.C. Sham, A note on the calculation of empirical P values from Monte Carlo procedures, *Am. J. Hum. Genet.*, 72:498–499, 2003.

Ochman, H., J.G. Lawrence, and E.A. Groisman, Lateral gene transfer and the nature of bacterial innovation, *Nature*, 405:299–304, 2000.

O'Connell, M., Differential expression, class discovery and class prediction using S-PLUS and S+ArrayAnalyzer, *SIGKDD Explorations*, 5:38–47, 2003.

O'Donovan, M.C., N.M. Williams, and M.J. Owen, Recent advances in the genetics of schizophrenia, *Hum. Mol. Genet.*, 12:R125–R133, 2003.

Ohno, S., *Evolution by Gene Duplication*, New York: Springer-Verlag, 1970.

Olson, J.M. et al., Linkage of chromosome 1 markers to alcoholism related phenotypes by sib pair linkage analysis of principal components, *Genet. Epidemiol.*, 17 Suppl 1:S271–S276, 1999.

Olson, J.M. and E. Wijsman, Linkage between quantitative trait and marker locus: methods using all relative pairs, *Genet. Epidemiol.*, 10:87–102, 1993.

Olson, M.V., When less is more: gene loss as an engine of evolutionary change, *Am. J. Hum. Genet.*, 64:18–23, 1999.

Ott, J., *Analysis of Human Genetic Linkage*, Baltimore, MD: Johns Hopkins University Press, 3rd ed., 1999.

Overbeek, R. et al., The use of gene clusters to infer functional coupling, *Proc. Natl. Acad. Sci. USA*, 96:2896–2901, 2000.

Page, R., GeneTree: Comparing gene and species phylogenies using reconciled trees, *Bioinformatics*, 14:819–820, 1998.

Page, R. and M.A. Charleston, From gene to organismal phylogeny: Reconciled trees and the gene tree/species tree problem, *Mol. Phylogenet. Evol.*, 7:231–240, 1997.

Page, R. and M.A. Charleston, Trees within trees: phylogeny and historical associations, *Trends Ecol. Evol.*, 13:356–359, 1998.

Pahl, R., H. Schäfer, and H.H. Müller, Optimal multistage designs – a general framework for efficient genome-wide association studies, *Biostatistics*, 2008, advance access at doi:10.1093/biostatistics/kxn036.

Pan, W., Incorporating gene functions as priors in model-based clustering of microarray gene expression data, *Bioinformatics*, 22:795, 2006.

Papaemmanuil, E. et al., Deciphering the genetics of hereditary non-syndromic colorectal cancer, *Eur. J. Hum. Genet.*, 82:1477–1486, 2008.

Pardi, F., D.F. Levinson, and C.M. Lewis, GSMA: software implementation of the genome search meta-analysis method, *Bioinformatics*, 21:4430–4431, 2005.

Parker, R.A. and R.B. Rothenberg, Identifying important results from multiple statistical tests, *Stat. Med.*, 7:1031–1043, 1988.

Parmigiani, G. et al., A statistical framework for molecular-based classification in cancer, *J. Roy. Stat. Soc. B*, 64:717–736, 2002.

Parmigiani, G. et al., A cross-study comparison of gene expression studies for the molecular classification of lung cancer, *Clin. Cancer Res.*, 10:2922–2927, 2004.

Pavlidis, P. and J. Weston, Gene functional classification from heterogeneous data, in Lengauer, T., ed., *Proceedings of the Fifth International Conference on Computational Molecular Biology (2003)*, pp. 249–255, New York: Association for Computing Machinery, 2001.

Pearson, K., On a method of determining whether a sample of size n supposed to have been drawn from a parent population having a known probability integral has probably been drawn at random, *Biometrika*, 25:379–410, 1933.

Pearson, T.A. and T.A. Maniolo, How to interpret a genome-wide association study, *JAMA*, 299:1335–1344, 2008.

Pellegrini, M. et al., Assigning protein functions by comparative genome analysis: protein phylogenetic profiles, *Proc. Natl. Acad. Sci. USA*, 96:4285–4288, 1999.

Peña-Castillo, L. et al., A critical assessment of *Mus musculus* gene function prediction using integrated genomic evidence, *Genome Biol.*, 9 Suppl 1:S2, 2008.

Peng, G. et al., Gene and pathway-based analysis – second wave of genome-wide association studies, 2008, URL http://hdl.handle.net/10101/npre.2008.2068.1, available from Nature Precedings.

Peri, S. et al., Development of human protein reference database as an initial platform for approaching systems biology in humans, *Genome Res.*, 13:2363–2371, 2003.

Pollard, D.A. et al., Widespread discordance of gene trees with species tree in *Drosophila*: evidence for incomplete lineage sorting, *PLoS Genet.*, 2:e173, 2006.

Posada, D. and K.A. Crandall, The effect of recombination on the accuracy of phylogeny estimation, *J. Mol. Evol.*, 54:396–402, 2002.

Posada, D., K.A. Crandall, and E.C. Holmes, Recombination in evolutionary genomics, *Annu. Rev. Genet.*, 36:75–97, 2002.

Pounds, S. and C. Cheng, Improving false discovery rate estimation, *Bioinformatics*, 20:1737–1745, 2004.

Pounds, S. and S. Morris, Estimating the occurrence of false positives and false negatives in microarray studies by approximating and partitioning the empirical distribution of p-values, *Bioinformatics*, 19:1236–1242, 2003.

Pratt, J.W., Bayesian interpretation of standard inference statements, *J. Roy. Stat. Soc. B*, 27:169–203, 1965.

Price, A.L. et al., Principal components analysis corrects for stratification in genome-wide association studies, *Nat. Genet.*, 38:904–909, 2006.

Pritchard, J.K. et al., Association mapping in structured populations, *Am. J. Hum. Genet.*, 67:170–181, 2000.

Prokopenko, I. et al., Variants in MTNR1B influence fasting glucose levels, *Nat. Genet.*, 41:77–81, 2008.

Province, M.A., The significance of not finding a gene, *Am. J. Hum. Genet.*, 69:660–663, 2001.

Province, M.A. et al., A meta-analysis of genome-wide linkage scans for hypertension: the National Heart, Lung and Blood Institute Family Blood Pressure Program, *Biometrika*, 16:144–147, 2003.

Putter, H., J. Lebrec, and J.C. van Houwelingen, Selection strategies for linkage studies using twins, *Twin Res.*, 6:377–382, 2003.

Putter, H., L.A. Sandkuijl, and J.C. van Houwelingen, Score test for detecting linkage to quantitative traits, *Genet. Epidemiol.*, 22:345–355, 2002.

R Development Core Team, *R: A Language and Environment for Statistical Computing*, R Foundation for Statistical Computing, Vienna, Austria, 2005, URL {http://www.R-project.org}, ISBN 3-900051-07-0.

Ramalho-Santos, M. et al., "stemness": Transcriptional profiling of embryonic and adult stem cells, *Science*, 298:587–600, 2002.

Ramasamy, A. et al., Key issues in conducting a meta-analysis of gene expression microarray datasets, *PLoS Med.*, 5:e184, 2008.

Rao, D.C., CAT scans, PET scans, and genomic scans, *Genet. Epidemiol.*, 15:1–18, 1998.

Rao, D.C. and C. Gu, False positives and false negatives in genome scans, *Adv. Genet.*, 42:487–498, 2001.

Rao, D.C. and M.A. Province, *Genetic Dissection of Complex Traits*, San Diego, CA: Academic Press, 2001.

Reiner, A., D. Yekutieli, and Y. Benjamini, Identifying differentially expressed genes using false discovery rate controlling procedures, *Bioinformatics*, 19:368–375, 2003.

Ren, B. et al., Genome-wide location and function of DNA binding proteins, *Science*, 290:2306–2309, 2000.

Rhodes, D.R. et al., Meta-analysis of microarrays: interstudy validation of gene expression profiles reveals pathway dysregulation in prostate cancer, *Cancer Res.*, 62:4427–4433, 2002.

Rhodes, D.R. et al., Mining for regulatory programs in the cancer transcriptome, *Nat. Genet.*, 37:579–583, 2005a.

Rhodes, D.R. et al., Probabilistic model of the human protein-protein interaction network, *Nat. Biotechnol.*, 23:951–959, 2005b.

Rhodes, D.R. et al., Large-scale meta-analysis of cancer microarray data identifies common transcriptional profiles of neoplastic transformation and progression, *Proc. Natl. Acad. Sci. USA*, 101:9309–9314, 2004a.

Rhodes, D.R. et al., ONCOMINE: a cancer microarray database and integrated data-mining platform, *Neoplasia*, 6:1–6, 2004b.

Rice, J.P., The role of meta-analysis in linkage studies of complex traits, *Am. J. Med. Genet.*, 74:112–114, 1997.

Rice, W.R., A consensus combined p-value test and the family-wide significance of component tests, *Biometrics*, 46:303–308, 1990.

Richardson, S., Measurent error modelling from a Bayesian perspective, in *Bulletin of the International Statistical Institute*, 1999, 52nd Session: Helsinki 1999.

Richardson, S. and W.R. Gilks, Conditional independence models for epidemiological studies with covariate measurement error, *Stat. Med.*, 12:1703–1722, 1993.

Rieseberg, L.H., S.J. Baird, and K.A. Gardner, Hybridization, introgression, and linkage evolution, *Plant Mol. Biol.*, 42:205–224, 2000.

Rieseberg, L.H. and S.E. Carney, Plant hybridization, *New Phytol.*, 140:599–624, 1998.

Risch, N., Linkage strategies for genetically complex traits. II. The power of affected relative pairs, *Am. J. Hum. Genet.*, 46:229–241, 1990.

Risch, N. and K. Merikangas, The future of genetic studies of complex human diseases, *Science*, 273:1516–1517, 1996.

Roberts, C.J. et al., Signaling and circuitry of multiple MAPK pathways revealed by a matrix of global gene expression profiles, *Science*, 287:873–880, 2000.

Rokas, A. et al., Genome-scale approaches to resolving incongruence in molecular phylogenies, *Nature*, 425:798–804, 2003.

Rosenberg, N., The probability of topological concordance of gene trees and species tree, *Theor. Popul. Biol.*, 61:225–247, 2002.

Rosenberg, N.A., Gene genealogies, in Fox, C.W. and J.B. Wolf, eds., *Evolutionary Genetics: Concepts and Case Studies*, pp. 55–67, Oxford: Oxford University Press, 2005.

Rosenthal, R., The "file drawer problem" and tolerance for null results, *Psychol. Bull.*, 86:638–641, 1979.

Rothstein, H.R., A.J. Sutton, and M. Borenstein, eds., *Publication Bias in Meta-Analysis: Prevention, Assessment and Adjustments*, New York: Wiley, 2005.

Roy, S.N., On a heuristic method of test construction and its use in multivariate analysis, *Ann. Math. Stat.*, 24:220–238, 1953.

Ruvolo, M., Molecular phylogeny of the hominoids: Inferences from multiple independent DNA sequence data sets, *Mol. Biol. Evol.*, 14:248–265, 1997.

Sagoo, G.S. et al., Meta-analysis of genome-wide studies of psoriasis susceptibility reveals linkage to chromosomes 6p21 and 4q28-q31 in Caucasian and Chinese Hans population, *J. Invest. Dermatol.*, 122:1401–1405, 2004.

Saito, R., H. Suzuki, and Y. Hayashizaki, Construction of reliable protein-protein interaction networks with a new interaction generality measure, *Bioinformatics*, 19:756–763, 2003.

Sasieni, P.D., From genotypes to genes: doubling the sample size, *Biometrics*, 53:1253–1261, 1997.

Sawcer, S. et al., Empirical genomewide significance levels established by whole genome simulations, *Genet. Epidemiol.*, 14:223–229, 1990.

Sax, K., The association of size differences with seed coat pattern and pigmentation in Phaseolus vulgaris, *Genetics*, 8:552–560, 1923.

Schabenberger, O. and C. Gotway, *Statistical Methods for Spatial Data Analysis*, Chapman & Hall/CRC, 2005.

Schadt, E. et al., Genetics of gene expression surveyed in maize, mouse and man, *Nature*, 422:297–302, 2003.

Schadt, E.E. et al., Mapping the genetic architecture of gene expression in human liver, *PLoS Biol.*, 6:e107, 2008.

Schicklmaier, P. and H. Schmieger, Frequency of generalized transducing phages in natural isolates of the *Salmonella typhimurium* complex, *Appl. Environ. Microbiol.*, 61:1637–1640, 1995.

Schwikowski, B., P. Uetz, and S. Fields, A network of protein-protein interactions in yeast, *Nat. Biotechnol.*, 18:1257–1261, 2000.

Segal, E. et al., A module map showing conditional activity of expression modules in cancer, *Nat. Genet.*, 36:1090–1098, 2004.

Segurado, R. et al., Genome scan meta-analysis of schizophrenia and bipolar disorder, Part III: Bipolar disorder, *Am. J. Hum. Genet.*, 73:49–62, 2003.

Self, S.G. and K.Y. Liang, Asymptotic properties of maximum likelihood estimators and likelihood ratio tests under nonstandard conditions, *J. Am. Stat. Assoc.*, 82:605–610, 1987.

Sham, P.C. and S. Purcell, Equivalence between Haseman-Elston and Variance-Components linkage analyses for sib-pairs, *Am. J. Hum. Genet.*, 68:1527–1532, 2001.

Sham, P.C. et al., Powerful regression based quantitative linkage analysis of general pedigrees, *Am. J. Hum. Genet.*, 71:238–253, 2002.

Shen, R., D. Ghosh, and A.M. Chinnaiyan, Prognostic meta-signature of breast cancer developed by two-stage mixture modeling of microarray data, *BMC Genomics*, 5:94, 2004.

Shmulevich, I. et al., Probabilistic Boolean networks: a rule-based uncertainty model for gene regulatory networks, *Bioinformatics*, 18:261–274, 2002.

Šidák, Z., Rectangular confidence regions for the means of multivariate normal distributions, *J. Am. Stat. Assoc.*, 62:626–633, 1967.

Simpson, E.H., The interpretation of interaction in contingency tables, *J. Roy. Stat. Soc. B*, 13:238–241, 1951.

Smyth, G.K., Linear models and empirical Bayes methods for assessing differential expression in microarray experiments, *Stat. Appl. Genet. Mol. Biol.*, 3:Article 3, 2004.

Smyth, G.K., Limma: linear models for microarray data, in Gentleman, R. et al., eds., *Bioinformatics and Computational Biology Solutions using R and Bioconductor*, pp. 397–420, New York: Springer, 2005.

Sørlie, T. et al., Repeated observation of breast tumor subtypes in independent gene expression data sets, *Proc. Natl. Acad. Sci. USA*, 100:8418–8423, 2003.

Spellman, P.T. et al., Comprehensive identification of cell cycle-regulated genes of the yeast Saccharomyces cerevisiae by microarray hybridization, *Mol. Biol. Cell*, 9:3273–3297, 1998.

Stawiki, E.W. et al., Progress in predicting protein function from structure: Unique features of O-Glycosidases, in Altman, R.B. et al., eds., *Proceedings of the Seventh Pacific Symposium on Biocomputing (PSB02)*, pp. 637–648, River Edge, NJ: World Scientific, 2002.

Stec, J. et al., Comparison of the predictive accuracy of DNA array based multigene classifiers across cDNA arrays and Affymetrix GeneChips, *J. Mol. Diagn.*, 7:357–367, 2005.

Stege, U., Gene trees and species trees: the gene-duplication problem is fixed-parameter tractable, in *Proceedings of the Sixth Workshop Algorithms and Data Structures (WADS99)*, vol. 1663 of *Lecture Notes in Computer Science*, pp. 288–293, 1999.

Sterne, J.A., M. Egger, and G.D. Smith, Investigating and dealing with publication and other biases in meta-analysis, *BMJ*, 323:101–105, 2001.

Stevens, J.R. and R.W. Doerge, Combining Affymetrix microarray results, *BMC Bioinformatics*, 6:57, 2005.

Stewart, G.J. and C.D. Sinigalliano, Exchange of chromosomal markers by natural transformation between the soil isolate, *Pseudomonas stutzeri* JM300, and the marine isolate, *Pseudomonas stutzeri* strain ZoBell., *Antonie Van Leeuwenhoek*, 59:19–25, 1991.

Stoehr, J. et al., Genetic obesity unmasks nonlinear interactions between murine type 2 diabetes susceptibility loci, *Diabetes*, 49:1946–1954, 2000.

Stoehr, J.P. et al., Identification of major quantitative trait loci controlling body weight variation in ob/ob mice, *Diabetes*, 53:245–249, 2004.

Stoesz, M.R. et al., Extension of the Haseman-Elston method to multiple alleles and multiple loci: theory and practice for candidate genes, *Ann. Hum. Genet.*, 61:263–274, 1997.

Storey, J. and R. Tibshirani, SAM thresholding and false discovery rates for detecting differential gene expression in DNA microarrays, in Parmigiani, G. et al., eds., *The Analysis of Gene Expression Data: Methods and Software*, pp. 272–290, New York: Springer-Verlag, 2003a.

Storey, J.D., A direct approach to false discovery rates, *J. Roy. Stat. Soc. B*, 64:479–498, 2002.

Storey, J.D. and R. Tibshirani, Statistical significance for genome-wide experiments, *Proc. Natl. Acad. Sci. USA*, 100:9440–9445, 2003b.

Stouffer, S.A. et al., *The American Soldier: Adjustment During Army Life, Vol. 1*, Princeton, New Jersey: Princeton University Press, 1949.

Stranger, B.E. et al., Relative impact of nucleotide and copy number variation on gene expression phenotypes, *Science*, 315:848–853, 2007.

Sturm, J.F., Using SeDuMi 1.02, a MATLAB toolbox for optimization over symmetric cones, *Optim. Method. Softw.*, 11-12:625–653, 1999.

Subramanian, A. et al., Gene set enrichment analysis: a knowledge-based approach for interpreting genome-wide expression profiles, *Proc. Natl. Acad. Sci. USA*, 102:15545–15550, 2005.

Sun, N. and H. Zhao, Genomic approaches in dissecting complex biological pathways, *Pharmacogenomics*, 5:163–179, 2004.

Sutton, A.J. et al., *Methods for Meta-Analysis in Medical Research*, New York: John Wiley & Sons, 2000.

Sweet-Cordero, A. et al., An oncogenic KRAS2 expression signature identified by cross-species gene-expression analysis, *Nat. Genet.*, 37:48–55, 2005.

Syring, J. et al., Evolutionary relationships among *Pinus* (Pinaceae) subsections inferred from multiple low-copy nuclear loci, *Am. J.Bot.*, 92:2086–2100, 2005.

Tajima, F., Evolutionary relationship of DNA sequences in finite populations, *Genetics*, 105:437–460, 1983.

Takahata, N., Gene genealogy in three related populations: Consistency probability between gene and population trees, *Genetics*, 122:957–966, 1989.

Takahata, N. and Y. Satta, Evolution of the primate lineage leading to modern humans: phylogenetic and demographic inferences from DNA sequences, *Proc. Natl. Acad. Sci. USA*, 94:4811–4815, 1997.

Takahata, N., Y. Satta, and J. Klein, Divergence time and population size in the lineage leading to modern humans, *Theor. Popul. Biol.*, 48:198–221, 1995.

Tan, P.K. et al., Evaluation of gene expression measurements from commercial microarray platforms, *Nucleic Acids Res.*, 31:5676–5684, 2003.

Tang, R., Fitting and evaluating certain two-level hierarchical models, Ph.D. thesis, Department of Statistics, Harvard University, 2002.

Tegnér, J. et al., Reverse engineering gene networks: integrating genetic perturbations with dynamical modeling, *Proc. Natl. Acad. Sci. USA*, 100:5944–5949, 2003.

Terwilliger, J.D. and J. Ott, *Handbook of Human Genetic Linkage*, Baltimore, Maryland: Johns Hopkins University Press, 1994.

Than, C., G. Jin, and L. Nakhleh, Integrating sequence and topology for efficient and accurate detection of horizontal gene transfer, in Nelson, C.E. and S. Vialette, eds., *Comparative Genomics. Proceedings of the Sixth RECOMB Comparative Genomics Satellite Workshop.*, vol. 5267 of *Lecture Notes in Computer Science*, pp. 113–127, New York: Springer, 2008a.

Than, C. and L. Nakhleh, SPR-based tree reconciliation: Non-binary trees and multiple solutions, in Brazma, A., S. Miyano, and T. Akutsu, eds., *Proceedings of the 6th Asia-Pacific Bioinformatics Conference, APBC 2008*, vol. 6 of *Advances in Bioinformatics and Computational Biology*, pp. 251–260, Imperial College Press, 2008.

Than, C. and L. Nakhleh, Efficient genome-scale inference of species trees by minimizing deep coalescences, 2009, submitted.

Than, C. et al., Confounding factors in HGT detection: Statistical error, coalescent effects, and multiple solutions, *J. Comput. Biol.*, 14:517–535, 2007.

Than, C., D. Ruths, and L. Nakhleh, PhyloNet: a software package for analyzing and reconstructing reticulate evolutionary relationships, *BMC Bioinformatics*, 9:322, 2008b.

Than, C. et al., Efficient inference of bacterial strain trees from genome-scale multi-locus data, *Bioinformatics*, 24:i123–i131, 2008c.

The Transatlantic Multiple Sclerosis Genetics Cooperative, A meta-analysis of genomic screens in multiple sclerosis, *Mult. Scler.*, 7:3–11, 2001.

Tippett, L.H.C., *The Methods of Statistics*, London: Williams & Norgate, 1st ed., 1931.

Tong, A.H.Y. et al., A combined experimental and computational strategy to define protein interaction networks for peptide recognition modules, *Science*, 295:321–324, 2002.

Troyanskaya, O.G. et al., A Bayesian framework for combining heterogeneous data sources for gene function prediction in *Saccharomyces cerevisiae*, *Proc. Natl. Acad. Sci. USA*, 100:8348–8353, 2003.

Tu, Z.D. et al., Understanding protein essentiality – linking genomic information with phenotype, Tech. rep., University of Southern California, Los Angeles, CA, 2005.

Tukey, J.W., *Exploratory Data Analysis*, Reading, MA: Addison-Wesley, 1977.

Tumbar, T. et al., Defining the epithelial stem cell niche in skin, *Science*, 303:359–363, 2004.

Tusher, V.G., R. Tibshirani, and G. Chu, Significance analysis of microarrays applied to the ionizing radiation response, *Proc. Natl. Acad. Sci. USA*, 98:5116–5121, 2001.

Uetz, P. et al., A comprehensive analysis of protein-protein interactions in saccharomyces cerevisiae, *Nature*, 403:623–627, 2000.

van der Laan, M.J., S. Dudoit, and K.S. Pollard, Augmentation procedures for control of the generalized family-wise error rate and tail probabilities for the proportion of false positives, *Stat. Appl. Genet. Mol. Biol.*, 3:Article 15, 2004a.

van der Laan, M.J., S. Dudoit, and K.S. Pollard, Multiple testing. Part II. Step-down procedures for control of the family-wise error rate, *Stat. Appl. Genet. Mol. Biol.*, 3:Article 14, 2004b.

van Heel, D.A. et al., Inflammatory bowel disease susceptibility loci defined by genome scan meta-analysis of 1952 affected relative pairs, *Hum. Mol. Genet.*, 13:763–770, 2004.

van Houwelingen, J.C., L.R. Arends, and T. Stijnen, Advanced methods in meta-analysis: multivariate approach and meta-regression, *Stat. Med.*, 21:589–624, 2002.

Varambally, S. et al., The polycomb group protein EZH2 is involved in progression of prostate cancer, *Nature*, 419:624–629, 2002.

Vazquez, A. et al., Global protein function prediction from protein-protein interaction networks, *Nat. Biotechnol.*, 21:697–700, 2003.

Velculescu, V.E. et al., Serial analysis of gene expression, *Science*, 270:484–487, 1995.

Venables, W.N. and B.D. Ripley, *Modern Applied Statistics with S*, New York: Springer-Verlag, 4th ed., 2002.

Venezia, T.A. et al., Molecular signatures of proliferation and quiescence in hematopoietic stem cells, *PLoS Biol.*, 2:e301, 2004.

Venter, J.C. et al., The sequence of the human genome, *Science*, 291:1304–1351, 2001.

Wang, J. et al., Differences in gene expression between B-cell chronic lymphocytic leukemia and normal B cells: a meta-analysis of three microarray studies, *Bioinformatics*, 20:3166–3178, 2004.

Wang, K., M. Li, and M. Bucan, Pathway-based approaches for analysis of genomewide association studies, *Am. J. Hum. Genet.*, 81:1278–1283, 2007.

Wang, W. et al., A systematic approach to reconstructing transcription networks in *Saccharomyces cerevisiae*, *Proc. Natl. Acad. Sci. USA*, 99:16893–16898, 2002.

Wasserman, S. and K. Faust, *Social Network Analysis: Methods and Applications*, Cambridge, UK: Cambridge University Press, 1994.

Welch, B.L., The significance of the difference between two means when the population variances are unequal, *Biometrika*, 29:350–362, 1938.

Welch, R.A., Extensive mosaic structure revealed by the complete genome sequence of uropathogenic *Escherichia coli*, *Proc. Natl. Acad. Sci. USA*, 99:17020–17024, 2002.

Westfall, P.H. and S.S. Young, *Resampling-based Multiple Testing: Examples and Methods for p-value Adjustment*, New York: Wiley, 1999.

Whitfield, M.L. et al., Identification of genes periodically expressed in the human cell cycle and their expression in tumors, *Mol. Biol. Cell*, 13:1977–2000, 2002.

Wilkinson, B., A statisical consideration in psychological research, *Psychol. Bull.*, 48:156–158, 1951.

Wirapati, P., D.R. Goldstein, and M. Delorenzi, Integrated analysis of gene expression profiling studies – examples in breast cancer, in Appel, R.D. and E. Feytmans, eds., *Bioinformatics: A Swiss Perspective*, Singapore: World Scientific, 2009.

Wirapati, P. et al., Meta-analysis of gene expression profiles in breast cancer: toward a unified understanding of breast cancer subtyping and prognosis signatures, *Breast Cancer Res.*, 10:R65, 2008.

Wise, L.H., J.S. Lanchbury, and C.M. Lewis, Meta-analysis of genome searches, *Ann. Hum. Genet.*, 63:263–272, 1999.

Witte, J.S., Genetic analysis with hierarchical models, *Genet. Epidemiol.*, 14:1137–1142, 1997.

Woegerbauer, M. et al., Natural genetic transformation of clinical isolates of *Escherichia coli* in urine and water, *Appl. Environ. Microbiol.*, 68:440–443, 2002.

Wolfinger, R.D. et al., Assessing gene significance from cDNA microarray expression data via mixed models, *J. Comput. Biol.*, 8:37–52, 2001.

Wright, G. et al., A gene expression-based method to diagnose clinically distinct subgroups of diffuse large B cell lymphoma, *Proc. Natl. Acad. Sci. USA*, 100:10585–10587, 2003.

Wright, G.W. and R.M. Simon, A random variance model for detection of differential gene expression in small microarray experiments, *Bioinformatics*, 19:2448–2455, 2003.

Wu, C. et al., A probe-to-transcripts mapping method for cross-platform comparisons of microarray data, Tech. rep., BEPress, 2005.

Wu, C.I., Inferences of species phylogeny in relation to segregation of ancient polymorphisms, *Genetics*, 127:429–435, 1991.

Wu, X. et al., A combined analysis of genomewide linkage scans for body mass index from the National Heart, Lung, and Blood Institute Family Blood Pressure Program, *Am. J. Hum. Genet.*, 70:1247–1256, 2002.

Xenarios, I. et al., DIP: The Database of Interacting Proteins. a research tool for studying cellular networks of protein interactions, *Nucleic Acids Res.*, 30:303–305, 2002.

Xu, X. et al., A unified Haseman-Elston method for testing linkage with quantitative traits, *Am. J. Hum. Genet.*, 67:1025–1028, 2000.

Yan, P.S. et al., Dissecting complex epigenetic alterations in breast cancer using CpG island microarrays, *Cancer Res.*, 61:8375–8380, 2001.

Yang, Y.H., M.J. Buckley, and T.P. Speed, Analysis of cDNA microarray images, *Brief. Bioinform.*, 2:341–349, 2001.

Yang, Y.H. et al., Normalization for cDNA microarray data: a robust composite method addressing single and multiple slide systematic variation, *Nucleic Acids*

Res., 30:e15, 2002.

Young, A. et al., OntologyTraverser: an R package for GO analysis, *Bioinformatics*, 21:275–276, 2005.

Yvert, G. et al., Trans-acting regulatory variation in *Saccharomyces cerevisiae* and the role of transcription factors, *Nat. Genet.*, 35:57–64, 2003.

Zeggini, E. et al., Meta-analysis of genome-wide association data and large-scale replication identifies additional susceptibility loci for type 2 diabetes, *Nat. Genet.*, 40:638–645, 2008.

Zhang, K. et al., An empirical Bayes method for updating inferences in analysis of quantitative trait loci using information from related genome scans, *Genetics*, 173:2283–2296, 2006.

Zhang, L., On a Mirkin-Muchnik-Smith conjecture for comparing molecular phylogenies, *J. Comput. Biol.*, 4:177–187, 1997.

Zhang, L., M.F. Miles, and K.D. Aldape, A model of molecular interactions on short oligonucleotide microarrays, *Nat. Biotechnol.*, 21:818–821, 2003.

Zhang, Q. et al., Mapping quantitative trait loci for milk production and health of dairy cattle in a large outbred pedigree, *Genetics*, 149:1959–1973, 1998.

Zhang, Z. and M. Gerstein, Reconstructing genetic networks in yeast, *Nat. Biotechnol.*, 21:1295–1297, 2003.

Zhao, H., B. Wu, and N. Sun, DNA-Protein binding and gene expression patterns, in Goldstein, D.R., ed., *Science and Statistics: A Festschrift for Terry Speed*, vol. 40 of *Lecture Notes Monograph Series*, pp. 259–274, Beachwood, Ohio: Institute of Mathematical Statistics, 2003.

Zheng, Y., R.J. Roberts, and S. Kasif, Genomic functional annotation using co-evolution profiles of gene clusters, *Genome Biol.*, 3:1–9, 2003.

Zhou, X., M. Kao, and W. Wong, Transitive functional annotation by shortest-path analysis of gene expression data, *Proc. Natl. Acad. Sci. USA*, 99:12783–12788, 2002.

Zhu, J. et al., Combining genotypic and expression data in segregating populations, *PLoS Comput. Biol.*, 3:e69, 2007.

Zintzaras, E. and J.P. Ioannidis, HEGESMA: genome search meta-analysis and heterogeneity testing, *Bioinformatics*, 21:3672–3673, 2005a.

Zintzaras, E. and J.P.A. Ioannidis, Heterogeneity testing in meta-analysis of genome searches, *Genet. Epidemiol.*, 28:123–137, 2005b.

Zmasek, C.M. and S.R. Eddy, A simple algorithm to infer gene duplication and speciation events on a gene tree, *Bioinformatics*, 17:821–828, 2001.

Index

Adaptive Interval (AI) algorithm, 101–102
affected sib pairs, *see* sib pairs
Affymetrix GeneChips, 14–16, 113, 157
  gene expression, 15, 164, 171
  HG-U133A, 157, 168
  HG-U95Av2, 157, 166, 168
  HuGeneFL, 109, 157, 166, 168
  MG-U74Av2, 99, 190
  MOE 430A, 136
  MOE430A, 222
  MOE430B, 222
  Mu 11K-A, 120
  probe set, 18
AmiGO, *see* software
ArrayExpress, *see* databases
ASP, *see* sib pairs
association mapping, 10
AUC, *see* Receiver Operating Characteristic
  curves

bacteria, 280, 290
bandwidth, 121
Bayes' rule, 70, 89
Bayesian analysis, 104, 117, 203, 243, 257,
  265, 283
Bayesian hierarchical model, 70, 104, 160,
  204
Bayesian networks, *see* networks
between-study heterogeneity, 4, 6, 23, 29,
  79, 140
between-study variance σ_B^2, 6, 27, 52, 58,
  64, 146
bias, 8, 10, 11, 25, 31, 43, 55, 56, 78, 97,
  122, 123, 125, 133, 134, 170, 203, 218
  publication bias, 8, 31, 43
BioConductor, 136, *see* software
BioGRID, *see* databases
biological distance, 225, 231–235, 237
biological replicates, 106, 108, 120, 133
biomarkers, 202

bipartite graph, 244
biweight estimator, 15
BLAST, *see* software
BOND, *see* databases
Bonferroni procedure, 35, 41, 99, 113–114
Boolean network, *see* networks
borrowing strength, 75, 96, 125
BUM, *see* mixture model

cancer
  biomarkers, 202
  data integration, 202–207
  tumorigenesis, 202, 209
cDNA microarrays, 13–14, 113
  dye swap experiment, 119
  gene expression, 14
  preprocessing, 14, 134
  self-self hybridization, 120, 122
cell cycle data, 253
central dogma of molecular biology, 13
chromatin immunoprecipitation microarrays
  (ChIP-chip), 201, 244
classifier, 204
cluster analysis, 141, 164, 204, 209, 225,
  227, 230, 236, 240, 241, 243, 260
  dendrogram, 141, 232
coalescent, 276, 281–283, 290–292
Cochran-Armitage trend test, 12
combining information, 4–8, 112, 114–119,
  157, 160, 201, 206, 226, 227, 255
  collaboration, 29, 34, 37, 38
  decisions, 8
  estimates, 5–6
  integrative correlation coefficient, 159,
    203
  mega-analysis, 4, 16–18, 23, 29, 33, 41,
    50, 136
  meta-analysis, *see* meta-analysis
  p-values, 6–7, 26, 68, 159

pooling data, 4, 18, 24, 34, 50, 153, 168, 174
ranks, 7
spectrum, 4, 158
composite hypothesis testing, 84–86
    significance, 86–87
confounding, 11
Cyber-T, 117, 121, 123

DAG, *see* directed acyclic graph
data pooling, *see* combining information
databases, 178
    ArrayExpress, 158
    Biomolecular Object Network Databank (BOND), 261
    Database of Interacting Proteins (DIP), 261, 263, 264, 273
    FlyBase, 178
    GenBank, 161, 205
    Gene Expression Omnibus (GEO), 135, 158, 175
    General Repository for Interaction Datasets (BioGRID), 261
    H-Invitational Database (H-InvDb), 170
    Human Protein Reference Database (HPRD), 207
    Kyoto Encyclopedia of Genes and Genomes (KEGG), 205, 226, 229
    Munich Information Center for Protein Sequences (MIPS), 261, 264, 272
    NCBI Reference Sequence Database (RefSeq), 161, 170
    ONCOMINE, 205, 207, 210
    Saccharomyces Genome Database (SGD), 178
    Swiss-Prot, 205
    SwissPfam, 262
    TRANSFAC, 244
    UniGene, 161, 203
DAVID, *see* software
dChip, *see* software
dendrogram, 141, 232
dictionary model, 207
differential gene expression, 14, 18, 95, 110, 123, 133, 137–138, 145–146, 204, 210, 214, 216
DIP, *see* databases
directed acyclic graph (DAG), 176, 179, 228–229, 286–289

edge, 176, 229, 286
indegree, 286
node, 176, 229, 286
outdegree, 286
subgraph, 185, 229
Drosophila experiments, 119
dye swaps, 119

EB, *see* empirical Bayes methods
EEEP, *see* software
EIGENSTRAT, 11
EM algorithm, 53
empirical Bayes (EB) methods, 17, 67, 88, 104, 114, 116–119, 216
    background study, 68, 72, 73
    heterogeneous error model (HEM), 96, 103–108, 110
        FDR evaluation, 108
    primary study, 68, 71–73
ETL, *see* expression trait loci
evidence codes, 181, 229
exchangeability, 131
expression trait loci (ETL), 19, 213
    mapping, 214–217, 226–228, 230
        analysis of variance (ANOVA), 230, 238
        hybrid method, 235, 238
        integrating gene ontology, 225, 227, 230–233, 241
        marker-based (MB), 216–217
        mixture over markers model (MOM), 217
        multivariate analysis of variance (MANOVA), 230, 239
        transcript-based (TB), 216
    method evaluation, 217–222
    power, 218, 224

F-test, 230, 239
false discovery rate (FDR), *see* multiple testing
family-wise error rate (FWER), *see* multiple testing
FatiGO, *see* software
FDR, *see* multiple testing
file drawer problem, 8, 43
Fisher's exact test, 188, 205

Fisher's method, 7, 16, 24–26, 30, 45, 68, 86, 140, 151–152, 203
    power, 24
fixed effects model, 5, 6, 16, 17, 51, 58, 140, 146, 160
fold change, 95, 137
forest plot, *see* plots
full-length transcript-based probe set, 158, 168–170
functional analysis, 204–206, 210
funnel plot, *see* plots
FWER, *see* multiple testing

Gaussian Regression Independent Multilevel Model (GRIMM), 71
GenBank, *see* databases
gene count (GO count), 182–187
    joint distribution, 184–187
        covariance, 186
        enrichment statistic, Q_p, 186
    standardized, 184
    unbranched path, 185–186
gene duplication, 275, 278, 284–286, 293
    duplication-loss model, 285
    duplication-only model, 285
gene expression, 14, 15
    differential expression, 14, 18
gene list, 175, 182–189
    content analysis, 183
    intersection, 159, 187
    multiple, 187–189
        combining information, 18
        distance between, 188
        enrichment, 188–189
        meta-analysis, 191–197
        multi-way exact test, 188–189, 193
        path enrichment analysis, 189, 193
    single, 183–184
gene loss, 275, 278, 284–286, 293
    duplication-loss model, 285
Gene Ontology (GO), 18, 175–182, 205, 206, 210, 225, 262, 272
    annotations, 180, 228–229
    application to stem cell data, 189–197
    data simulation, 236–238
    directed acyclic graph, 179, 228–229
    evidence codes, 181, 229
    formats, 182
    is-a, 180

part-of, 180
    software tools, 181–182
        DAVID, 182
        FatiGO, 182
        Ontology Traverser, 182, 191
        XML, 182
    structured vocabularies, 178–179
        biological process, 178, 229, 236
        cellular component, 178–179, 229
        molecular function, 178, 229
    true-path rule, 180, 185
gene ontology distance, 230, 234
    longest path, 231, 233
    union-intersection, 231, 233, 236, 240
gene set enrichment analysis (GSEA), 20
gene tree, *see* trees
genetic heterogeneity, 11, 29, 45
genetic marker, 214
genetic network, *see* networks
genetic recombination, 8, 9, 276, 277, 280
Genome Search Meta-Analysis, 17, 26, 30, 34–37, 50
    extensions, 42–43
    heterogeneity, 26, 43
    limitations, 43–44
    power, 38, 40–41
    software, 46
    weighted analysis, 38–40
genome-wide association study, 12, 17
genome-wide scan, 11, 23, 67, 71
    evidence for linkage
        genome-wide, 35, 41, 45, 216
        nominal, 37, 41, 45
        suggestive, 28, 35, 41, 45
genome-wide significance levels, 16
genomic control, 11
GEO, *see* databases
Gibbs sampling, 107, 248, 270
GO, *see* Gene Ontology
graphical model, 245
graphical models, *see* networks
GRIMM, *see* Gaussian Regression Independent Multilevel Model
GSEA, *see* gene set enrichment analysis
GSMA, *see* Genome Search Meta-Analysis
GWAS, *see* genome-wide association study

H-InvDB, *see* databases

Haseman-Elston regression, 10, 27, 56, 68, 69
HEM, *see* empirical Bayes (EB) methods
heterogeneity, 4–5, 18, 26, 43, 49, 67, 72, 140, 146–148, 152
    between-study heterogeneity, 4, 6, 23, 29, 79, 140
    genetic heterogeneity, 11, 29, 45
    locus heterogeneity, 49, 50, 53–54, 64
    size heterogeneity, 49, 50, 52–53, 64
    sources of, 4, 29, 31, 44, 159–160
heterogeneity lod score, *see* lod score
HGT, *see* horizontal gene transfer
hierarchical model, 52, 70, 204, 245
HLOD, *see* lod score
homogeneity test, 5, 43, 51–52, 61, 64, 140, 146
homology, 177–178
horizontal gene transfer, 275, 276, 278, 287
HorizStory, *see* software
HPRD, *see* databases
Human Genome Project, 201
hybrid speciation, 276, 278, 287, 289
hypergeometric distribution, 183–184, 205, 220
    binomial approximation, 184
    Fisher's exact test, 205
    multivariate hypergeometric distribution, 185

IBD, *see* identity by descent
IDD, *see* integration-driven discovery
identity by descent, 9, 16, 55, 64, 69
ILP, *see* integer linear programming
IM, *see* interval mapping
image analysis, 14
in vivo/in vitro data integration, 209–210
incomplete information, *see* missing data
Individual Patient Data, 29
integer linear programming, 284
integration-driven discovery, 3
integrative correlation coefficient, *see* combining information
intersection-union test, 84, 90
interval mapping, 69–70
    Interval Mapping-Empirical Bayes method, 71, 73–77
inverse regression, 55–56
inverse variance weighting, 6

IUT, *see* intersection-union test

KEGG, *see* databases
kernel-based Markov random field, 260, 270–272
Kolmogorov-Smirnov statistic, 172, 209, 210

LD, *see* linkage disequilibrium
likelihood function, 89
likelihood ratio test, 54, 167, 239
limma, 117, 123, 139, 217
lineage sorting, 275, 278, 281–284
linear model, 139, 140, 245, 257
linkage analysis, 8–10, 33
    genome-wide scan, 11, 23, 67, 71
    power, 24, 26, 34, 38, 241
linkage disequilibrium, 10
local pooled error, 18
local pooled error (LPE), 96–99, 116, 121, 123, 133
locus heterogeneity, 49, 50, 53–54, 64
lod score, 9, 16, 25, 33, 51, 58, 61, 216, 221
    heterogeneity lod score, 28, 32
    multipoint lod score, 34
    nonparametric lod score, 26, 28
loess, 101, 116, 121, 123, 133
LPE, *see* local pooled error

MA plot, *see* plots
MAGS, *see* Meta-Analysis for Genome Studies
Markov Chain Monte Carlo (MCMC), 107, 204, 248–252, 272
Markov random field (MRF), 260, 266
    partition function, 268
    potential function, 267, 269
MAS 5.0, 15, 99, 109, 134, 171
MATLAB, *see* software
MCMC, *see* Markov Chain Monte Carlo
mega-analysis, *see* combining information
MENDEL, *see* software
MERLIN, *see* software
meta-analysis, 3–4, 16, 18, 23, 43, 50, 64, 67, 146, 160, 161, 175, 201
    bias, 8, 31, 43
    combining *p*-values

Fisher's method, 7, 16, 24–26, 30, 45,
     68, 86, 140, 151–152, 203
  effect size, 204
  of microarray experiments, 17, 135,
     139–140, 203–204
  of microarray studies, 148–152
  power, 26, 34
Meta-Analysis for Genome Studies, 28
MGI, *see* databases
microarray technologies, 13–15
  preprocessing, 134
MIPS, *see* databases
mismatch probes, 15
missing data, 36, 50, 288
mixture model, 53, 58, 59, 61, 64, 90, 92,
     204, 217, 236, 241, 257
  beta mixture, 88, 167
MM, *see* mismatch probes
moderated t-test, 117, 137–138, 146, 151,
     216
MOM, *see* mixture over markers model
Moran model, 290
most recent common ancestor, 276, 283
mouse experiments
  Huntington's disease, 120, 136
  immune system, 98–99, 110
  lung cancer, 209
MRCA, *see* most recent common ancestor
MRF, *see* Markov Random Field
MSP, *see* Multiple Scan Probability method
Multiple Scan Probability method, 24–25, 45
  power, 25, 45–46
multiple testing, 24, 31, 35, 43, 83, 95, 99,
     114, 133, 138–139, 189, 214, 224, 227
  adjusted p-value, 85, 216
  Bonferroni procedure, 99, 113–114
  composite hypothesis testing, 83
  false discovery rate (FDR), 99, 102–104,
     131, 138–139, 167, 214, 216–218, 224,
     230
  family-wise error rate (FWER), 95, 99
  q-value, 138–139, 143, 205, 217, 221
  Significance Analysis of Microarrays
     (SAM), 98, 101, 103, 110, 114, 217
  step-down procedure, 99
multipoint lod score, *see* lod score
Mx, *see* software

networks, 275

Bayesian, 206–207, 243
  Boolean, 243
  model, 286
  phylogenetic, 286–290
  protein interaction, 260
  reconstructible, 288
  reconstruction, 289–290
  regulatory, 259
  transcriptional regulatory (TRN), 243,
     245–247
nonparametric lod score, *see* lod score
normalization, 14, 15, 135, 141, 160
  Affymetrix GeneChips, 134
  cDNA arrays, 134
NPL, *see* lod score

oligonucleotide arrays, *see* Affymetrix
     GeneChips
ONCOMINE, *see* databases
ontology, 175–176
Ontology Traverser, *see* software
ordered rank, 45
  p-value, 35, 37, 46

p-values
  combining, 6–7, 26, 159
     average, 86
     Fisher's method, 7, 16, 24–26, 30, 45,
        68, 86, 140, 151–152, 203
     maximum, 86
     Pearson's method, 7, 86
     Stouffer's method, 7, 25, 86
     Tippett's method, 7, 86
partial probe set, 157, 162–163
pathway analysis, 20, 204–206
PDNN, *see* position-dependent nearest
     neighbor
perfect match probes, 15
Perl, *see* software
permutation, 35, 43, 98, 101, 103, 112, 113,
     115, 118–119, 122, 125–131, 133, 167,
     216
phylogenetic network, *see* networks
phylogenetic tree, *see* trees
phylogeny, 19, 275
PhyloNet, *see* software
pleiotropy, 91, 227, 234, 235, 241
plots

dendrogram, 141, 232
forest plot, 5, 140
funnel plot, 8
MA plot, 97, 98, 100, 141
volcano plot, 99
pooled t-statistic, 118, 123
pooling data, *see* combining information
population substructure, 11, 67
position-dependent nearest neighbor
(PDNN), 164, 171, 172
posterior distribution, 53, 107, 204, 216,
266, 270
posterior probability, 87, 89, 91, 221, 266
power, 29, 33, 34, 46, 49, 64, 69, 79, 98,
120, 121, 131, 158, 159, 161, 187–188,
227, 241
combined Haseman-Elston estimates, 27
expression trait loci (ETL) mapping, 218,
224
Fisher's method, 24
Genome Search Meta-Analysis, 38, 40–41
Interval Mapping-Empirical Bayes
method, 73–77
Multiple Scan Probability method, 25,
45–46
pooling raw data, 24
weighted nonparametric lod scores, 26
preprocessing, 134
prior distribution, 105–107
probe set, 15, 18
full-length transcript-based probe set,
158, 168–170
partial probe set, 157, 162–163
protein function prediction, 260, 266–272
computational approaches, 261
statistical approaches
kernel-based Markov random field,
260, 270–272
Markov random field (MRF), 260,
266–269
support vector machine, 271–272
support vector machine (SVM), 260
protein interaction network, *see* networks
protein synthesis, 12
transcription, 13
translation, 13
protein-DNA interaction, 18, 19, 243, 244,
247, 249, 250, 254, 255

protein-protein interactions, 18, 19, 201,
206, 210, 226, 242, 244, 259
coverage, 260
databases, 261
reliability, 260
assessment, 262–264
publication bias, 8, 31, 43
file drawer problem, 8, 43

Q-statistic, 5, 52, 64, 140
q-value, 138–139, 143, 205, 217, 221
QTL, *see* quantitative trait loci
quantile normalization, 15, 143
quantitative trait loci (QTL), 10, 26, 49, 67,
91, 213, 214, 226

R, *see* software
random effects model, 5, 6, 16, 52, 58, 61,
64, 140, 146, 160
rank-invariant resampling (RIR), 101–102
Receiver Operating Characteristic (ROC)
curves, 226, 273
area under the curve (AUC), 231, 238
recombination fraction, 9, 34, 69
RefSeq, *see* databases
regulatory network, *see* networks
replicates
biological, 96, 106, 108, 120, 133
technical, 96, 106, 108, 120, 133
reticulate evolution, 275, 276, 278, 286–290,
293
RIATA-HGT, *see* software
RMA, *see* robust multi-array analysis
robust multi-array average (RMA), 15, 134,
171

S-PLUS, *see* software
SAM, *see* multiple testing
scree slope, 37
sensitivity analysis, 8
sequence data, 19, 275, 283
SGD, *see* databases
shrinkage, 70, 118, 122, 123
sib pairs, 29, 38, 55, 58, 68, 69, 71
Significance Analysis of Microarrays
(SAM), *see* multiple testing
Simpson's paradox, 4, 11

simulation studies, 25–27, 34, 35, 37, 38, 40,
    41, 55–57, 68, 72, 90, 96, 98, 116, 133,
    155, 217, 233, 249
size heterogeneity, 49, 50, 52–53, 64
software
    BioConductor, 96, 115, 170, 171, 210
    BLAST, 170
    DAVID, 182
    dChip, 171
    EEEP, 290
    FatiGO, 182
    GSMA, 46
    HEGESMA, 26
    HorizStory, 290
    MATLAB, 210
    MENDEL, 54
    MERLIN, 56, 58, 64
    Mx, 54
    Ontology Traverser, 182, 191
    PerfectMatch, 171
    Perl, 210
    PhyloNet, 290
    R, 96, 115, 121, 136, 191, 210
    RIATA-HGT, 290
    S-PLUS, 96, 270
    SeDuMi, 271
    T-REX, 290
species tree, see trees
stem cells, 189–191
Stouffer's method, 7, 25, 86
structured association, 11
study designs, 10, 29, 171
summed rank, 34
    distribution function, 35
    p-value, 35, 37, 46
supervised learning, 204
support vector machine (SVM), 260,
    271–272
    1-norm soft margin SVM, 271
SVM, see support vector machine
Swiss-Prot, see databases
SwissPfam, see databases
systems biology, 201

T-REX, see software
t-statistic, 69, 95, 99, 113–117, 137, 203,
    216
technical replicates, 106, 108, 120, 133
transcription factor, 19, 243

cooperativity, 244
    transcriptional synergy, 244
transcriptional regulatory network, see
    networks
TRANSFAC, see databases
trees, 275, 286, 288
    gene tree, 277–284
        reconciliation problem, 285
    leaf set, 286
    phylogenetic tree, 277
    root, 286
    species tree, 277, 282–284
        reconciliation problem, 285
TRN, see networks
tumorigenesis, 202, 209
two-way table, 183, 188
Type I error, 25, 73, 74, 86, 99, 120, 121, 216

UIT, see union-intersection test
UniGene, see databases
union-intersection test, 84
unsupervised learning, 204

variance components, 50–52, 54, 55, 69
volcano plot, see plots

weighting, 5, 6, 25–27, 57, 80
Wright-Fisher model, 276, 290

XML, see software

yeast experiments
    cell cycle data, 253
    sequence data, 283
    two-hybrid assay, 201, 206, 259, 261, 262

Printed and bound by CPI Group (UK) Ltd, Croydon, CR0 4YY

21/10/2024

01777042-0014